CRC Handbook of Agricultural Energy Potential of Developing Countries

Volume III

Author
James A. Duke
Germplasm Introduction and Evaluation Laboratory
Plant Genetics and Germplasm Institute
Agricultural Research Service
United States Department of Agriculture
Beltsville, Maryland

With Special Contributors

Climate
Alan A. Atchley
Germplasm Introduction and Evaluation Laboratory
Plant Genetics and Germplasm Institute
Agricultural Research Service
United States Department of Agriculture
Beltsville, Maryland

Soils
Kenneth T. Ackerson
Soil Conservation Service
United States Department of Agriculture
Lanham, Maryland

Soil and Vegetation Maps
Peggy K. Duke
Botany Department
University of Maryland
College Park, Maryland

CRC Press, Inc.
Boca Raton, Florida

Library of Congress Cataloging-in-Publication Data

Duke, James A., 1929-
CRC handbook of agricultural energy potential of developing countries.

1. Biomass energy—Developing countries.
I. Title. II. Title: Handbook of agricultural energy potential of developing countries.
TP360.D85 1987 333.79'38 85-31392
ISBN 0-8493-3640-6 (Set)

Direct all inquiries to CRC Press, Inc., 2000 Corporate Blvd., N.W., Boca Raton, Florida, 33431.

International Standard Book Number 0-8493-3640-6 (Set)
International Standard Book Number 0-8493-3641-4 (Volume I)
International Standard Book Number 0-8493-3642-2 (Volume II)
International Standard Book Number 0-8493-3643-0 (Volume III)
International Standard Book Number 0-8493-3644-9 (Volume IV)

Library of Congress Card Number 85-31392
Printed in the United States

PREFACE

I project that in the future, at least half of the world's energy requirements will be furnished by biomass . . . For this generation, association with the development of bio-energy is one of the most challenging existing opportunities.

P. H. Abelson
EPA Journal (July/August, 1980)

Over the years, the Germplasm Resources Laboratory and the Economic Botany Laboratory of the USDA have prepared background information summaries on the agroenergetic potential of several developing countries. Rescuing them from anonymity, I have edited them to reduce redundancy and updated them with a few recent notes on energetics.

Finding these summaries useful to travelers visiting a developing country for the first time, I am pleased that CRC, after review, elected to publish them, rescuing them from obscurity in a dusty file cabinet. I hope that readers and travelers will find them as useful as I have for capsular views of factors important in agroenergetic appraisals of developing countries, with notes on agricultural and human demography, energy, ecological and agricultural productivity, climate, and soils.

THE AUTHOR

Born in Birmingham, Alabama, in 1929, James A. Duke is a Phi Beta Kappa graduate of the University of North Carolina, where he took his AB (1952) and BS (1955) in Botany. Following a 2½ year tour of military duty, which included a mycological assignment at Fort Dietrick, Maryland, he took his PhD in Botany, in 1961, at the University of North Carolina, moving on to postdoctoral activities at Washington University and the Missouri Botanical Garden, in St. Louis, Missouri, where he assumed professorial and curatorial duties, respectively. At the Missouri Botanical Garden, Duke began intensive studies of neotropical ethnobotany, which is his overriding interest to this day. From 1963 to 1965, Duke served with the U.S. Department of Agriculture at Beltsville, Maryland, devoting much of his time to neotropical ecology, especially seedling ecology. In 1965, he joined Battelle Columbus Laboratories, for ecological and ethnological studies in Panama and Colombia, returning to the USDA in 1971 for crop diversification and medicinal plant studies in developing countries. As a collaborator with the Smithsonian Institution, he has lectured and taught there, with emphasis on neotropcal ethnobotany and folk medicine. Considered a key figure in the "Herbal Renaissance", Duke received the Cutty Sark Science Award in 1981. In 1985, he received the Maharishi University Award and was awarded Honorary Membership in the Oriental Healing Arts Institute. Currently he is an economic botanist with the Germplasm Introduction and Evaluation Laboratory, Agricultural Research Service, USDA, Beltsville, Maryland.

Dr. Duke is an active member in the American Herb Association (Life), American Society of Pharmacognosy, Association for Tropical Biology, Herb Research Foundation, International Association of Plant Taxonomists (Life), International Society for Tropical Ecology, International Weed Science Society (Life), National Association for Professional Bureaucrats, Organization for Tropical Studies (Life), Society for Economic Botany (Life), Southern Appalachian Botanical Club (Life), Tri-State Bluegrass Association (Life), Weed Science Society of America, International Platform Association, Herb Research Foundation, and Oriental Healing Arts Institute.

Widely travelled in his ethnobotanical studies, Dr. Duke has more than 100 scientific publications to his credit. His books include *Isthmian Ethnobotanical Dictionary,* Harrod and Company, Baltimore, 1972 (being reissued with illustrations by Scientific Publishers, Jodphur, India); *A Handbook of Legumes of World Economic Importance,* Plenum Press, New York, 1981; *Medicinal Plants of the Bible,* Trado-Medic Books, Buffalo, New York, 1983; with E. S. Ayensu, *Medicinal Plants of China,* 2 volumes, Reference Publications, Algonac, Michigan, 1984; and *A Culinary Herbal,* Trado-Medic Books, Buffalo, New York, 1985. Also in press for 1986: *Gilding of the Ginseng,* Reference Publications, Algonac, Michigan, *Handbook of Boreal Amerindian Herbs,* Quarterman Publications, Lawrence, Massachusetts and, with A. A. Atchley, *Handbook of Proximate Analysis Tables of Higher Plants,* CRC Press, Boca Raton, Florida. In early 1986, Dr. Duke's first LP album was produced, entitled "Herbalbum" with herbal country music.

ACKNOWLEDGMENTS

As always, I had a lot of good help in assembling this book. Margaret Vinton and Judy duCellier helped pull data from dry statistical tables and insert them in our format. Then Judy entered everything in the word processor. Ruth Nash typed the agroecological appendix (Appendix I), Judy typed the phytomass appendix (Appendix II). Dr. A. A. Atchley and his staff helped with the computerized climatic data, averaging out all our data for each country and correcting older data and translating from Fahrenheit to Centigrade, feet to meters, and inches to millimeters. Dr. K. T. Ackerson provided soil maps and descriptions for several countries. Mrs. Peggy K. Duke redrew these maps and CIA "vegetation" maps in black and white so that they could be reproduced here. I did a little map-making myself in Panama. How well I remember the dimly lit room in the Hotel Ideal in Panama, where I transposed data from the FAO (1976)[1] into what was paragraph one of the Ag Summary — only to have it superceded when FAO (1980)[2] came out.

TABLE OF CONTENTS

Volume I

Volume II

Volume III

Volume IV

KENYA*

Agriculture

Of 58,265,000 ha in total area, Kenya was estimated in 1978 to have 2,270,000 ha of arable or permanent cropland, 3,770,000 ha in permanent pasture, 1,874,000 ha in forest and woodland, and 49,011,000 ha in other terrain categories. In 1978, approximately 44,000 ha were irrigated. Of the 1979 population, 78.1% or 4,682,000 individuals of the economically active population were engaged in agriculture. The total population in 1979 was 15,780,000. According to FAO statistics for 1979, there were 2000 horses, 10,470,000 cattle, 550,000 camels, 65,000 pigs, 4,000,000 sheep, 4,500,000 goats, and 17,500,000 chickens.[2] Agricultural production is summarized in Table 1. Conventional and energetic yield data on most of these will be found in Appendixes I and II.

Climate

Data from Wernstedt[3] are presented in the Climatological Data Table (see Introduction for explanation). Based on the means for data points cited in Wernstedt, the mean annual precipitation is 8.8 dm (mean of 77 sites) and the mean annual temperature 21.2°C (mean of 34 sites). The averages would suggest that much of the country could be classified as Subtropical Dry Forest Life Zone.[4]

Productivity

Major vegetation types are depicted in the vegetation map. Annual productivities of the mapping units are estimated as follows:

Vegetation type	MT/ha/yr
Montane, riverine, and coastal forest	12—30
Coastal brush	6—12
Savanna	6—12
Desert shrub and grass	3—12
Desert	0—3
Marsh and swamp	6—36

Economics

Food consumption per capita was 2095 cal/day, energy consumption 135 kg coal equivalent per capita in 1977. In 1970, the caloric supply was 98.0% of requirements. The total population growth rate was 3.4% in 1970. Population density was 19 persons per square kilometer in 1970, with 10.2% of the total population urban. Life expectancy at birth of those born in 1970 was 52 years. In 1977, with GNP at $290 per capita, agriculture accounted for 38.3% of GDP, manufacturing for 12.7%, other industry for 7.1%, and services for 41.9%. In 1977, food and beverage constituted 64.6% of exports, nonfood agriculture for 5.4%, fuels, minerals, and metal for 17.8%, machines for 1.9%, and other manufacturing for 10.3% of total merchandise exports. Food and beverage constituted 5.8% of imports, nonfood agriculture 2.7%, fuel and lubricants 22.3%, nonfuel minerals and metals 1.6%, machinery and equipment 33.9%, and other manufactured goods constituted 33.7% of total merchandise imports.[10]

Energetics

Kenya has a hydroelectric potential of 13,440 MW. Where surplus molasses is available from existing sugar production in remote mills, or where sugar has a low value,

* Maps and table follow text.

Table 1
AGRICULTURAL PRODUCTION IN KENYA

Crop	Area planted (ha)	Yield (kg/ha)	Total production (MT)
Wheat	117,000	923	108,000
Rice paddy	7,000	6,143	43,000
Barley	80,000	938	75,000
Corn	1,400,000	1,286	1,800,000
Oats	7,000	1,000	7,000
Millet	81,000	1,358	110,000
Sorghum	210,000	886	186,000
Potatoes	48,000	7,500	360,000
Sweet potatoes	38,000	8,947	340,000
Cassava	80,000	7,875	630,000
Peanuts in shell	14,000	589	8,000
Castor beans	9,000	250	2,000
Sunflower seed	14,000	1,103	15,000
Sesame seed	19,000	411	8,000
Linseed		1,000	
Seed cotton	120,000	278	33,000
Cottonseed			22,000
Coconuts	80,000		
Copra			2,000
Tomatoes	4,000	14,444	52,000
Dry onions	2,000	2,343	4,000
Sugar cane	34,000	106,823	3,621,000
Sugar (centrifugal, raw)			250,000
Sugar (noncentrifugal)			28,000
Oranges			14,000
Grapefruit and pomelo			7,000
Pineapples			8,000
Bananas			130,000
Plantains			225,000
Cashew nuts			58,000
Green coffee	105,000	762	80,000
Tea	83,000	1,196	99,000
Tobacco leaves	2,000	900	2,000
Sisal	44,000	828	36,000
Cotton lint			11,000

From FAO 1979 Production Yearbook, Vol. 33, Food and Agriculture Organization, Rome, 1980.

ethanol production seems economically justified. Kenya is expected to be producing geothermal energy in 1990.[10]

Soils

The soils of Kenya are shown on the Soils Map in eight associations: (1) Reddish-Brown Lateritic soils, (2) Latosols with laterite, Noncalcic Brown soils, and silty Regosols, (3) Grumusols, Planosols, and Solonchak soils, (4) sandy Regosols and Grumusols, (5) Reddish-Brown soils, (6) Red Desert soils and Lithosols, (7) Lithosols, and (8) Ando and Alpine Meadow soils.

Reddish-Brown Lateritic soils, which are among the most productive soils of the country, occur on nearly level to moderately sloping areas of the highlands of southwest Kenya. They are deep, well-drained, moderately acid, friable, reddish-brown clays that become redder and more acid in reaction with depth; they are derived from vol-

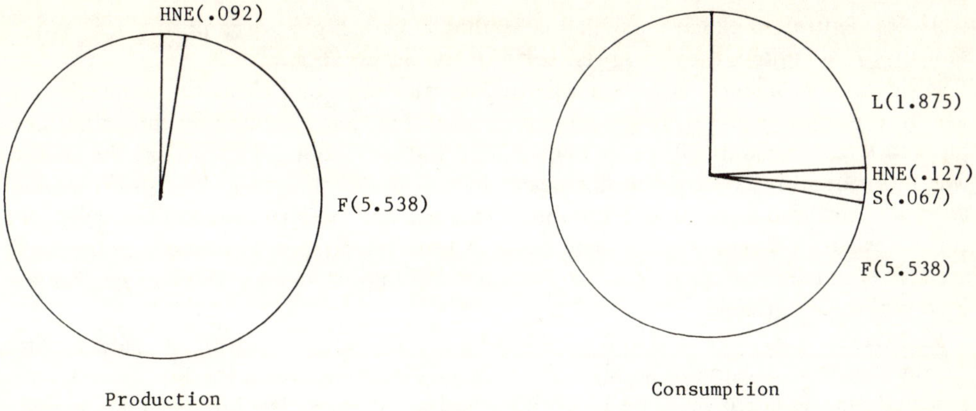

1978 Energy production and consumption in Kenya; fuel types listed in million metric tons coal equivalents (MMTCE). 1 MMTCE = 27.78×10^{12} Btu. L = liquid fuel, S = solid fuel, HNE = hydro, nuclear, and electric fuel, F = firewood, NG = natural gas. (Data from United Nations, World Energy Supplies, 1973—1978, Statistical Papers Series J, 22, Department of International Economic and Social Affairs, Statistical Office, New York, 1979.)

canic ash and basic rocks. These soils are fairly well supplied with plant nutrients and produce a wide variety of crops. Among the Reddish-Brown Lateritic soils are some Red Latosols, which are more leached and have a lower supply of plant nutrients than the Reddish-Brown Lateritic soils. Consequently, crop yields on the Red Latosols are lower, and extensive areas are left in savanna.

Latosols with laterite, Noncalcic Brown soils, and silty Regosols occur in the southwestern highlands and also near the coast. The Latosols have a grayish-brown friable clay surface over a reddish-brown to brown subsoil containing numerous iron concretions, which in places form a layer of hard laterite. Noncalcic Brown soils are much less leached than the Latosols. They have a dark brown loam surface, which becomes hard when dry, and a more clayey subsoil that has a few iron concretions. The silty Regosols occur on recent volcanic ash. They have a dark brown silt loam surface and a loamy fine sand subsoil. The soils of this association are less productive than the Reddish-Brown Lateritic soils. Much of this land is used for grazing, particularly the areas of Regosols.

Grumusols, Planosols, and Solonchak soils are predominantly clayey soils occurring in depressional areas of Kenya, mainly in the southwest and along the Tana. Grumusols are dark to nearly black clay loams having clayey, generally calcareous subsoils. Grumusols are difficult to work because they are sticky when wet and hard when dry. The Planosols in Kenya have a gritty loam or silt loam surface, and a gray mottled silt loam or compacted silt loam subsoil, which overlies nearly impervious clay. Solonchak soils are saline soils that occur not only among the Grumusols and Planosols but also among the Reddish-Brown and Red Desert soils. Solonchak soils are primarily in grass; where they are not too saline and water is available, they have a potential for other crops.

Grumusols occur also in low areas along the coast, where they are associated with sandy Regosols. Near the coast, the Regosols are grayish-brown to brown loose sands that become somewhat reddish with depth; most of these can be used for crop cultivation. Near Lake Rudolf, the Regosols are brown sands that become calcareous with depth; these have no agricultural value.

Alluvial soils occur throughout the country, but cannot be shown at the small scale of the soil map. The largest areas are along the Tana and Turkwel rivers. The Alluvial soils are neutral to calcareous silt loams that have a highly organic surface. Where moisture is adequate, they comprise some of the best crop land.

Reddish-Brown soils occur especially in eastern Kenya and along the Tanzania border. In the south, these soils are principally dark red sandy loams overlying a permeable well-structured subsoil, probably a buried Latosol from the past when the climate was more humid. Toward the northeast these soils become dark reddish-brown and overlie a calcareous subsoil, although where the soil overlies a soft limy layer, the surface is a dark grayish-brown clay loam. A large proportion of the soils in the north are somewhat shallow. The Reddish Brown soils support some dryland crops, but are used mainly for grazing.

Red Desert soils of central and northern Kenya are dark reddish-brown sandy loams with a compact, calcareous sandy loam subsoil. Associated with the Red Desert soils, especially in the northwest, are Lithosols (shallow, stony soils). Lithosols not in association with Red Desert soils cover much of northern Kenya, as well as smaller areas in the south. Many of the nearly level areas of Lithosols are also strewn with large volcanic boulders. Elsewhere the Lithosols are steep and barren. Red Desert soils are unsuited for agriculture, primarily because of a lack of soil moisture, although they do provide some sparse grazing. Lithosols in Kenya have no agricultural value.

Ando soils and Alpine Meadow soils occur on the ash-mantled mountains of southwest Kenya and on their lower slopes. Ando soils are high in organic matter and are very permeable to roots and water. Because of the steep slope of the land, however, these soils are principally in forest reserves. The Alpine Meadow soils, which occupy the highest elevations, have a dark peaty surface and a thin subsoil; rock outcrops are numerous. Alpine Meadow soils have a patchy grass cover that is of little agricultural value.

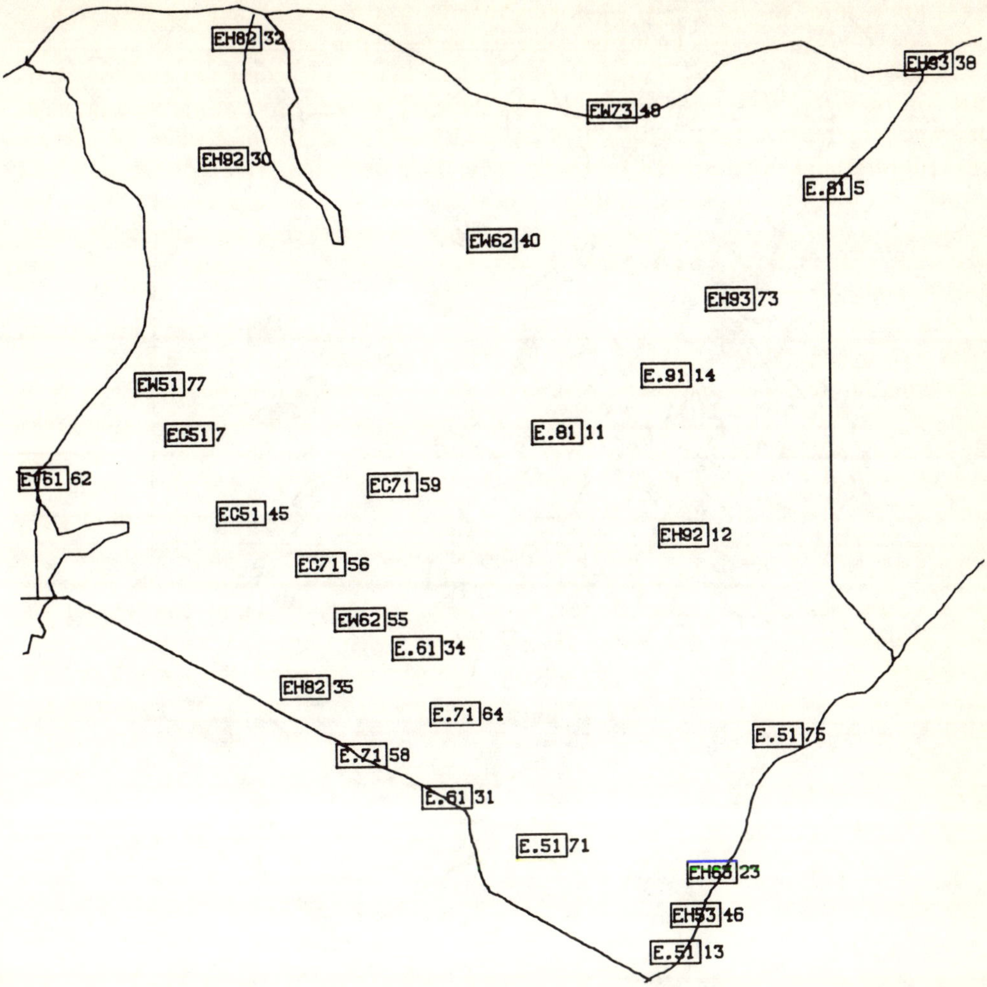

Selected weather stations for Kenya.

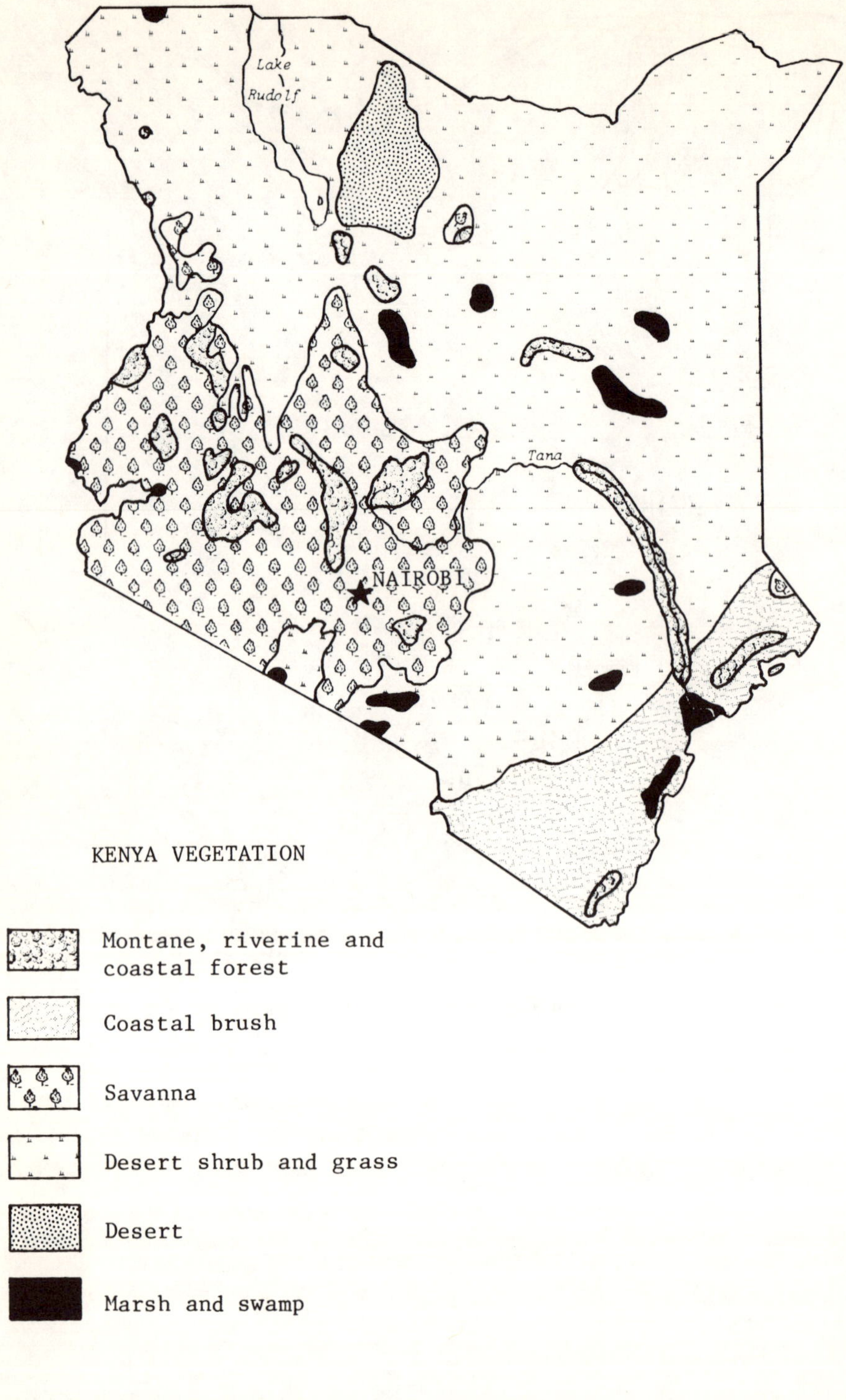
Lake
Rudolf
Tana
NAIROBI
KENYA VEGETATION
Montane, riverine and coastal forest
Coastal brush
Savanna
Desert shrub and grass
Desert
Marsh and swamp

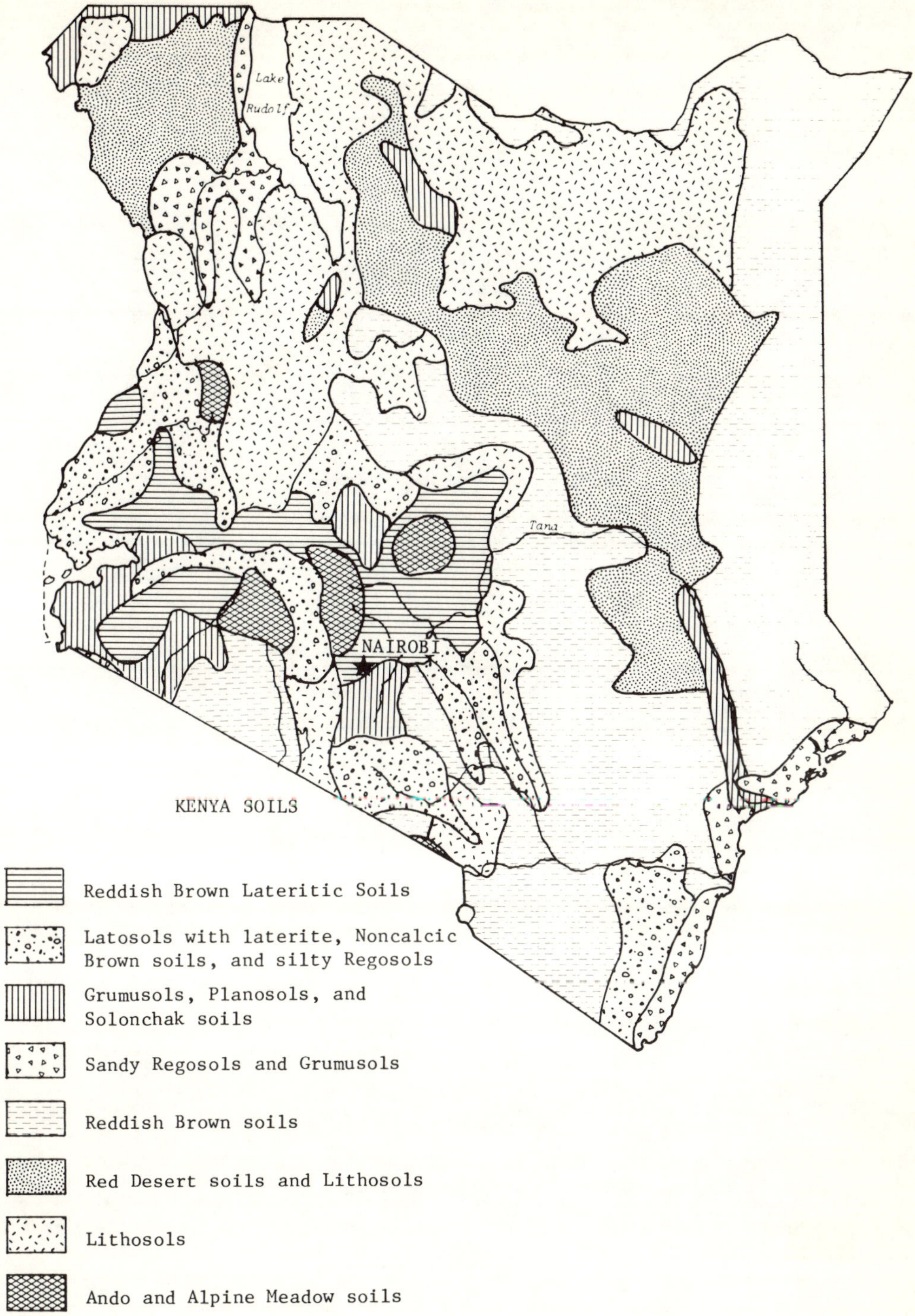
Lake
Rudolf
Tana
NAIROBI
KENYA SOILS
Reddish Brown Lateritic Soils
Latosols with laterite, Noncalcic Brown soils, and silty Regosols
Grumusols, Planosols, and Solonchak soils
Sandy Regosols and Grumusols
Reddish Brown soils
Red Desert soils and Lithosols
Lithosols
Ando and Alpine Meadow soils

CLIMATOLOGICAL DATA FOR KENYA

GEOGRAPHY

OBS.	LOCATION	YEARS	LATITUDE	LONGITUDE	ELEVATION -M-	LIFEZONE	CCTX	ID	ONMAP
1	BUKURIA	17	1.22 S	34.62 E	1524.0	.	.	1	NO
2	CHERANGANI	30	1.00 N	35.27 E	1920.2	.	.	2	NO
3	DONYO SABUK	30	1.00 N	37.47 E	1295.4	.	.	3	NO
4	EL ROBA	18	3.93 N	40.03 E	1005.8	.	.	4	NO
5	EL WAK	15	2.78 N	40.95 E	365.8	.	.	5	YES
6	ELDAMA RAVINE	41	0.08 N	35.75 E	2194.6	WM	EC51	6	NO
7	ELDORET	16	0.52 N	35.28 E	2083.3	WM	EC51	7	YES
8	EQUATOR	25	0.02 S	35.55 E	2761.2	WM	EC51	8	NO
9	FAZA	41	2.05 S	41.10 E	6.1	.	.	9	NO
10	FORT HALL	62	0.72 S	37.17 E	1280.2	SM	EW53	10	NO
11	GARBA TULLA	13	0.50 N	38.53 E	457.2	.	.	11	YES
12	GARISSA	32	0.48 S	39.63 E	127.7	TT	EH92	12	YES

AVERAGE PRECIPITATION (MM)

OBS.	JAN.	FEB.	MAR.	APR.	MAY	JUN.	JUL.	AUG.	SEP.	OCT.	NOV.	DEC.	ANNUAL
1	70	94	155	198	204	101	53	64	105	113	106	127	1390
2	26	36	66	131	142	115	171	171	96	62	49	37	1102
3	46	45	102	204	75	5	3	5	8	50	188	101	832
4	3	4	38	122	43	1	1	5	1	59	49	8	334
5	0	0	37	82	20	0	2	4	3	51	95	16	310
6	25	39	86	170	152	103	132	134	70	67	77	46	1101
7	35	26	53	145	124	99	157	189	90	51	45	49	1063
8	33	34	72	168	142	123	163	205	111	53	63	55	1222
9	5	2	17	116	282	168	86	48	42	43	45	27	881
10	33	36	112	306	198	45	20	24	27	116	189	74	1180
11	6	6	28	112	13	0	0	0	0	25	115	56	361
12	10	6	27	59	16	5	1	6	6	21	77	64	298

AVERAGE TEMPERATURE (DEGREES CENTIGRADE)

OBS.	JAN.	FEB.	MAR.	APR.	MAY	JUN.	JUL.	AUG.	SEP.	OCT.	NOV.	DEC.	ANNUAL
1	.	.	.	.	.	.	.	.	.	.	.	.	.
2	.	.	.	.	.	.	.	.	.	.	.	.	.
3	.	.	.	.	.	.	.	.	.	.	.	.	.
4	.	.	.	.	.	.	.	.	.	.	.	.	.
5	.	.	.	.	.	.	.	.	.	.	.	.	.
6	16.6	17.0	15.9	17.3	16.7	16.2	15.4	15.4	15.9	16.4	16.1	16.9	16.3
7	16.9	17.3	17.9	17.7	16.9	15.7	15.3	15.5	15.9	16.6	16.9	16.7	16.6
8	13.7	14.3	14.5	14.0	13.5	12.6	11.8	11.9	12.7	13.1	13.3	13.0	13.1
9	.	.	.	.	.	.	.	.	.	.	.	.	.
10	19.7	20.8	21.2	20.4	19.7	18.9	18.3	17.9	19.3	20.4	22.6	19.6	19.7
11	.	.	.	.	.	.	.	.	.	.	.	.	.
12	28.6	29.4	30.3	30.1	28.9	27.3	26.5	26.7	27.5	28.7	28.9	28.3	28.4

CLIMATOLOGICAL DATA FOR KENYA

GEOGRAPHY

OBS.	LOCATION	YEARS	LATITUDE	LONGITUDE	ELEVATION -M-	LIFEZONE	CCTX	ID	ONMAP
13	GAZI	30	4.42 S	39.52 E	45.7	.	.	13	YES
14	HABBASWEIN	12	1.03 N	39.50 E	198.1	.	.	14	YES
15	HOEYS BRIDGE	27	0.87 N	35.12 E	1828.8	.	.	15	NO
16	IKUTHA	15	2.05 S	38.17 E	731.5	.	.	16	NO
17	ISIOLO	34	0.35 N	37.58 E	1103.7	SD	EW72	17	NO
18	KAKMEGA	67	0.28 N	34.75 E	1554.5	.	.	18	NO
19	KAPENGURIA	32	1.23 N	35.10 E	2133.3	WM	EC51	19	NO
20	KAPSABET	59	0.20 N	35.12 E	1996.4	WM	EC41	20	NO
21	KERICHO	59	0.58 S	35.12 E	1889.5	SM	EC41	21	NO
22	KIBWEZI	30	2.45 S	37.93 E	914.7	.	.	22	NO
23	KILIFI	45	3.67 S	39.85 E	3.0	TV	EH63	23	YES
24	KISII	52	0.68 S	34.78 E	1767.8	.	.	24	NO

AVERAGE PRECIPITATION (MM)

OBS.	JAN.	FEB.	MAR.	APR.	MAY	JUN.	JUL.	AUG.	SEP.	OCT.	NOV.	DEC.	ANNUAL
13	27	19	78	250	401	147	99	80	75	82	78	59	1395
14	5	5	17	91	8	0	1	0	2	24	42	23	218
15	29	26	65	154	180	137	174	194	97	50	53	39	1198
16	23	12	65	85	13	0	2	2	2	30	189	117	540
17	33	33	87	127	34	3	7	5	5	65	139	75	613
18	57	87	151	249	265	191	175	230	187	125	107	87	1911
19	15	29	72	150	199	131	191	200	109	80	59	30	1265
20	51	70	116	196	215	161	167	205	153	90	83	56	1563
21	67	93	154	267	252	162	152	185	157	138	127	93	1847
22	38	28	75	111	29	5	0	2	3	20	148	139	598
23	18	14	36	124	277	110	79	54	59	72	65	44	952
24	61	94	164	253	214	148	102	149	164	137	153	103	1742

AVERAGE TEMPERATURE (DEGREES CENTIGRADE)

OBS.	JAN.	FEB.	MAR.	APR.	MAY	JUN.	JUL.	AUG.	SEP.	OCT.	NOV.	DEC.	ANNUAL
13	.	.	.	.	.	.	.	.	.	.	.	.	.
14	.	.	.	.	.	.	.	.	.	.	.	.	.
15	.	.	.	.	.	.	.	.	.	.	.	.	.
16	.	.	.	.	.	.	.	.	.	.	.	.	.
17	23.3	24.3	24.7	23.9	23.9	23.2	22.6	22.9	23.9	24.3	22.6	22.3	23.5
18	.	.	.	.	.	.	.	.	.	.	.	.	.
19	16.5	17.2	17.5	17.3	16.8	15.9	15.5	15.9	15.9	16.3	16.1	16.1	16.5
20	18.1	18.2	18.6	18.8	16.6	16.6	16.2	16.1	16.1	17.5	17.8	17.5	17.3
21	18.8	18.9	18.9	18.2	17.3	17.0	16.3	16.5	16.7	17.1	17.3	17.7	17.6
22	.	.	.	.	.	.	.	.	.	.	.	.	.
23	26.9	27.3	28.0	27.6	25.9	25.1	24.3	24.5	24.8	25.9	26.5	27.0	26.1
24	.	.	.	.	.	.	.	.	.	.	.	.	.

CLIMATOLOGICAL DATA FOR KENYA

GEOGRAPHY

OBS.	LOCATION	YEARS	LATITUDE	LONGITUDE	ELEVATION -M-	LIFEZONE	CCTX	ID	ONMAP
25	KISUMU	25	0.10 S	34.75 E	1145.4	SM	EW51	25	NO
26	KITALE	29	1.00 N	35.02 E	1895.2	SM	EW51	77	YES
27	KITUI	21	0.35 S	38.05 E	1176.5	.	.	26	NO
28	KONZA	30	1.73 S	37.13 E	1654.5	.	.	27	NO
29	KWALE	55	4.18 S	39.45 E	393.2	.	.	28	NO
30	LAMU	57	2.27 S	40.90 E	29.9	TV	EH63	29	NO
31	LODWAR	44	3.12 N	35.62 E	505.7	TX	EH92	30	YES
32	LOITOKITOK	30	2.93 S	37.52 E	1981.2	.	.	31	YES
33	LOKITAUNG	30	4.25 N	35.75 E	731.5	TT	EH82	32	YES
34	LUGARI	31	0.68 N	34.88 E	1615.4	.	.	33	NO
35	MACHAKOS	69	1.52 S	37.27 E	1645.9	.	.	34	YES
36	MAGADI	32	1.88 S	36.28 E	612.6	TT	EH82	35	YES

AVERAGE PRECIPITATION (MM)

OBS.	JAN.	FEB.	MAR.	APR.	MAY	JUN.	JUL.	AUG.	SEP.	OCT.	NOV.	DEC.	ANNUAL
25	57	70	160	195	177	101	68	96	79	64	106	105	1278
26	20	38	71	146	165	132	159	171	106	73	59	40	1180
27	24	13	64	187	33	1	0	1	1	42	148	119	633
28	21	27	58	119	58	8	1	1	3	19	65	46	426
29	33	19	59	155	230	95	81	61	68	90	94	82	1067
30	6	3	19	127	344	154	68	40	37	37	32	28	895
31	8	5	20	41	27	8	13	9	3	8	10	13	165
32	53	58	112	124	35	3	2	1	3	52	177	147	767
33	12	17	53	119	47	22	32	13	7	13	32	28	395
34	37	54	91	173	197	185	169	217	151	73	67	52	1466
35	49	53	124	210	76	12	5	6	9	53	189	122	908
36	32	43	66	89	55	8	1	3	5	15	34	47	398

AVERAGE TEMPERATURE (DEGREES CENTIGRADE)

OBS.	JAN.	FEB.	MAR.	APR.	MAY	JUN.	JUL.	AUG.	SEP.	OCT.	NOV.	DEC.	ANNUAL
25	23.9	24.3	24.2	23.4	22.8	22.3	21.9	22.1	22.3	23.8	23.9	23.7	23.3
26	18.7	19.1	19.4	19.3	18.7	17.7	17.2	17.3	17.7	18.3	18.0	18.1	18.3
27	.	.	.	.	.	.	.	.	.	.	.	.	.
28	.	.	.	.	.	.	.	.	.	.	.	.	.
29	.	.	.	.	.	.	.	.	.	.	.	.	.
30	26.9	27.3	28.1	28.1	26.6	25.8	25.1	25.1	25.5	26.5	27.2	27.3	26.7
31	28.9	29.9	30.3	29.7	29.8	29.1	28.3	28.5	29.4	29.9	29.1	28.7	29.3
32	.	.	.	.	.	.	.	.	.	.	.	.	.
33	27.9	28.4	28.3	26.1	26.6	26.6	26.0	25.8	26.9	26.9	27.0	27.1	26.9
34	.	.	.	.	.	.	.	.	.	.	.	.	.
35	.	.	.	.	.	.	.	.	.	.	.	.	.
36	30.3	30.7	30.7	29.1	28.1	27.5	27.1	27.7	28.9	30.1	29.6	29.3	29.1

CLIMATOLOGICAL DATA FOR KENYA

GEOGRAPHY

OBS.	LOCATION	YEARS	LATITUDE	LONGITUDE	ELEVATION -M-	LIFEZONE	CCTX	ID	ONMAP
37	MAKINDU	59	2.28 S	37.83 E	997.6	SD	EW73	36	NO
38	MALINDI	15	3.22 S	40.12 E	19.8	TD	EH53	37	NO
39	MANDERA	27	3.95 N	41.87 E	330.7	TX	EH93	38	YES
40	MARALAL	28	1.08 N	36.70 E	1950.7	WD	EC72	39	NO
41	MARSABIT	45	2.32 N	37.98 E	1344.2	SD	EW62	40	YES
42	MAZERAS	59	3.97 S	39.55 E	161.5	.	.	41	NO
43	MERU	53	0.05 N	37.65 E	1569.7	.	.	42	NO
44	MIVUKONI	21	0.42 S	38.25 E	883.9	.	.	43	NO
45	MOHURU	20	1.00 S	34.12 E	1219.2	.	.	44	NO
46	MOLO	17	0.23 S	35.73 E	2498.1	WM	EC51	45	YES
47	MOMBASA	73	4.07 S	39.70 E	15.8	TD	EH53	46	YES
48	MOUNT ELGON FOREST	34	1.13 N	34.75 E	2225.0	.	.	47	NO

AVERAGE PRECIPITATION (MM)

OBS.	JAN.	FEB.	MAR.	APR.	MAY	JUN.	JUL.	AUG.	SEP.	OCT.	NOV.	DEC.	ANNUAL
37	40	28	79	111	31	2	1	1	1	29	168	120	611
38	3	3	23	152	455	137	93	81	65	82	24	30	1148
39	1	4	23	86	27	0	0	1	2	42	34	8	228
40	11	19	38	85	69	54	109	92	30	39	52	22	620
41	35	17	79	207	105	8	17	20	13	120	142	71	834
42	41	20	61	144	232	80	77	51	60	93	96	79	1034
43	54	25	117	277	145	9	12	10	21	232	311	145	1368
44	12	6	48	121	22	0	0	1	1	24	130	93	458
45	37	31	80	111	108	28	12	25	19	44	57	46	598
46	39	28	58	170	138	101	132	182	100	57	87	82	1174
47	26	15	61	200	319	112	89	65	68	83	93	60	1191
48	25	37	74	153	156	124	174	178	102	88	70	39	1220

AVERAGE TEMPERATURE (DEGREES CENTIGRADE)

OBS.	JAN.	FEB.	MAR.	APR.	MAY	JUN.	JUL.	AUG.	SEP.	OCT.	NOV.	DEC.	ANNUAL
37	23.2	24.5	24.7	24.0	22.7	21.1	20.1	20.5	21.9	23.3	23.3	22.6	22.7
38	26.7	27.2	28.1	27.9	26.4	25.6	24.9	25.1	25.5	26.1	26.9	27.2	26.5
39	28.9	29.9	31.1	29.9	29.1	28.6	27.9	28.1	28.9	28.7	28.4	28.6	29.0
40	16.1	16.7	17.4	17.6	17.0	15.7	15.3	15.0	15.0	16.0	16.1	15.6	16.1
41	20.7	21.2	21.4	21.1	20.5	19.5	18.7	18.6	19.3	20.0	19.9	19.6	20.0
42	.	.	.	.	.	.	.	.	.	.	.	.	.
43	.	.	.	.	.	.	.	.	.	.	.	.	.
44	.	.	.	.	.	.	.	.	.	.	.	.	.
45	.	.	.	.	.	.	.	.	.	.	.	.	.
46	14.1	14.1	14.8	14.7	14.3	13.2	12.6	12.9	13.3	13.7	13.7	13.8	13.7
47	27.9	28.5	28.9	27.9	26.2	25.5	24.7	24.7	25.2	26.5	27.3	27.7	26.7
48	.	.	.	.	.	.	.	.	.	.	.	.	.

CLIMATOLOGICAL DATA FOR KENYA

GEOGRAPHY

OBS.	LOCATION	YEARS	LATITUDE	LONGITUDE	ELEVATION -M-	LIFEZONE	CCTX	ID	ONMAP
49	MOYALE	48	3.53 N	39.05 E	1112.5	SD	EW73	48	YES
50	MTITO ANDEI	12	2.75 S	38.12 E	914.4	.	.	49	NO
51	MUGUGA	12	1.22 S	36.63 E	2095.2	WD	EC62	50	NO
52	MUMIAS	30	0.32 N	34.50 E	1339.6	.	.	51	NO
53	MURRI	15	4.27 N	40.10 E	1005.8	.	.	52	NO
54	MUTONO	22	1.85 S	38.20 E	914.4	.	.	53	NO
55	MWATATE	30	3.52 S	38.40 E	822.7	.	.	54	NO
56	NAIROBI	40	1.25 S	36.77 E	1737.4	SD	EW62	55	YES
57	NAIVASHA	53	0.72 S	36.43 E	1899.2	WD	EC71	56	YES
58	NAKURU	30	0.28 S	36.07 E	1835.5	SD	EW61	57	NO
59	NAMANGA	21	2.53 S	36.77 E	1371.6	.	.	58	YES
60	NANYUKI	24	0.02 N	37.07 E	1945.2	WD	EC71	59	YES

AVERAGE PRECIPITATION (MM)

OBS.	JAN.	FEB.	MAR.	APR.	MAY	JUN.	JUL.	AUG.	SEP.	OCT.	NOV.	DEC.	ANNUAL
49	11	17	55	182	118	17	17	17	25	96	86	41	682
50	39	24	65	113	29	3	1	2	3	23	173	141	616
51	61	40	60	219	203	49	22	22	26	59	141	89	991
52	45	91	138	239	270	167	143	159	160	128	103	87	1730
53	0	1	28	95	48	0	0	0	0	54	79	29	334
54	27	16	65	94	25	3	2	3	5	33	209	165	647
55	29	35	91	116	58	9	6	11	17	34	94	111	611
56	50	47	109	211	161	43	15	25	24	55	130	84	954
57	22	35	59	107	86	41	32	44	44	47	58	36	611
58	15	33	59	135	114	86	109	109	69	56	60	32	877
59	73	63	107	159	72	9	1	1	9	26	78	106	704
60	22	21	54	121	84	42	64	68	51	78	90	44	739

AVERAGE TEMPERATURE (DEGREES CENTIGRADE)

OBS.	JAN.	FEB.	MAR.	APR.	MAY	JUN.	JUL.	AUG.	SEP.	OCT.	NOV.	DEC.	ANNUAL
49	24.4	25.0	24.7	22.7	21.2	20.3	19.6	20.1	21.0	21.7	22.3	23.1	22.1
50	.	.	.	.	.	.	.	.	.	.	.	.	.
51	17.0	17.3	17.7	17.1	15.9	14.4	13.4	14.2	15.3	16.6	16.1	16.3	15.9
52	.	.	.	.	.	.	.	.	.	.	.	.	.
53	.	.	.	.	.	.	.	.	.	.	.	.	.
54	.	.	.	.	.	.	.	.	.	.	.	.	.
55	.	.	.	.	.	.	.	.	.	.	.	.	.
56	19.9	20.5	20.9	20.5	19.6	18.3	17.1	17.5	19.0	20.1	19.6	19.4	19.4
57	17.9	18.3	18.5	18.3	17.5	16.5	15.9	16.1	16.7	17.3	17.0	17.3	17.3
58	18.5	19.3	19.5	19.1	18.5	17.8	17.3	17.3	17.5	17.7	17.7	17.9	18.2
59	.	.	.	.	.	.	.	.	.	.	.	.	.
60	16.1	16.6	17.1	17.0	16.5	15.9	15.5	15.5	15.9	16.1	15.7	15.5	16.1

CLIMATOLOGICAL DATA FOR KENYA

GEOGRAPHY

OBS.	LOCATION	YEARS	LATITUDE	LONGITUDE	ELEVATION -M-	LIFEZONE	CCTX	ID	ONMAP
61	NAROK	50	1.13 S	35.83 E	1889.5	WD	EC72	60	NO
62	NYERI	59	0.43 S	36.95 E	1828.8	.	.	61	NO
63	PORT VICTORIA	24	0.12 N	33.98 E	1249.7	.	.	62	YES
64	RUMURUTI	56	0.27 N	36.53 E	1856.2	WD	EC71	63	NO
65	SIMBA	30	2.15 S	37.60 E	1036.6	.	.	64	YES
66	SUBUKIA	40	0.12 N	36.25 E	1676.4	.	.	65	NO
67	TAMBACH	30	0.60 N	35.53 E	1828.5	.	.	66	NO
68	TAVETA	58	3.43 S	37.67 E	722.4	.	.	67	NO
69	THIKA	22	1.02 S	37.10 E	1462.4	SD	EW62	68	NO
70	THOMSONS FALLS	42	0.08 N	36.42 E	2194.6	.	.	69	NO
71	TIMAU	30	0.12 N	37.47 E	2575.3	.	.	70	NO
72	TSAVO STATION	30	3.40 S	38.35 E	1462.4	.	.	71	YES

AVERAGE PRECIPITATION (MM)

OBS.	JAN.	FEB.	MAR.	APR.	MAY	JUN.	JUL.	AUG.	SEP.	OCT.	NOV.	DEC.	ANNUAL
61	69	75	102	140	91	27	14	21	21	25	62	71	718
62	45	43	67	182	172	32	36	38	31	94	114	76	930
63	47	56	126	174	131	51	32	54	41	58	77	28	875
64	20	24	47	93	66	48	92	90	38	44	78	33	673
65	41	43	70	120	41	4	0	2	4	21	134	118	598
66	20	28	55	141	138	81	134	153	71	53	65	41	980
67	33	46	88	188	173	103	134	125	65	76	93	73	1197
68	37	42	93	141	75	10	6	7	7	18	88	61	585
69	33	30	96	213	130	25	15	18	17	60	145	76	858
70	30	40	60	111	74	76	145	137	58	47	85	63	926
71	41	24	87	142	65	16	27	21	12	122	174	107	838
72	45	48	129	257	165	40	32	51	62	90	225	185	1329

AVERAGE TEMPERATURE (DEGREES CENTIGRADE)

OBS.	JAN.	FEB.	MAR.	APR.	MAY	JUN.	JUL.	AUG.	SEP.	OCT.	NOV.	DEC.	ANNUAL
61	17.4	17.3	17.5	17.7	17.0	15.5	14.7	15.2	15.9	16.7	16.6	16.7	16.5
62	.	.	.	.	.	.	.	.	.	.	.	.	.
63	.	.	.	.	.	.	.	.	.	.	.	.	.
64	16.8	17.4	18.2	18.6	17.9	17.4	16.8	16.6	17.2	17.6	17.4	17.7	17.5
65	.	.	.	.	.	.	.	.	.	.	.	.	.
66	.	.	.	.	.	.	.	.	.	.	.	.	.
67	.	.	.	.	.	.	.	.	.	.	.	.	.
68	.	.	.	.	.	.	.	.	.	.	.	.	.
69	19.8	20.3	21.2	21.3	20.3	18.8	17.7	18.2	19.7	20.7	20.1	19.7	19.9
70	.	.	.	.	.	.	.	.	.	.	.	.	.
71	.	.	.	.	.	.	.	.	.	.	.	.	.
72	.	.	.	.	.	.	.	.	.	.	.	.	.

CLIMATOLOGICAL DATA FOR KENYA

GEOGRAPHY

OBS.	LOCATION	YEARS	LATITUDE	LONGITUDE	ELEVATION -M-	LIFEZONE	CCTX	ID	ONMAP
73	VOI	59	3.40 S	38.57 E	559.6	TV	EW73	72	NO
74	WAJIR	46	1.75 N	40.07 E	243.8	TT	EH93	73	YES
75	WAYU	10	1.53 S	39.57 E	152.4	.	.	74	NO
76	WITU	23	2.38 S	40.45 E	3.0	.	.	75	YES
77	WUNDANYI	34	3.40 S	38.35 E	1463.0	.	.	76	NO

AVERAGE PRECIPITATION (MM)

OBS.	JAN.	FEB.	MAR.	APR.	MAY	JUN.	JUL.	AUG.	SEP.	OCT.	NOV.	DEC.	ANNUAL
73	32	30	73	92	29	7	3	8	15	27	96	126	538
74	5	5	23	68	33	1	3	2	6	26	61	21	254
75	13	4	62	80	28	20	51	4	50	42	84	88	526
76	21	6	36	104	330	147	84	63	50	60	75	109	1085
77	55	45	145	268	141	41	33	50	60	101	213	177	1329

AVERAGE TEMPERATURE (DEGREES CENTIGRADE)

OBS.	JAN.	FEB.	MAR.	APR.	MAY	JUN.	JUL.	AUG.	SEP.	OCT.	NOV.	DEC.	ANNUAL
73	26.1	26.9	27.2	25.9	24.9	23.7	22.7	22.7	23.4	24.9	25.7	25.7	24.9
74	28.7	29.3	29.9	28.9	27.9	26.9	26.1	26.4	27.1	27.7	27.5	27.9	27.9
75	.	.	.	.	.	.	.	.	.	.	.	.	.
76	.	.	.	.	.	.	.	.	.	.	.	.	.
77	.	.	.	.	.	.	.	.	.	.	.	.	.

SOUTH KOREA*

Agriculture

Of 9,848,000 ha in total area, the Republic of South Korea was estimated in 1978 to have 2,222,000 ha of arable or permanent cropland, 41,000 ha in permanent pasture, 6,578,000 ha in forest and woodland, and 978,000 ha in other terrain categories. In 1978, approximately 1,122,000 ha were irrigated. Of the 1979 population, 39.9% or 5,650,000 individuals of the economically active population were engaged in agriculture. The total population in 1979 was 37,313,000. According to FAO statistics for 1979, there were 5000 horses, 1,651,000 cattle, 1,719,000 pigs, 8000 sheep, 224,000 goats, 40,753,000 chickens, 560,000 ducks, and 8000 turkeys.[2] Agricultural production is summarized in Table 1. Conventional and energetic yield data on most of these will be found in Appendixes I and II.

Climate

Data from Wernstedt[3] are presented in the Climatological Data Table (see Introduction for explanation). Based on the means for data points cited in Wernstedt, the mean annual precipitation is 11.7 dm (mean of 98 sites) and the mean annual temperature 12.3°C (mean of 82 sites). The averages would suggest that much of the country could be classified as Warm Temperate Moist Forest Life Zone.[4]

Productivity

The vegetation map shows the distribution of some catchall vegetation categories. Temperate forests are not as productive as tropical forests in general, since they may be "working half-time". My estimates of productivity are tempered by limited field work in the Republic of Korea, which seems rather analagous to southeastern U.S.

Vegetation type	MT/ha/yr
Paddy crops	6—12
Dry crops	3—12
Forest	6—24
Scrub, brush, and barren land	0—12

Economics

Food consumption per capita was 2785 cal/day, energy consumption 1240 kg coal equivalent per capita in 1977. In 1970, the caloric supply was 114.0% of requirements. The total population growth rate was 2.4% in 1970. Population density was 319 persons per square kilometer in 1970, with 41.2% of the total population urban. Life expectancy at birth of those born in 1977 was 59 years. In 1977, with GNP at $980 per capita, agriculture accounted for 24.0% of GDP, manufacturing for 24.2%, other industry for 10.5%, and services for 41.3%. In 1977, food and beverage constituted 10.6% of exports, nonfood agriculture for 2.2%, fuels, minerals, and metal for 2.3%, machines for 17.4%, and other manufacturing for 67.6% of total merchandise exports. Food and beverage constituted 8.2% of imports, nonfood agriculture 13.6%, fuel and lubricants 20.2%, nonfuel minerals and metals 5.6%, machinery and equipment 26.9%, and other manufactured goods constituted 25.5% of total merchandise imports.[10]

Energetics

Korea has 2000 million ton coal (480 technically and economically recoverable), and a hydroelectric potential of 5514 MW. With a growing demand for oil, Korea could

* Maps and table follow text.

Table 1
AGRICULTURAL PRODUCTION IN THE REPUBLIC OF SOUTH KOREA

Crop	Area planted (ha)	Yield (kg/ha)	Total production (MT)
Wheat	13,000	3,206	42,000
Rice paddy	1,228,000	6,556	8,051,000
Barley	473,000	3,186	1,508,000
Rye	3,000	1,963	5,000
Millet	3,000	1,200	3,000
Sorghum	5,000	800	4,000
Potatoes	32,000	11,125	356,000
Sweet potatoes	70,000	19,814	1,387,000
Dry beans	39,000	865	34,000
Dry peas	3,000	880	2,000
Soybeans	260,000	988	257,000
Peanuts in shell	11,000	1,455	16,000
Castor beans	1,000	340	
Rapeseed	17,000	1,612	27,000
Sesame seed	68,000	393	27,000
Seed cotton	5,000	1,000	5,000
Cottonseed			6,000
Cabbages	83,000	12,409	1,034,000
Tomatoes	2,000	16,667	35,000
Pumpkins, squash, gourds	11,000	15,307	175,000
Cucumbers and gherkins	9,000	16,941	144,000
Eggplants	3,000	17,231	45,000
Chilies and peppers, green	70,000		
Dry onions	15,000	26,200	393,000
Garlic	50,000	7,000	350,000
Carrots	1,000	10,000	9,000
Watermelons	9,000	22,389	202,000
Cantaloupes and other melons	11,000	14,943	158,000
Grapes	8,000	7,829	60,000
Apples			444,000
Pears			65,000
Peaches and nectarines			92,000
Plums			7,000
Oranges			108,000
Strawberries			35,100
Chestnuts			19,000
Walnuts			130
Tea	1,000	300	
Hops		621	
Tobacco leaves	56,000	2,210	123,000
Hemp fiber and tow	1,000	2,157	1,000
Cotton lint			3,000

From FAO 1979 Production Yearbook, Vol. 33, Food and Agriculture Organization, Rome, 1980.

become increasingly dependent on imports unless they develop alternative sources. Coal is used for household fuel in some of Korea's cities. Korea is already generating some nuclear power. Importing less than 26% of its commercial energy demands as oil, Korea is not expected to be experiencing fuelwood problems by the year 2000.[10]

Soils

Much Korean soil is unsuited for cultivation because it is too steep, shallow, and

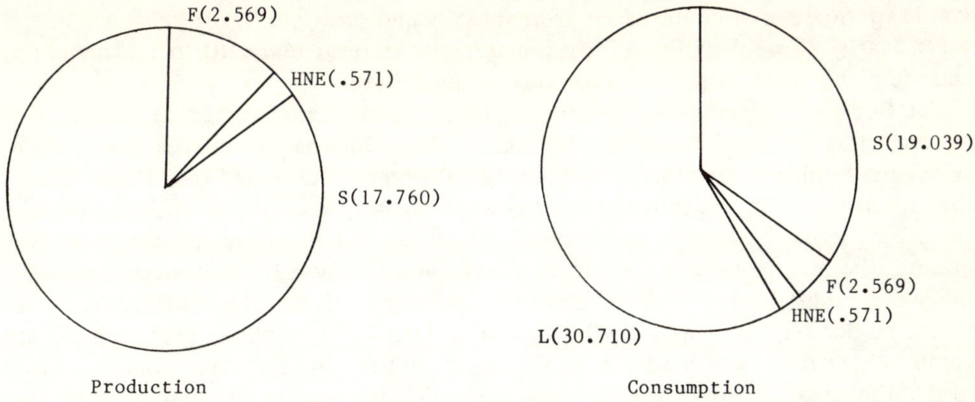

1978 Energy production and consumption in the Republic of South Korea; fuel types listed in million metric tons coal equivalents (MMTCE). 1 MMTCE = 27.78 × 10^{12} Btu. L = liquid fuel, S = solid fuel, HNE = hydro, nuclear, and electric fuel, F = firewood, NG = natural gas. (Data from United Nations, World Energy Supplies, 1973—1978, Statistical Papers Series J, 22, Department of International Economic and Social Affairs, Statistical Office, New York, 1979.)

stony. Lithosols are the most extensive of the five soil associations mapped. They cover the mountainous and hilly terrain that comprises about 80% of the country's area. Soils suitable for cultivation occur on valley floors (clayey Alluvial and sandy Alluvial soils), on the coastal lowlands, and in the interior basins (Podzolic and Alluvial soils), and on the plains of Quelpart Island (Reddish-Brown Latosols and Ando soils).

The clayey Alluvial soils are the most productive and most intensively cultivated soils in the Republic of Korea. They consist of fine-grained sediments that cover the heavily populated plains along the coasts and the broader inland valleys. Each year new areas of these important soils are reclaimed from the extensive mud flats along the western coast, but these small areas do not add significantly to the total cultivated area. The clayey Alluvial soils have a clay loam surface layer and an equally heavy subsoil; but areas shown on the map as clayey Alluvial soils include silt loams and loams with correspondingly lighter textured subsoils. All these soils are now fertile because they receive nearly all of the manure and other organic amendments available on the populated plains. They are flooded during the warm months for growing rice and drained the rest of the year for winter grains. This double-cropping system is commonly practiced on most of the Alluvial soils in the Republic of Korea.

Less productive or extensive than the clayey Alluvial soils, the sandy Alluvial soils occur on coarse sediments covering many of the flat-bottomed interior valleys. Most of these soils have a sandy loam surface layer, but in some places they are gravelly at the surface as well as in the subsoil. Although they are generally well-drained, permeable, and easy to work, their low moisture-holding capacity makes crops susceptible to drought. These soils are generally strongly acid, and low in organic matter and plant nutrients. High fertility levels can be maintained on these highly porous soils only by frequent applications of fertilizer, lime, and organic material. The sandy Alluvial soils are devoted chiefly to dryland crops such as millet, wheat, barley, potatoes, and soybeans.

In a few of the valleys, e.g., near Pusan and Seoul, rice is grown exclusively because the water table is high and water is available for irrigation. In other places, these soils are not generally cultivated because they are subject to destructive floods; with proper flood control, much of their small area could be brought into production.

Podzolic soils occur in a discontinuous belt along the coast of the Republic of Korea, but include many areas of Alluvial soils (usually rice paddies) too small to show on the

Soil Map. Some are like the sandy Alluvial soils and some are like the clayey Alluvial soils; nearly all are used intensively for agriculture, particularly for rice production, and are highly important to the economy of the country.

The Podzolic soils, typical of rolling and hilly uplands, are leached but are variable in color, texture, depth, and natural fertility, depending on parent rock and climate. Along the southern and western coasts, where the temperatures and rainfall are higher, the soils are pale red and yellow; here they are more deeply weathered and leached than the brown Podzolic soils of the interior and north. Sandy textures predominate over granite, the most extensive rock type. Clayey and loamy textures are common over other rocks, chiefly shales, shists, and limestones. In general, Podzolic soils are low in bases, moderately acid, and low in organic matter. In the rolling areas the soils are commonly terraced and used chiefly for wheat, millet, soybeans, and potatoes. Most of the hilly areas are covered with brush that serves for pasture and fuel.

The extensive Lithosols on hilly and mountainous terrain are shallow and stony. Only small scattered patches are cultivated. For the most part, these soils now support brush, but could support forests under proper management.

Reddish-Brown Latosols and Ando soils occur only on Quelpart Island. These soils cover the plains that slope gently from the central volcanic cone. This island, composed entirely of volcanic rock, is milder and more humid than the mainland. The Reddish-Brown Latosols, the more extensive soil of the association, are moderately deep, reddish silt loams, which, in many places are excessively stony. Except in a small area in the south, rice is not grown on these soils, primarily because of the limited water supply. Even with adequate water, the subsoil and underlying rock are generally too porous for wetland rice. The Reddish-Brown Latosols are only slightly acid and moderately supplied with some nutrients, but they are low in organic matter. Cultivation is restricted chiefly to the coastal belt; small grains, sweet potatoes, and soybeans are the chief crops. Farther inland, livestock grazing is the chief use of these soils. They are not highly productive, however, because they are generally stony and have low moisture retention.

Of higher productivity are the Ando soils, which cover the northeastern part of Quelpart Island. These soils have a black loamy surface layer that grades into silty volcanic ash with depth. Little is known about these soils, but the few existing reports indicate they are similar to the highly productive Ando soils of nearby Japan.

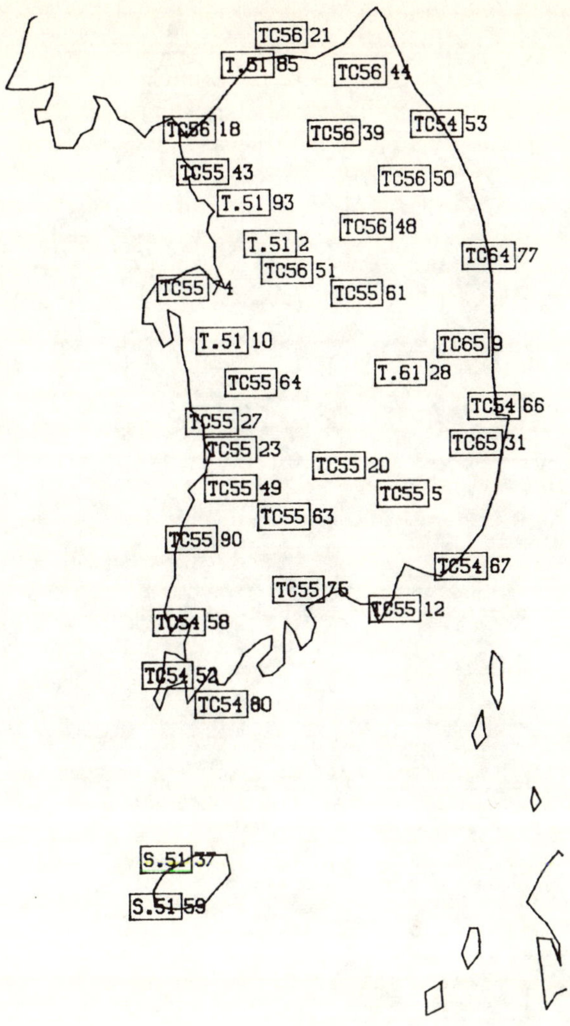

Selected weather stations for South Korea.

SOUTH KOREA VEGETATION

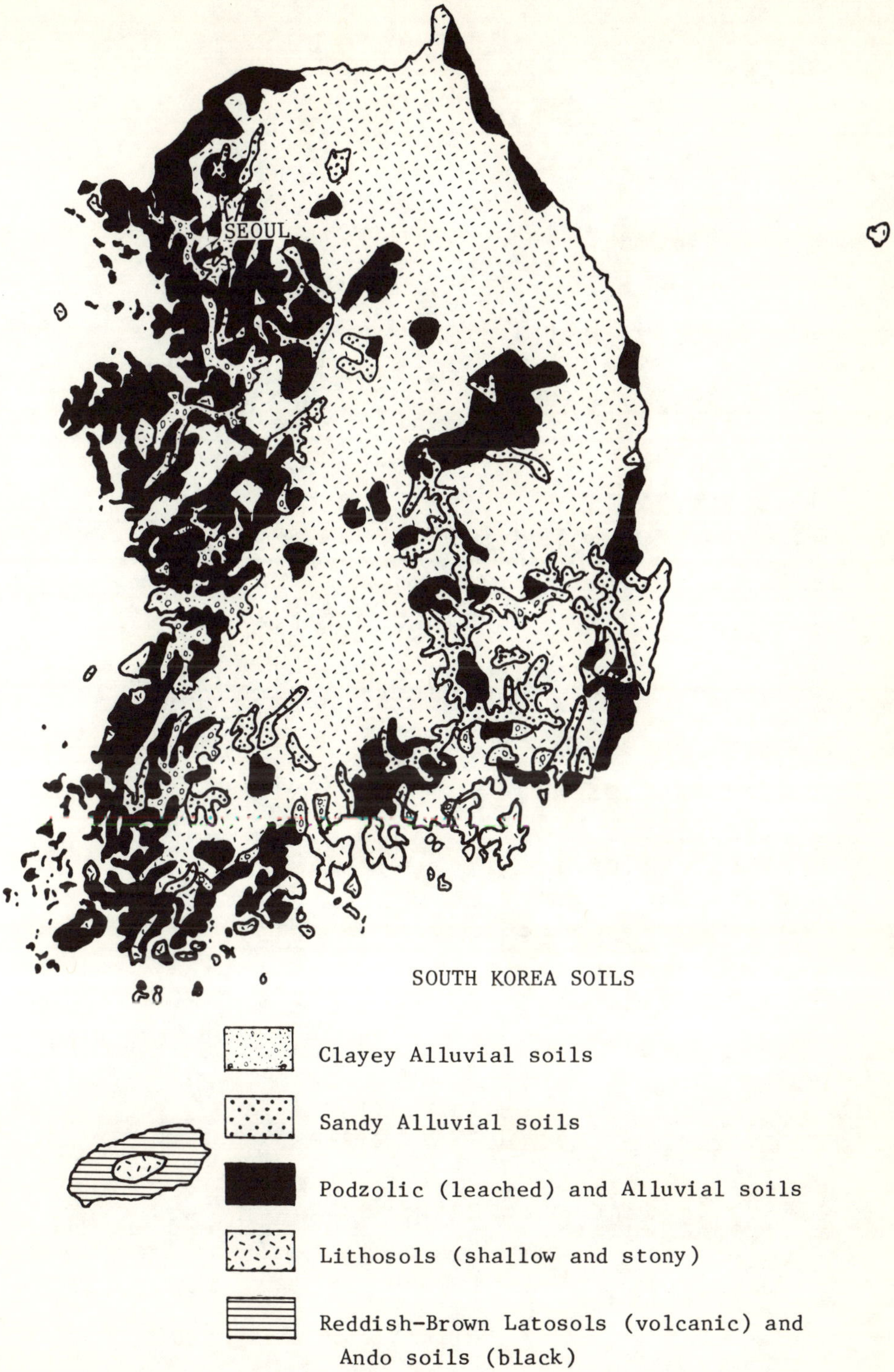

SOUTH KOREA SOILS

CLIMATOLOGICAL DATA FOR KOREA(SOUTH)

GEOGRAPHY

OBS.	LOCATION	YEARS	LATITUDE	LONGITUDE	ELEVATION -M-	LIFEZONE	CCTX	ID	ONMAP
1	ANDONG	30	36.57 N	128.73 E	.	.	.	1	NO
2	ANSEONG	30	37.03 N	127.27 E	.	.	.	2	YES
3	BOEUN	30	36.47 N	127.70 E	.	WM	TC55	3	NO
4	BOSEONG	30	34.75 N	127.08 E	.	WM	TC55	4	NO
5	CHANGNYEONG	30	35.53 N	128.50 E	.	WM	TC55	5	YES
6	CHEJU	30	33.52 N	126.53 E	.	WM	SC54	6	NO
7	CHENGJU	30	36.97 N	127.97 E	.	.	.	7	NO
8	CHEOLWEON	30	38.23 N	127.20 E	.	CW	TC56	8	NO
9	CHEONGSONG	30	36.42 N	129.10 E	.	CM	TC65	9	YES
10	CHEONGYANG	30	36.45 N	126.80 E	.	.	.	10	YES
11	CHONJU	30	35.82 N	127.15 E	.	WM	TC55	11	NO
12	CHUNGMU	30	34.85 N	128.40 E	.	WM	TC55	12	YES

AVERAGE PRECIPITATION (MM)

OBS.	JAN.	FEB.	MAR.	APR.	MAY	JUN.	JUL.	AUG.	SEP.	OCT.	NOV.	DEC.	ANNUAL
1	17	23	34	70	84	132	218	151	111	36	26	21	924
2	12	25	38	73	74	131	297	189	127	33	44	33	1077
3	23	26	40	85	84	158	305	205	125	43	44	30	1167
4	24	41	69	128	141	203	254	267	174	64	43	31	1442
5	21	29	39	75	84	157	204	189	151	52	30	19	1053
6	59	76	73	82	89	158	210	227	250	88	69	60	1440
7	13	29	42	74	87	122	269	191	129	43	37	20	1050
8	14	18	83	70	99	127	404	294	133	57	40	29	1321
9	18	22	32	64	67	114	186	146	111	33	25	22	840
10	16	26	49	82	87	145	264	255	146	41	53	32	1196
11	27	33	61	76	85	155	280	240	156	52	42	36	1241
12	34	34	68	119	154	194	275	159	167	71	47	36	1358

AVERAGE TEMPERATURE (DEGREES CENTIGRADE)

OBS.	JAN.	FEB.	MAR.	APR.	MAY	JUN.	JUL.	AUG.	SEP.	OCT.	NOV.	DEC.	ANNUAL
1	.	.	.	.	.	.	.	.	.	.	.	.	.
2	.	.	.	.	.	.	.	.	.	.	.	.	.
3	-4.1	-2.0	3.8	10.8	16.4	21.2	25.3	25.5	19.7	12.6	6.0	-0.6	11.2
4	-0.5	0.9	5.6	11.3	16.6	20.6	24.8	25.7	20.7	14.3	8.1	2.5	12.6
5	-1.1	1.0	6.0	12.6	18.0	22.2	26.0	26.3	21.2	15.2	8.3	2.0	13.1
6	4.8	5.2	8.0	12.3	16.2	20.4	25.1	25.8	21.7	16.8	12.1	7.6	14.7
7	.	.	.	.	.	.	.	.	.	.	.	.	.
8	-7.7	-4.7	2.1	9.4	15.4	20.2	24.0	24.4	18.8	11.6	4.3	-3.6	9.5
9	-4.2	-1.5	3.7	10.5	16.0	20.3	24.5	24.9	19.1	12.6	6.0	-0.5	11.0
10	.	.	.	.	.	.	.	.	.	.	.	.	.
11	-1.7	0.2	5.0	11.3	16.8	21.3	25.7	25.9	20.6	13.9	7.8	1.7	12.4
12	2.1	3.7	7.9	13.3	18.0	21.6	25.4	27.3	23.0	17.7	11.2	5.3	14.7

CLIMATOLOGICAL DATA FOR KOREA(SOUTH)

GEOGRAPHY

OBS.	LOCATION	YEARS	LATITUDE	LONGITUDE	ELEVATION -M-	LIFEZONE	CCTX	ID	ONMAP
13	CHUPUNGNYONG	30	36.22 N	128.00 E	.	WM	TC55	13	NO
14	DAECHON	30	36.33 N	126.60 E	.	WM	TC55	14	NO
15	DANJIN	30	36.90 N	126.63 E	.	WM	TC55	15	NO
16	DANYANG	30	36.93 N	128.33 E	.	WM	TC55	16	NO
17	EUIJEONGBU	30	37.73 N	127.05 E	.	.	.	17	NO
18	GANGHWA	30	37.73 N	126.50 E	.	WM	TC56	18	YES
19	GAPYEONG	30	37.85 N	127.47 E	.	WM	TC56	19	NO
20	GEOCHANG	30	35.70 N	127.90 E	.	WM	TC55	20	YES
21	GEUMHWA	30	38.30 N	127.40 E	.	CW	TC56	21	YES
22	GEUMSAN	30	36.12 N	127.45 E	.	WM	TC55	22	NO
23	GIMJE	30	35.80 N	126.87 E	.	WM	TC55	23	YES
24	GOCHANG	30	35.45 N	126.67 E	.	WM	TC55	24	NO

AVERAGE PRECIPITATION (MM)

OBS.	JAN.	FEB.	MAR.	APR.	MAY	JUN.	JUL.	AUG.	SEP.	OCT.	NOV.	DEC.	ANNUAL
13	25	30	57	72	75	167	268	191	155	40	37	30	1147
14	19	22	33	69	89	141	295	201	138	43	45	33	1129
15	17	18	39	76	81	109	325	243	162	42	41	27	1181
16	24	28	39	76	83	124	295	194	116	35	30	29	1073
17	21	21	36	78	104	134	410	273	143	43	43	28	1332
18	16	22	34	68	86	104	349	230	123	50	41	25	1143
19	14	21	42	81	96	145	411	280	145	42	39	23	1336
20	21	27	37	80	82	173	232	211	147	43	37	25	1114
21	15	23	34	61	95	116	390	264	109	57	41	25	1229
22	22	28	44	77	83	174	275	235	137	43	39	32	1187
23	25	26	36	77	84	143	235	197	129	47	44	29	1070
24	34	34	39	74	75	141	240	208	132	46	45	42	1107

AVERAGE TEMPERATURE (DEGREES CENTIGRADE)

OBS.	JAN.	FEB.	MAR.	APR.	MAY	JUN.	JUL.	AUG.	SEP.	OCT.	NOV.	DEC.	ANNUAL
13	-3.1	-0.7	4.5	11.0	16.7	20.6	24.5	24.7	19.4	13.0	6.7	0.2	11.5
14	-2.2	-0.2	4.5	10.6	16.1	20.8	25.2	26.2	20.8	14.5	7.9	1.7	12.1
15	-3.6	-1.2	4.2	10.4	15.7	20.7	24.9	25.6	20.2	13.4	6.4	0.1	11.4
16	-4.7	-1.6	4.6	11.3	17.3	21.8	25.6	26.1	20.3	13.2	6.1	-0.7	11.6
17	.	.	.	.	.	.	.	.	.	.	.	.	.
18	-4.7	-2.0	3.7	10.5	16.2	20.7	24.8	25.9	20.4	14.3	6.8	-1.6	11.2
19	-6.6	-3.1	3.8	10.5	16.2	21.0	24.7	25.3	19.4	12.5	5.1	-2.6	10.5
20	-2.5	-0.2	4.9	11.4	16.9	21.6	25.9	26.1	20.4	13.5	7.0	0.8	12.2
21	-7.7	-5.1	2.0	9.3	15.4	20.0	24.2	24.8	18.8	11.5	4.3	-4.0	9.4
22	-3.6	-1.1	4.2	11.2	17.0	21.6	25.8	26.2	20.3	12.9	6.3	-0.6	11.6
23	-1.6	0.1	5.1	11.3	17.0	21.7	26.1	26.7	21.2	13.2	7.8	1.9	14.0
24	-1.7	0.5	5.5	11.6	16.9	21.6	25.9	26.5	21.0	14.7	8.3	2.2	12.8

CLIMATOLOGICAL DATA FOR KOREA(SOUTH)

GEOGRAPHY

OBS.	LOCATION	YEARS	LATITUDE	LONGITUDE	ELEVATION -M-	LIFEZONE	CCTX	ID	ONMAP
25	GOHEUNG	30	34.60 N	127.30 E	.	WM	TC55	25	NO
26	GONGJU	30	36.47 N	127.13 E	.	WM	TC55	26	NO
27	GUNSAN	30	35.97 N	126.70 E	.	WM	TC55	27	YES
28	GUNWI	30	36.25 N	128.50 E	.	.	.	28	YES
29	GURYE	30	35.23 N	127.45 E	.	WM	TC55	29	NO
30	GWANGYANG	30	35.97 N	127.60 E	.	WM	TC55	30	NO
31	GYEONGJU	30	35.83 N	129.20 E	.	WD	TC65	31	YES
32	HABCHEON	30	35.57 N	128.17 E	.	WM	TC55	32	NO
33	HADONG	30	35.07 N	127.73 E	.	WM	TC45	33	NO
34	HAENAM	30	34.57 N	126.60 E	.	WM	TC54	34	NO
35	HAMAN	30	35.23 N	128.43 E	.	.	.	35	NO
36	HAMYANG	30	35.53 N	127.70 E	.	WM	TC55	36	NO

AVERAGE PRECIPITATION (MM)

OBS.	JAN.	FEB.	MAR.	APR.	MAY	JUN.	JUL.	AUG.	SEP.	OCT.	NOV.	DEC.	ANNUAL
25	28	31	60	117	130	191	241	212	162	66	37	32	1308
26	27	30	43	95	92	173	332	260	155	49	44	32	1330
27	23	22	34	78	77	154	228	202	143	47	48	34	1089
28	14	27	39	66	68	142	175	198	131	29	33	25	947
29	22	27	46	87	87	176	273	229	156	52	36	27	1315
30	16	26	63	101	130	204	213	241	179	58	37	29	1296
31	23	34	43	65	75	133	178	185	154	50	34	24	998
32	22	29	41	80	81	164	242	205	155	45	25	20	1110
33	29	36	65	135	149	226	294	270	196	62	42	34	1539
34	27	34	50	106	99	158	199	202	161	52	43	39	1133
35	16	28	57	94	118	157	234	236	214	75	27	30	1286
36	25	33	48	79	83	168	253	219	153	47	41	30	1179

AVERAGE TEMPERATURE (DEGREES CENTIGRADE)

OBS.	JAN.	FEB.	MAR.	APR.	MAY	JUN.	JUL.	AUG.	SEP.	OCT.	NOV.	DEC.	ANNUAL
25	1.5	2.4	6.8	12.3	17.4	21.1	25.0	26.3	21.9	15.9	9.7	4.1	13.7
26	-3.5	-0.6	4.6	11.3	17.2	22.0	25.9	26.2	20.4	13.6	6.8	0.1	12.0
27	-1.6	0.2	4.8	11.0	16.5	21.2	25.8	26.5	21.2	14.7	8.1	1.6	12.4
28	.	.	.	.	.	.	.	.	.	.	.	.	.
29	-0.2	1.4	6.2	12.4	18.2	22.4	26.3	26.7	21.6	14.8	8.6	2.5	13.4
30	0.1	2.1	7.3	11.9	17.3	21.9	26.7	27.8	21.9	14.9	9.6	3.0	13.7
31	-0.1	1.5	6.5	12.6	17.7	21.6	25.8	26.6	21.1	15.2	8.9	2.6	13.3
32	-1.1	1.0	6.0	12.5	18.0	22.8	26.6	27.1	21.5	14.6	8.6	2.1	13.3
33	0.9	2.6	7.1	12.8	17.7	21.6	25.6	26.6	21.8	15.8	9.6	3.7	13.8
34	0.9	2.1	6.0	11.8	16.7	21.1	25.5	26.5	21.9	15.8	9.7	3.9	13.5
35	.	.	.	.	.	.	.	.	.	.	.	.	.
36	-1.2	0.9	5.3	11.8	17.4	21.6	25.8	26.2	20.6	13.8	7.5	2.0	12.6

CLIMATOLOGICAL DATA FOR KOREA(SOUTH)

GEOGRAPHY

OBS.	LOCATION	YEARS	LATITUDE	LONGITUDE	ELEVATION -M-	LIFEZONE	CCTX	ID	ONMAP
37	HANRIM	30	33.40 N	126.25 E	.	.	.	37	YES
38	HOENGSEONG	30	37.48 N	127.95 E	.	WM	TC56	38	NO
39	HONGCHEON	30	37.70 N	127.90 E	.	WM	TC56	39	YES
40	HONGSEONG	30	36.60 N	126.67 E	.	WM	TC55	40	NO
41	HWACHEON	30	38.10 N	127.70 E	.	WM	TC56	41	NO
42	ICHEON	30	37.28 N	127.45 E	.	.	.	42	NO
43	INCHON	30	37.47 N	126.63 E	.	WM	TC55	43	YES
44	INJE	30	38.07 N	128.17 E	.	WM	TC56	44	YES
45	IRI	30	35.93 N	126.93 E	.	WM	TC55	45	NO
46	JANGHOWEON	30	37.13 N	127.65 E	.	.	.	46	NO
47	JANGSEONG	30	35.33 N	126.80 E	.	WM	TC55	47	NO
48	JECHEON	30	37.13 N	128.20 E	.	CW	TC56	48	YES

AVERAGE PRECIPITATION (MM)

OBS.	JAN.	FEB.	MAR.	APR.	MAY	JUN.	JUL.	AUG.	SEP.	OCT.	NOV.	DEC.	ANNUAL
37	35	46	71	72	98	136	171	183	193	88	70	43	1205
38	19	17	31	71	82	125	358	226	126	32	31	23	1140
39	18	22	31	66	86	113	364	282	115	34	37	24	1142
40	21	24	36	78	82	131	307	217	147	47	45	33	1167
41	12	25	26	56	79	111	336	228	101	36	31	24	1065
42	22	26	41	83	87	135	378	217	123	36	42	28	1219
43	16	18	50	66	73	139	304	180	137	45	35	30	1093
44	18	18	28	65	84	107	291	220	106	35	33	18	1021
45	28	27	59	77	82	163	253	204	157	48	41	34	1171
46	16	29	33	68	62	108	311	178	102	31	34	24	997
47	38	38	46	96	90	181	281	229	148	59	48	44	1296
48	19	21	33	71	79	130	339	191	113	35	34	28	1094

AVERAGE TEMPERATURE (DEGREES CENTIGRADE)

OBS.	JAN.	FEB.	MAR.	APR.	MAY	JUN.	JUL.	AUG.	SEP.	OCT.	NOV.	DEC.	ANNUAL
37	.	.	.	.	.	.	.	.	.	.	.	.	.
38	-6.1	-3.2	3.5	10.6	16.6	21.1	25.0	25.5	19.6	12.6	5.4	-2.0	10.7
39	-7.1	-4.6	4.5	11.7	16.5	21.3	25.4	26.0	20.2	14.2	6.8	-1.6	11.1
40	-3.6	-1.1	4.5	10.9	16.7	21.4	25.5	26.1	20.2	13.4	6.7	0.2	11.7
41	-7.2	-3.7	5.5	11.0	16.7	21.3	25.9	26.4	20.3	14.0	6.3	-2.1	11.2
42	.	.	.	.	.	.	.	.	.	.	.	.	.
43	-4.0	-1.6	3.4	9.7	15.3	19.5	23.9	25.1	20.6	14.2	7.2	-0.2	11.1
44	-7.7	-4.0	4.2	9.6	16.8	21.0	24.9	25.6	19.5	13.4	5.4	-2.2	10.5
45	-1.7	0.2	5.1	11.3	17.1	21.5	26.1	26.5	21.0	14.3	8.1	1.7	12.6
46	.	.	.	.	.	.	.	.	.	.	.	.	.
47	-1.5	0.3	5.2	11.0	17.0	21.2	26.3	26.3	20.8	14.1	8.3	1.8	12.6
48	-6.2	-3.6	2.5	9.6	15.8	20.7	24.7	24.9	19.2	11.7	4.7	-2.7	10.1

CLIMATOLOGICAL DATA FOR KOREA(SOUTH)

GEOGRAPHY

OBS.	LOCATION	YEARS	LATITUDE	LONGITUDE	ELEVATION -M-	LIFEZONE	CCTX	ID	ONMAP
49	JEONGEUB	30	35.57 N	126.87 E	.	WM	TC55	49	YES
50	JEONGSEON	30	37.42 N	128.58 E	.	CW	TC56	50	YES
51	JINCHEON	30	36.87 N	127.43 E	.	WM	TC56	51	YES
52	JINDO	30	34.47 N	126.27 E	.	WM	TC54	52	YES
53	JINJU	30	35.20 N	128.07 E	.	WM	TC55	95	NO
54	KANGNUNG	30	37.75 N	128.90 E	.	WM	TC54	53	YES
55	KIMCHEON	30	36.12 N	128.10 E	.	WD	TC65	54	NO
56	KWANGJU	30	35.15 N	126.90 E	.	WM	TC55	55	NO
57	MASAN	30	35.17 N	128.57 E	.	WM	TC55	56	NO
58	MIRYANG	30	35.50 N	128.77 E	.	WM	TC55	57	NO
59	MOKPO	30	34.78 N	126.38 E	.	WM	TC54	58	YES
60	MORADO	30	33.13 N	126.15 E	.	.	.	59	YES

AVERAGE PRECIPITATION (MM)

OBS.	JAN.	FEB.	MAR.	APR.	MAY	JUN.	JUL.	AUG.	SEP.	OCT.	NOV.	DEC.	ANNUAL
49	32	31	40	76	73	129	256	243	147	54	49	45	1173
50	15	21	36	76	76	130	291	174	121	33	35	23	1029
51	19	22	35	79	91	136	367	210	132	40	37	33	1199
52	34	37	58	100	102	132	183	189	145	58	46	38	1121
53	20	41	73	174	130	180	243	205	203	61	37	33	1345
54	37	73	73	70	64	135	212	191	198	88	88	53	1282
55	15	24	36	69	77	147	223	191	122	39	28	23	995
56	32	34	69	82	92	169	223	201	190	52	43	37	1223
57	26	28	60	130	141	210	287	213	180	70	37	28	1410
58	20	27	47	88	103	173	228	181	142	55	29	23	1115
59	37	40	58	83	102	136	183	188	156	55	44	43	1126
60	41	47	66	106	113	170	173	144	122	64	56	71	1173

AVERAGE TEMPERATURE (DEGREES CENTIGRADE)

OBS.	JAN.	FEB.	MAR.	APR.	MAY	JUN.	JUL.	AUG.	SEP.	OCT.	NOV.	DEC.	ANNUAL
49	-1.6	0.1	5.4	11.7	16.9	21.7	26.3	26.5	21.0	14.5	8.3	1.8	12.9
50	-5.7	-3.2	3.0	10.2	15.8	20.3	24.3	24.6	18.7	12.0	5.3	-2.1	10.3
51	-5.0	-1.6	4.1	11.0	16.7	21.3	25.4	25.8	20.1	12.9	5.9	-1.1	11.3
52	1.7	2.8	6.2	11.8	16.8	20.9	25.3	26.4	21.8	16.2	10.5	4.9	13.8
53	0.2	2.1	6.5	12.4	17.7	21.8	25.8	26.7	21.6	15.2	9.0	3.2	13.5
54	-1.0	0.3	4.7	11.5	16.7	19.7	23.5	24.3	19.7	14.4	8.8	2.4	12.1
55	-2.2	0.3	5.7	12.4	18.2	22.8	26.6	26.9	21.0	14.4	7.5	1.2	12.9
56	-0.6	1.1	5.7	11.4	16.8	21.4	25.6	26.1	20.9	14.0	8.2	2.4	12.8
57	1.2	2.9	7.3	12.8	17.7	21.1	25.2	26.4	22.1	16.6	10.0	4.5	14.0
58	-0.7	1.8	6.0	12.6	18.1	22.7	26.8	27.0	22.3	15.4	9.3	2.8	13.7
59	1.0	2.1	5.9	11.5	16.5	20.6	24.8	26.1	21.7	16.1	10.3	4.3	13.4
60	.	.	.	.	.	.	.	.	.	.	.	.	.

CLIMATOLOGICAL DATA FOR KOREA(SOUTH)

GEOGRAPHY

OBS.	LOCATION	YEARS	LATITUDE	LONGITUDE	ELEVATION -M-	LIFEZONE	CCTX	ID	ONMAP
61	MUJU	30	36.00 N	127.67 E	.	WM	TC55	60	NO
62	MUNGYEONG	30	36.73 N	128.10 E	.	WM	TC55	61	YES
63	NAJU	30	35.03 N	126.73 E	.	WM	TC55	62	NO
64	NAMWEON	30	35.40 N	127.37 E	.	WM	TC55	63	YES
65	NONSAN	30	36.20 N	127.07 E	.	WM	TC55	64	YES
66	ONYANG	30	36.77 N	127.03 E	.	WM	TC55	65	NO
67	POHANG	30	36.05 N	129.38 E	.	WM	TC54	66	YES
68	PUSAN	30	35.10 N	129.03 E	.	WM	TC54	67	YES
69	PYEONGCHANG	30	37.37 N	128.40 E	.	CW	TC56	68	NO
70	PYEONGTAEG	30	37.00 N	127.08 E	.	WM	TC55	69	NO
71	SACHEONPO	30	34.93 N	128.07 E	.	.	.	70	NO
72	SEOGIPO	30	33.23 N	126.57 E	.	.	.	71	NO

AVERAGE PRECIPITATION (MM)

OBS.	JAN.	FEB.	MAR.	APR.	MAY	JUN.	JUL.	AUG.	SEP.	OCT.	NOV.	DEC.	ANNUAL
61	24	29	40	77	76	160	245	202	136	42	41	33	1105
62	22	26	41	86	92	156	315	214	134	39	39	27	1189
63	37	31	44	88	100	188	238	207	157	63	47	45	1245
64	24	25	41	84	82	174	277	243	138	50	42	29	1212
65	21	20	35	80	80	158	280	242	134	44	38	30	1161
66	23	24	39	72	76	120	303	215	145	37	46	34	1133
67	30	41	57	67	75	139	158	134	173	59	60	36	1028
68	25	44	89	114	139	198	248	165	205	73	44	39	1382
69	20	23	37	73	82	129	321	185	120	30	36	26	1082
70	21	18	38	64	77	115	283	207	122	44	39	26	1049
71	17	24	76	85	135	168	210	168	163	94	38	27	1204
72	32	59	146	151	232	262	225	197	260	98	120	36	1817

AVERAGE TEMPERATURE (DEGREES CENTIGRADE)

OBS.	JAN.	FEB.	MAR.	APR.	MAY	JUN.	JUL.	AUG.	SEP.	OCT.	NOV.	DEC.	ANNUAL
61	-4.0	-1.5	4.3	10.7	16.8	21.4	25.6	25.5	19.4	12.8	6.1	0.3	11.5
62	-4.0	-1.2	4.3	10.9	17.0	21.3	25.3	25.4	19.8	13.3	6.0	-0.1	11.5
63	-0.5	1.1	5.9	11.7	17.4	21.8	26.2	26.9	21.7	15.0	8.8	2.7	13.2
64	-1.7	0.5	5.3	12.1	17.6	22.0	26.3	26.6	21.2	14.4	7.7	1.8	12.8
65	-2.6	-0.2	5.3	11.7	17.5	22.3	26.2	26.9	21.1	14.1	10.8	1.1	12.8
66	-4.0	-1.2	4.6	11.0	17.0	21.7	25.0	26.3	20.1	13.5	6.8	-0.1	11.8
67	0.6	2.2	6.1	12.1	16.8	20.1	23.6	25.0	20.6	15.2	9.9	3.4	13.0
68	1.8	3.5	7.3	12.5	16.7	19.8	23.7	25.4	21.6	16.6	11.1	5.0	13.8
69	-6.1	-3.1	3.1	10.3	16.2	20.4	24.2	24.5	18.9	12.2	5.1	-2.1	10.3
70	-4.5	-2.0	3.8	10.8	17.0	21.7	26.0	26.4	20.5	13.5	6.4	-0.5	11.6
71	.	.	.	.	.	.	.	.	.	.	.	.	.
72	.	.	.	.	.	.	.	.	.	.	.	.	.

CLIMATOLOGICAL DATA FOR KOREA(SOUTH)

GEOGRAPHY

OBS.	LOCATION	YEARS	LATITUDE	LONGITUDE	ELEVATION -M-	LIFEZONE	CCTX	ID	ONMAP
73	SEONGHWAN	30	36.60 N	127.28 E	.	WM	TC56	72	NO
74	SEONSAN	30	36.23 N	128.30 E	.	WD	TC65	73	NO
75	SEOSAN	30	36.77 N	126.43 E	.	WM	TC55	74	YES
76	SEOUL	30	37.57 N	126.97 E	.	WM	TC56	94	NO
77	SUNCHEON	30	34.97 N	127.50 E	.	WM	TC55	75	YES
78	SUWON	30	37.28 N	127.02 E	.	.	.	93	YES
79	TAEGU	30	35.88 N	128.62 E	.	WD	TC65	76	NO
80	ULJIN	30	36.95 N	129.37 E	.	WD	TC64	77	YES
81	ULLUNG-DO	30	37.48 N	130.90 E	.	WM	TC53	78	NO
82	ULSAN	30	35.55 N	129.32 E	.	WM	TC54	79	NO
83	WANDO	30	34.30 N	126.77 E	.	WM	TC54	80	YES
84	WEONJU	30	37.33 N	127.93 E	.	WM	TC56	81	NO

AVERAGE PRECIPITATION (MM)

OBS.	JAN.	FEB.	MAR.	APR.	MAY	JUN.	JUL.	AUG.	SEP.	OCT.	NOV.	DEC.	ANNUAL
73	19	23	30	64	72	110	321	198	124	34	35	27	1055
74	17	24	29	68	75	151	226	165	108	31	27	23	942
75	18	25	32	77	80	123	298	226	154	38	48	36	1155
76	17	21	56	68	86	169	358	224	142	49	36	32	1259
77	32	34	58	126	139	220	270	276	187	57	41	29	1469
78	21	19	54	72	82	169	334	219	157	50	38	34	1247
79	16	27	46	64	67	133	200	166	162	44	30	25	979
80	34	44	45	61	60	90	176	164	154	66	50	32	975
81	177	107	89	80	70	129	146	98	190	112	121	166	1485
82	24	46	68	88	106	154	204	167	209	65	46	40	1218
83	33	27	59	111	118	173	204	175	166	64	47	31	1207
84	18	20	37	75	84	130	377	216	115	37	38	24	1171

AVERAGE TEMPERATURE (DEGREES CENTIGRADE)

OBS.	JAN.	FEB.	MAR.	APR.	MAY	JUN.	JUL.	AUG.	SEP.	OCT.	NOV.	DEC.	ANNUAL
73	-4.7	-1.6	3.8	10.8	16.6	21.1	25.2	25.8	20.1	13.4	6.3	-0.7	11.3
74	-2.2	0.3	5.4	12.8	18.2	22.6	26.6	26.8	21.2	14.7	7.4	1.2	12.9
75	-2.7	-0.7	4.2	10.5	16.1	20.4	24.7	25.6	20.6	14.5	7.7	0.9	11.8
76	-4.7	-1.7	3.6	10.5	16.3	20.8	24.5	25.4	20.3	13.4	6.3	-1.2	11.1
77	1.1	2.8	7.0	13.0	18.3	22.4	26.1	27.1	22.2	16.6	10.1	4.0	13.9
78	.	.	.	.	.	.	.	.	.	.	.	.	.
79	-1.6	0.6	5.7	12.1	17.6	21.6	25.3	25.9	20.5	14.2	7.8	1.4	12.6
80	-0.1	1.4	6.1	11.5	16.5	19.9	24.5	25.5	20.5	15.1	9.3	3.4	12.8
81	0.6	1.1	4.5	10.2	15.0	18.4	22.1	23.9	19.9	15.0	9.7	3.9	12.0
82	0.4	2.1	6.0	11.5	16.3	20.0	24.3	25.1	20.5	14.8	9.0	3.2	12.8
83	2.2	2.8	6.7	12.0	16.8	20.5	24.8	26.3	22.3	16.9	10.9	5.0	13.9
84	-6.2	-2.7	3.5	10.6	16.7	21.3	25.5	25.9	19.7	12.5	5.2	-2.2	10.8

CLIMATOLOGICAL DATA FOR KOREA(SOUTH)

GEOGRAPHY

OBS.	LOCATION	YEARS	LATITUDE	LONGITUDE	ELEVATION -M-	LIFEZONE	CCTX	ID	ONMAP
85	YANGPYEONG	30	37.50 N	127.50 E	.	WM	TC56	82	NO
86	YANGYANG	30	38.07 N	128.63 E	.	WM	TC55	83	NO
87	YEOJU	30	37.30 N	127.62 E	.	.	.	84	NO
88	YEONCHEON	30	38.12 N	127.07 E	.	.	.	85	YES
89	YEONGCHEON	30	35.92 N	128.92 E	.	WD	TC65	86	NO
90	YEONGDEOG	30	36.42 N	129.33 E	.	WD	TC65	87	NO
91	YEONGDONG	30	36.17 N	127.77 E	.	WM	TC55	88	NO
92	YEONGJU	30	36.83 N	128.58 E	.	WD	TC65	89	NO
93	YEONGKWANG	30	35.27 N	126.50 E	.	WM	TC55	90	YES
94	YEONGWEOL	30	37.17 N	128.47 E	.	WM	TC56	91	NO
95	YOSU	30	34.73 N	127.73 E	.	WM	TC55	92	NO

AVERAGE PRECIPITATION (MM)

OBS.	JAN.	FEB.	MAR.	APR.	MAY	JUN.	JUL.	AUG.	SEP.	OCT.	NOV.	DEC.	ANNUAL
85	27	17	26	70	88	128	406	266	121	33	37	21	1239
86	19	15	38	75	71	102	232	180	178	75	101	45	1132
87	23	21	34	77	85	122	375	223	118	31	40	24	1174
88	17	21	36	67	92	117	383	277	124	46	38	27	1245
89	17	21	37	63	77	118	184	149	122	38	28	20	874
90	22	45	51	61	69	104	179	175	141	46	38	38	968
91	19	26	38	68	81	132	233	193	119	37	38	29	1012
92	16	21	37	75	92	139	246	175	108	33	27	23	994
93	37	34	35	82	82	106	221	220	151	53	48	46	1174
94	17	12	31	70	78	118	294	189	104	34	27	26	1003
95	17	40	80	124	150	180	263	157	188	45	39	30	1314

AVERAGE TEMPERATURE (DEGREES CENTIGRADE)

OBS.	JAN.	FEB.	MAR.	APR.	MAY	JUN.	JUL.	AUG.	SEP.	OCT.	NOV.	DEC.	ANNUAL
85	-6.2	-2.7	3.7	10.9	16.6	21.6	25.5	26.0	20.2	12.5	5.1	-2.2	10.9
86	-2.2	0.3	6.1	11.8	17.5	21.4	24.4	24.3	19.6	14.4	8.7	1.3	12.3
87	.	.	.	.	.	.	.	.	.	.	.	.	.
88	.	.	.	.	.	.	.	.	.	.	.	.	.
89	-1.7	0.6	5.6	11.9	17.8	21.7	26.1	26.8	21.0	14.5	7.7	1.5	12.8
90	0.1	2.0	6.4	12.3	17.3	21.0	25.1	25.8	20.9	15.5	9.1	3.2	13.2
91	-3.5	-1.1	4.5	11.4	17.1	21.5	25.7	25.9	20.2	13.2	6.3	0.1	11.8
92	-4.1	-1.2	4.2	11.0	17.0	21.4	25.3	25.8	20.0	13.2	6.3	-0.6	11.6
93	-0.5	1.4	5.6	11.8	17.0	21.5	26.0	26.5	21.5	15.2	8.8	2.9	13.1
94	-5.7	-2.7	3.8	10.7	16.4	21.1	25.2	25.5	19.5	12.4	5.3	-1.7	10.8
95	1.5	3.0	6.8	12.3	16.8	20.2	23.9	25.8	21.9	16.6	10.9	4.5	13.7

LESOTHO*

Agriculture

Of 3,035,000 ha in total area, Lesotho was estimated in 1978 to have 310,000 ha of arable or permanent cropland, 2,000,000 ha in permanent pasture, and 725,000 ha in other terrain categories. Of the 1979 population, 84.4% or 583,000 individuals of the economically active population were engaged in agriculture. The total population in 1979 was 1,309,000. According to FAO statistics for 1979, there were 110,000 horses, 74,000 mules, 85,000 asses, 550,000 cattle, 80,000 pigs, 1,300,000 sheep, 730,000 goats, and 1,150,000 chickens.[2] Agricultural production is summarized in Table 1. Conventional and energetic yield data on most of these will be found in Appendixes I and II.

Climate

Data from Wernstedt[3] are presented in the Climatological Data Table (see Introduction for explanation). Based on the means for data points cited in Wernstedt, the mean annual precipitation is 7.3 dm (mean of 32 sites) and the mean annual temperature 14.2°C (mean of 7 sites). The averages would suggest that much of the climate is equivalent to Warm Temperate Dry Forest Life Zone.[4]

According to charts presented in Griffith,[9] Lesotho has more days with hail than any other part of southern Africa. However, much of Lesotho has an annual precipitation >1000 mm suggesting Warm Temperate Moist Forest Life Zone. There is thunder from 70 to >90 days, with the higher values to the northeast of Lesotho. In mean interdiurnal temperature, variation varies from 1.9 to >2.2 but <2.5°C. The annual frequency of cold spells (temperature fall >5°C) is 20 to 40. The higher numbers to the southeast, those with temperatures falling >10°C is <5 to <10, the higher numbers to the southeast. Most of Lesotho has absolute maximum temperatures at 35°C or less, but in the extreme northwest, and southwest, they may exceed 35°C. The lowest of daily maximum temperatures is 2.5°C or lower, except for the extreme northeast that may exceed 2.5°C slightly. The highest of daily minimum temperatures are mostly 20 to 22.5°C, the lower values to the southeast. There are usually 10 or fewer days with maximum temperatures >30°C, but to the west and northwest, there may be as many as 30. These are 60 to >90 days with the minimum temperature below 0°C, the greater number of frosty nights being at the edge of the great escarpment. For most of Lesotho, the frost period is 180 days or more. Average annual evaporation from Class A pans is 1500 to more than 2250 mm.[9]

Productivity

A very generalized map, extrapolated from Griffiths[9] and Wernstedt,[3] is presented as the vegetation map. The long pastoral uses of the lands of Lesotho suggest to me that pastoral degradation would lower the productivity of these lands, although with good management practices to upgrade the soils, we could expect productivities of 6 to 18 MT/ha/yr.

Economics

Food consumption per capita was 2245 cal/day. In 1970, the caloric supply was 95.0% of requirements. The total population growth rate was 2.2% in 1970. Population density was 36 persons per square kilometer in 1970, with 2.6% of the total population urban. Life expectancy at birth of those born in 1970 was 47 years. In 1977, with GNP at $250 per capita, agriculture accounted for 36.4% of GDP, manufacturing for 1.8%, other industry for 12.7%, and services for 49.0%.[10]

Table 1
AGRICULTURAL PRODUCTION IN LESOTHO

Crop	Area planted (ha)	Yield (kg/ha)	Total production (MT)
Wheat	38,000	798	30,000
Barley	1,000	308	
Corn	120,000	1,076	129,000
Oats	1,000	1,429	1,000
Sorghum	59,000	1,447	86,000
Dry beans	13,000	1,154	15,000
Dry peas	13,000	538	7,000

From FAO 1979 Production Yearbook, Vol. 33, Food and Agriculture Organization, Rome, 1980.

Energetics

Lesotho has a hydroelectric potential of 490 MW. Importing more than 75% of its commercial energy demands as oil, Lesotho is expected to be experiencing fuelwood problems by the year 2000.[10]

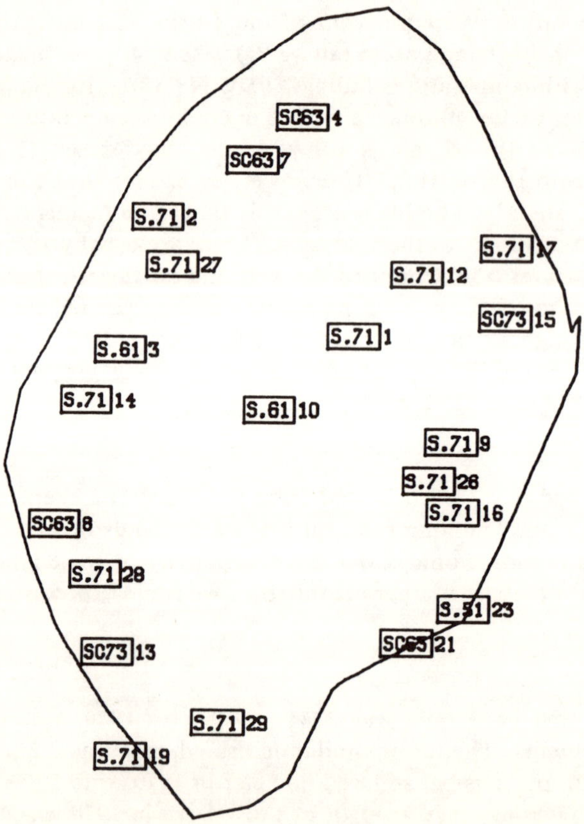

Selected weather stations for Lesotho.

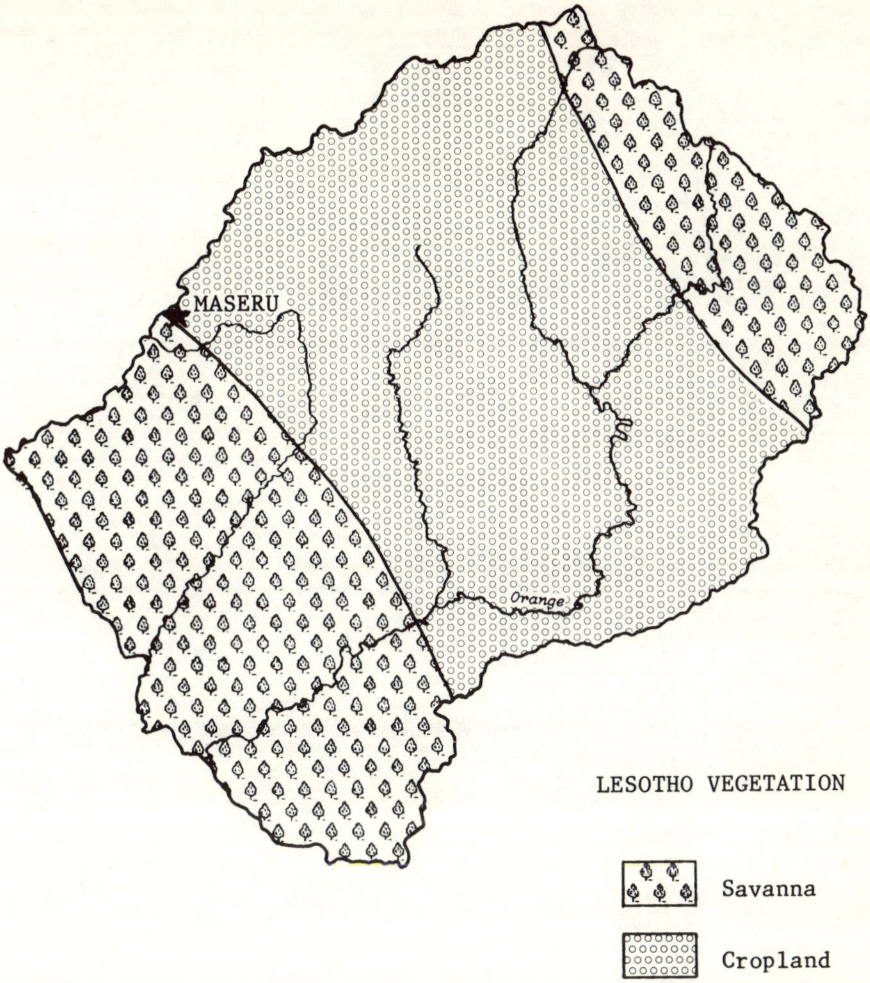

LESOTHO VEGETATION

CLIMATOLOGICAL DATA FOR LESOTHO

GEOGRAPHY

OBS.	LOCATION	YEARS	LATITUDE	LONGITUDE	ELEVATION -M-	LIFEZONE	CCTX	ID	ONMAP
1	BOKONG	22	29.33 S	28.47 E	1980.3	.	.	1	YES
2	BOLITHO	28	29.03 S	27.67 E	1554.5	.	.	2	YES
3	BOTSABELO	22	29.37 S	27.52 E	1645.3	.	.	3	YES
4	BUTHA BUTHE	42	28.77 S	28.25 E	1767.5	WD	SC63	4	YES
5	GIANT'S CASTLE	12	29.28 S	29.50 E	1980.3	.	.	5	NO
6	LELINGOANG	22	29.27 S	28.87 E	2011.4	.	.	6	NO
7	LERIBE	61	28.88 S	28.05 E	1736.4	WD	SC63	7	YES
8	MAFETENG	54	29.82 S	27.25 E	1614.5	WD	SC63	8	YES
9	MAKHEKA	18	29.60 S	28.87 E	2056.2	.	.	9	YES
10	MARAKABIES	17	29.52 S	28.13 E	1980.3	.	.	10	YES
11	MASERU	58	29.32 S	27.48 E	1527.4	WD	SC73	32	NO
12	MASHAI	24	29.68 S	28.80 E	1828.5	.	.	11	NO

AVERAGE PRECIPITATION (MM)

OBS.	JAN.	FEB.	MAR.	APR.	MAY	JUN.	JUL.	AUG.	SEP.	OCT.	NOV.	DEC.	ANNUAL
1	100	95	67	38	25	7	14	21	27	64	92	84	633
2	101	93	86	57	32	7	14	14	15	62	89	91	660
3	134	118	117	81	47	10	19	22	28	89	99	119	883
4	134	117	104	55	29	11	14	17	32	79	107	118	818
5	178	176	135	62	28	5	9	25	37	68	137	178	1038
6	98	89	58	30	24	6	10	13	21	53	63	71	537
7	143	122	118	57	34	12	13	16	31	74	106	118	843
8	118	119	104	64	36	14	14	18	33	58	78	101	755
9	114	92	81	36	25	5	10	17	26	71	84	93	655
10	124	121	92	71	49	16	23	27	44	95	99	115	874
11	100	95	94	55	26	12	13	15	20	57	79	81	646
12	94	79	63	25	29	3	8	12	21	50	67	92	543

AVERAGE TEMPERATURE (DEGREES CENTIGRADE)

OBS.	JAN.	FEB.	MAR.	APR.	MAY	JUN.	JUL.	AUG.	SEP.	OCT.	NOV.	DEC.	ANNUAL
1	.	.	.	.	.	.	.	.	.	.	.	.	.
2	.	.	.	.	.	.	.	.	.	.	.	.	.
3	.	.	.	.	.	.	.	.	.	.	.	.	.
4	19.2	19.1	17.4	14.4	10.1	6.9	7.1	9.6	13.1	16.2	17.9	19.4	14.3
5	.	.	.	.	.	.	.	.	.	.	.	.	.
6	.	.	.	.	.	.	.	.	.	.	.	.	.
7	20.2	19.2	17.7	14.4	10.5	7.3	7.1	9.9	13.6	16.9	18.3	19.8	14.6
8	20.4	19.5	17.5	14.4	10.9	7.9	7.4	10.3	13.2	16.4	17.6	19.9	14.6
9	.	.	.	.	.	.	.	.	.	.	.	.	.
10	.	.	.	.	.	.	.	.	.	.	.	.	.
11	21.1	20.5	18.6	14.9	10.7	7.7	7.7	10.6	14.1	17.1	18.7	20.7	15.2
12	.	.	.	.	.	.	.	.	.	.	.	.	.

CLIMATOLOGICAL DATA FOR LESOTHO

GEOGRAPHY

OBS.	LOCATION	YEARS	LATITUDE	LONGITUDE	ELEVATION -M-	LIFEZONE	CCTX	ID	ONMAP
13	MENO	16	29.17 S	28.72 E	2346.4	.	.	12	YES
14	MOHALE'S HOEK	63	30.15 S	27.47 E	1599.3	WD	SC73	13	YES
15	MOKHALINYANA	15	29.50 S	27.38 E	1553.6	.	.	14	YES
16	MOKHOTLONG	30	29.28 S	29.08 E	2374.1	CM	SC73	15	YES
17	MOKUINIKI	16	29.78 S	28.88 E	2285.4	.	.	16	YES
18	MOTHEBESOANE	17	29.10 S	29.08 E	2254.3	.	.	17	YES
19	MPHARANE	28	30.00 S	27.58 E	1751.4	.	.	18	NO
20	PALMIETFONTEIN	10	30.42 S	27.53 E	1325.6	.	.	19	YES
21	PONTMAIN	23	29.00 S	28.17 E	1828.5	.	.	20	NO
22	QACHA'S NEK	62	30.12 S	28.70 E	1971.4	WD	SC63	21	YES
23	QUTHING	53	30.42 S	27.72 E	1736.4	.	.	22	NO
24	RAMATSELISO	16	30.03 S	28.93 E	2193.3	.	.	23	YES

AVERAGE PRECIPITATION (MM)

OBS.	JAN.	FEB.	MAR.	APR.	MAY	JUN.	JUL.	AUG.	SEP.	OCT.	NOV.	DEC.	ANNUAL
13	119	104	82	43	40	11	16	15	29	76	97	109	740
14	107	106	107	56	39	13	15	16	30	58	78	99	723
15	84	113	101	50	28	12	10	12	19	44	56	74	601
16	97	85	64	34	23	5	10	14	22	58	78	85	575
17	113	86	71	36	27	9	7	14	23	61	70	88	605
18	101	96	71	49	23	9	14	18	29	60	77	105	650
19	101	115	112	65	36	10	15	19	31	72	84	111	770
20	123	100	94	64	34	20	12	14	41	56	62	97	717
21	150	142	105	76	45	12	14	18	40	124	114	151	991
22	163	157	124	50	28	14	11	20	42	69	103	146	928
23	117	104	112	67	40	19	15	22	36	70	81	116	798
24	193	174	124	55	29	13	11	16	39	77	132	159	1020

AVERAGE TEMPERATURE (DEGREES CENTIGRADE)

OBS.	JAN.	FEB.	MAR.	APR.	MAY	JUN.	JUL.	AUG.	SEP.	OCT.	NOV.	DEC.	ANNUAL
13	.	.	.	.	.	.	.	.	.	.	.	.	.
14	20.7	20.1	18.5	14.9	10.8	8.3	8.1	11.1	13.8	16.9	18.3	20.0	15.3
15	.	.	.	.	.	.	.	.	.	.	.	.	.
16	16.6	16.2	14.4	11.6	7.3	4.9	4.8	7.3	10.9	13.6	14.4	15.9	11.5
17	.	.	.	.	.	.	.	.	.	.	.	.	.
18	.	.	.	.	.	.	.	.	.	.	.	.	.
19	.	.	.	.	.	.	.	.	.	.	.	.	.
20	.	.	.	.	.	.	.	.	.	.	.	.	.
21	.	.	.	.	.	.	.	.	.	.	.	.	.
22	18.5	18.0	16.5	13.9	10.6	7.6	7.4	9.8	12.7	14.9	15.9	17.6	13.6
23	.	.	.	.	.	.	.	.	.	.	.	.	.
24	.	.	.	.	.	.	.	.	.	.	.	.	.

CLIMATOLOGICAL DATA FOR LESOTHO

GEOGRAPHY

OBS.	LOCATION	YEARS	LATITUDE	LONGITUDE	ELEVATION -M-	LIFEZONE	CCTX	ID	ONMAP
25	RAPASE	27	30.05 S	28.55 E	1828.5	.	.	24	NO
26	ROMA	21	29.45 S	27.73 E	1645.3	.	.	25	NO
27	SEHONG-HONG	19	29.70 S	28.78 E	1904.4	.	.	26	YES
28	TEYATEYANENG	62	29.15 S	27.73 E	1752.3	.	.	27	YES
29	THABA MARENA	16	29.95 S	27.42 E	1675.5	.	.	28	YES
30	TOSING	25	30.33 S	27.93 E	1767.5	.	.	29	YES
31	TSOELIKE	22	30.02 S	28.73 E	1767.5	.	.	30	NO
32	WHITE HILL ESTATE	27	30.05 S	28.48 E	1675.5	.	.	31	NO

AVERAGE PRECIPITATION (MM)

OBS.	JAN.	FEB.	MAR.	APR.	MAY	JUN.	JUL.	AUG.	SEP.	OCT.	NOV.	DEC.	ANNUAL
25	86	77	60	28	21	9	10	13	26	42	62	84	518
26	115	124	109	69	41	12	17	20	28	77	94	119	825
27	96	78	70	27	22	5	9	13	22	52	65	78	537
28	125	96	102	54	30	11	12	15	27	67	84	100	723
29	95	105	94	65	34	10	16	21	39	65	69	85	697
30	105	99	101	54	42	12	13	16	38	69	85	102	736
31	87	87	72	33	21	10	7	12	22	49	75	84	559
32	88	78	67	34	24	9	6	14	28	53	69	98	568

AVERAGE TEMPERATURE (DEGREES CENTIGRADE)

OBS.	JAN.	FEB.	MAR.	APR.	MAY	JUN.	JUL.	AUG.	SEP.	OCT.	NOV.	DEC.	ANNUAL
25	.	.	.	.	.	.	.	.	.	.	.	.	.
26	.	.	.	.	.	.	.	.	.	.	.	.	.
27	.	.	.	.	.	.	.	.	.	.	.	.	.
28	.	.	.	.	.	.	.	.	.	.	.	.	.
29	.	.	.	.	.	.	.	.	.	.	.	.	.
30	.	.	.	.	.	.	.	.	.	.	.	.	.
31	.	.	.	.	.	.	.	.	.	.	.	.	.
32	.	.	.	.	.	.	.	.	.	.	.	.	.

LIBERIA

Agriculture

Of 11,137,000 ha in total area, Liberia was estimated in 1978 to have 371,000 ha of arable or permanent cropland, 240,000 ha in permanent pasture, 250,000 ha in forest and woodland, and 6,521,000 ha in other terrain categories. In 1978, approximately 2000 ha were irrigated. Of the 1979 population, 70.3% or 469,000 individuals of the economically active population were engaged in agriculture. The total population in 1979 was 1,802,000. According to FAO statistics for 1979, there were 38,000 cattle, 100,000 pigs, 190,000 sheep, 190,000 goats, 2,200,000 chickens, and 212,000 ducks.[2] Agricultural production is summarized in Table 1. Conventional and energetic yield data on most of these will be found in Appendixes I and II.

Climate

Data for Wernstedt[3] are presented in the Climatological Table (see Introduction for explanation). Based on the means for data points cited in Wernstedt, the mean annual precipitation is 27.6 dm (mean of 24 sites) and the mean annual temperature 25.6°C (mean of 24 sites). The averages would suggest that much of the country could be classified as Tropical Moist Forest Life Zone.[4]

Productivity

In the vegetation map, general vegetation zones of Liberia are depicted. Because of its humid tropical climate, Liberia's projected productivity rates are rather high.

Vegetation type	MT/ha/yr
Broadleaf evergreen forest	18—36
Broadleaf deciduous and evergreen forest	12—36
Brush, grassland, cultivated and tree crops	6—18
Swamp	18—36

Economics

Food consumption per capita was 2404 cal/day, energy consumption 402 kg coal equivalent per capita in 1977. In 1970, the caloric supply was 84.0% of requirements. The total population growth rate was 3.1% in 1970. Population density was 12 persons per square kilometer in 1970, with 26.2% of the total population urban. Life expectancy at birth of those born in 1970 was 47 years. In 1977, with GNP at $410 per capita, agriculture accounted for 27.7% of GDP, manufacturing for 5.0%, other industry for 31.2%, and services for 36.0%. In 1970, food and beverage constituted 3.8% of exports, nonfood agriculture for 19.4%, fuels, minerals, and metal for 72.1%, machines for 1.3%, and other manufacturing for 3.4% of total merchandise exports. Food and beverage constituted 16.9% of imports, nonfood agriculture 0.3%, fuel and lubricants 6.4%, nonfuel minerals and metals 1.1%, machinery and equipment 33.5%, and other manufactured goods constituted 41.8% of total merchandise imports.[10]

Energetics

Liberia has a hydroelectric potential of 6000 MW. Importing more than 75% of its commercial energy demands as oil, Liberia is expected to be experiencing fuelwood problems by the year 2000.[10]

Table 1
AGRICULTURAL PRODUCTION IN LIBERIA

Crop	Area planted (ha)	Yield (kg/ha)	Total production (MT)
Rice paddy	200,000	1,300	260,000
Sweet potatoes	2,000	10,000	17,000
Cassava	60,000	3,000	180,000
Soybeans	5,000	362	2,000
Peanuts in shell	5,000	622	3,000
Coconuts	7,000		
Palm kernels			14,000
Palm oil			28,000
Sugar (centrifugal, raw)			4,000
Oranges			7,000
Pineapples			8,000
Bananas			69,000
Plantains			30,000
Green coffee	25,000	380	10,000
Cocoa beans	7,000	500	4,000
Natural rubber			75,000

From FAO 1979 Production Yearbook, Vol. 33, Food and Agriculture Organization, Rome, 1980.

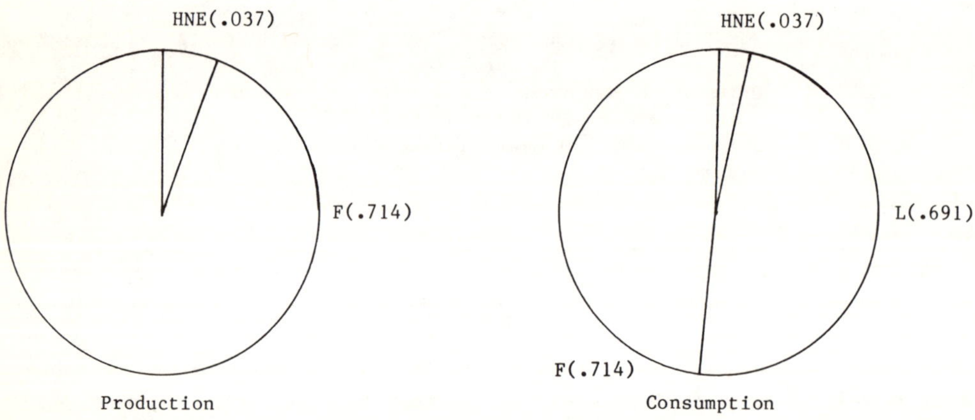

1978 Energy production and consumption in Liberia; fuel types listed in million metric tons coal equivalents (MMTCE). 1 MMTCE = 27.78×10^{12} Btu. L = liquid fuel, S = solid fuel, HNE = hydro, nuclear, and electric fuel, F = firewood, NG = natural gas. (Data from United Nations, World Energy Supplies, 1973—1978, Statistical Papers Series J, 22, Department of International Economic and Social Affairs, Statistical Office, New York, 1979.)

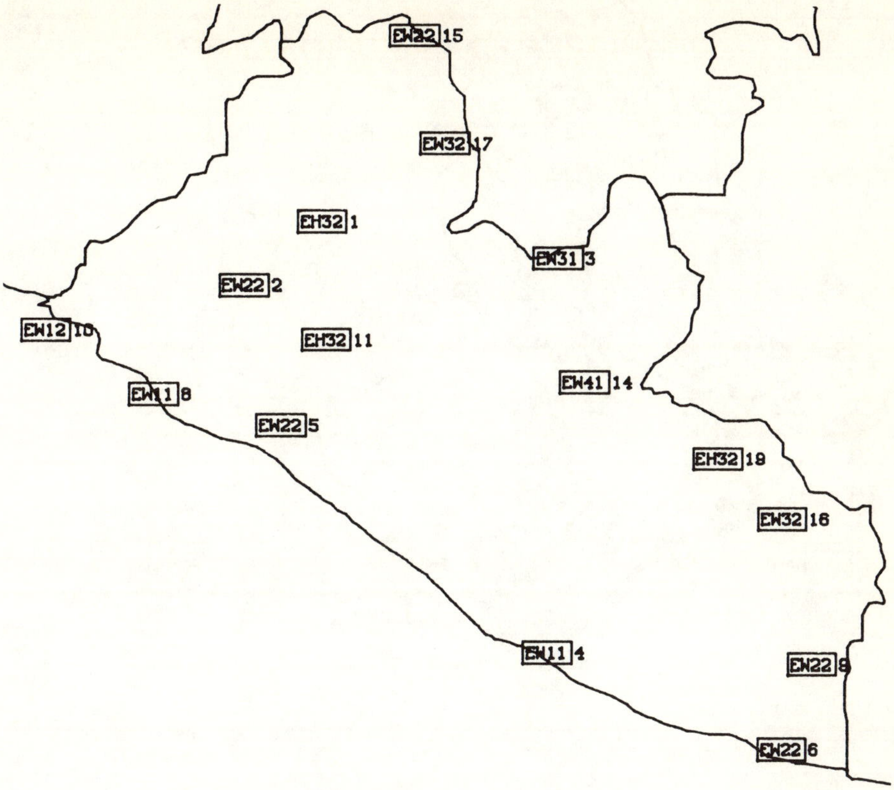

Selected weather stations for Liberia.

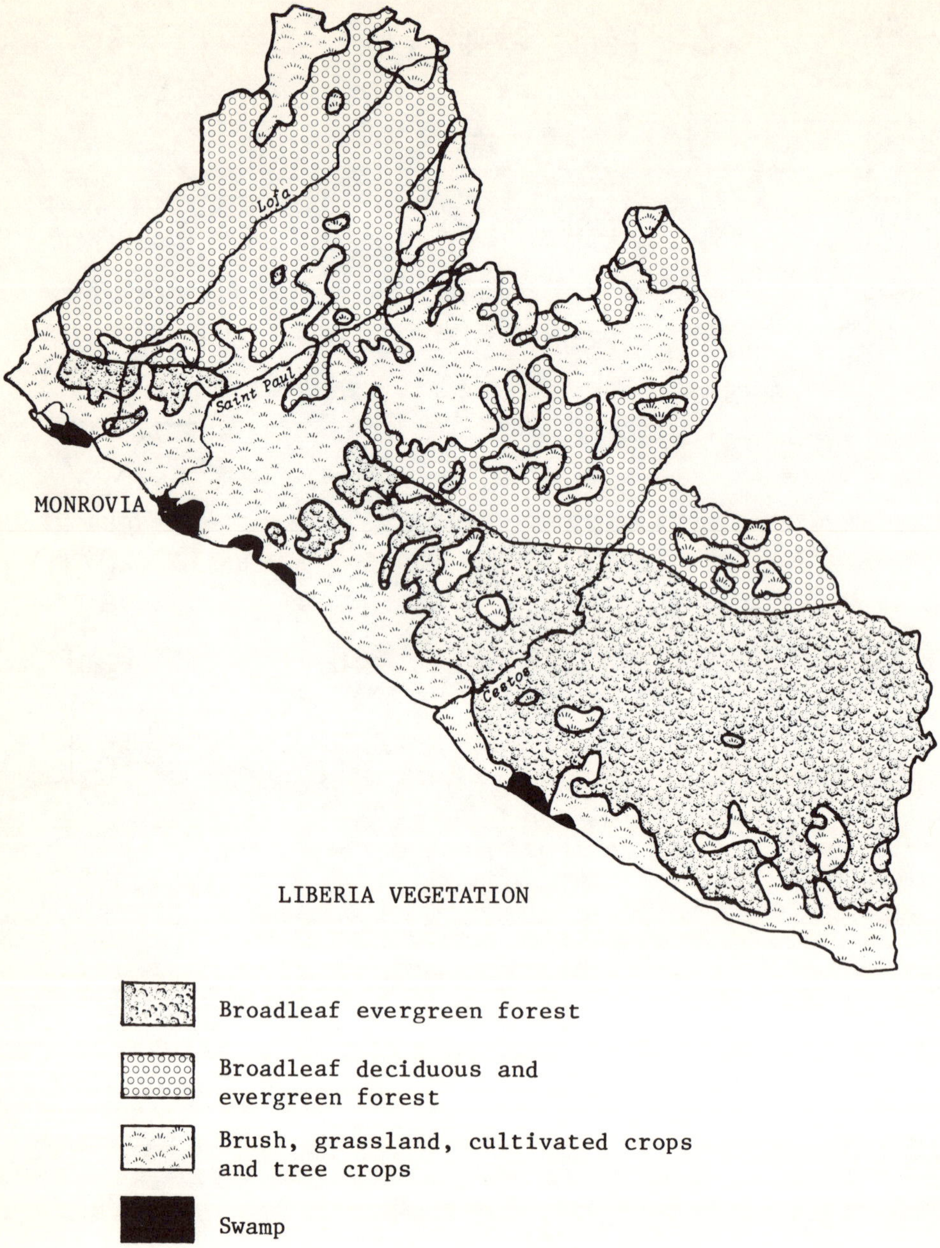
Lofa
Saint Paul
MONROVIA
Cestos
LIBERIA VEGETATION
Broadleaf evergreen forest
Broadleaf deciduous and evergreen forest
Brush, grassland, cultivated crops and tree crops
Swamp

CLIMATOLOGICAL DATA FOR LIBERIA

GEOGRAPHY

OBS.	LOCATION	YEARS	LATITUDE	LONGITUDE	ELEVATION -M-	LIFEZONE	CCTX	ID	ONMAP
1	BELLAYELLA	10	7.40 N	10.15 W	.	TM	EH32	1	YES
2	BOPOLU	10	7.05 N	10.53 W	.	TM	EW22	2	YES
3	GANTA	10	7.20 N	9.00 W	.	TM	EW31	3	YES
4	GREENVILLE	10	5.03 N	9.05 W	.	TW	EW11	4	YES
5	HARBEL	10	6.28 N	10.35 W	.	TM	EW22	5	YES
6	HARPER	10	4.50 N	7.90 W	.	TM	EW22	6	YES
7	KOLAHUN	10	8.50 N	10.13 W	.	TM	EW23	7	NO
8	MONROVIA	20	6.45 N	10.97 W	.	TW	EW11	8	YES
9	NYAAKE	10	4.97 N	7.75 W	.	TM	EW22	9	YES
10	ROBERTPORT	10	6.80 N	11.50 W	.	TW	EW12	10	YES
11	SALALA	10	6.75 N	10.13 W	.	TM	EH32	11	YES
12	SANNIQUELLIE	10	7.40 N	8.75 W	.	TM	EW32	12	NO

AVERAGE PRECIPITATION (MM)

OBS.	JAN.	FEB.	MAR.	AFR.	MAY	JUN.	JUL.	AUG.	SEP.	OCT.	NOV.	DEC.	ANNUAL
1	17	66	140	229	166	241	235	355	336	263	126	39	2212
2	25	63	127	155	201	444	328	324	455	313	147	42	2626
3	18	66	143	172	191	217	248	224	419	203	145	41	2087
4	129	139	223	193	555	721	264	193	597	503	323	292	4132
5	43	78	182	157	263	426	566	374	612	422	178	96	3397
6	165	129	128	182	500	556	85	128	245	307	266	307	2998
7	14	46	118	115	226	302	355	341	509	325	241	51	2644
8	43	62	97	161	394	942	890	474	740	594	224	113	4735
9	181	207	344	274	312	433	216	201	310	332	537	250	3601
10	33	46	109	164	416	835	910	436	577	404	156	99	4187
11	29	56	153	177	202	248	327	287	450	317	117	76	2439
12	21	58	120	116	151	360	415	339	448	274	126	42	2470

AVERAGE TEMPERATURE (DEGREES CENTIGRADE)

OBS.	JAN.	FEB.	MAR.	APR.	MAY	JUN.	JUL.	AUG.	SEP.	OCT.	NOV.	DEC.	ANNUAL
1	28.6	28.2	29.0	28.4	26.1	26.4	27.6	26.1	25.8	29.3	28.4	29.7	27.8
2	25.2	26.4	25.7	25.4	26.2	24.4	23.9	23.9	24.9	25.6	26.0	25.7	25.3
3	24.8	26.1	25.4	26.3	25.7	25.8	25.0	24.7	25.6	26.6	25.9	24.8	25.6
4	25.7	25.8	25.9	25.5	25.7	24.9	24.7	24.4	24.6	24.8	25.2	24.9	25.2
5	26.0	26.5	26.6	26.7	26.2	25.3	24.2	24.2	24.9	25.7	26.1	25.8	25.7
6	22.1	25.4	26.6	26.5	25.6	22.7	22.5	24.2	23.3	25.8	25.1	23.9	24.5
7	21.9	24.1	26.5	25.4	27.0	24.6	23.6	21.1	25.6	28.8	26.3	25.4	25.1
8	26.2	26.4	26.7	26.7	26.3	25.[illegible]	24.7	25.0	24.8	25.3	26.2	26.3	25.8
9	25.7	26.2	26.3	25.8	25.5	24.[illegible]	23.1	24.4	25.4	27.1	24.7	28.1	25.5
10	26.0	26.2	26.7	26.6	26.2	25.4	23.9	23.6	24.8	25.7	23.4	24.1	25.2
11	26.7	27.1	27.4	27.9	27.0	26.6	25.6	25.0	26.0	27.2	28.2	25.6	27.2
12	25.0	24.7	24.2	24.8	24.6	23.9	23.6	22.5	25.0	25.4	25.0	24.3	24.4

CLIMATOLOGICAL DATA FOR LIBERIA

GEOGRAPHY

OBS.	LOCATION	YEARS	LATITUDE	LONGITUDE	ELEVATION -M-	LIFEZONE	CCTX	ID	ONMAP
13	SUAKOKO	10	7.08 N	9.58 W	.	TD	EW42	13	NO
14	TAPPITA	10	6.52 N	8.87 W	.	TD	EW41	14	YES
15	VOINJANA	10	8.42 N	9.70 W	.	TM	EW22	15	YES
16	ZIATOWN	10	5.77 N	7.90 W	.	TM	EW32	16	YES
17	ZORZOR	10	7.83 N	9.55 W	.	SW	EW32	17	YES
18	ZUIE	10	7.15 N	10.47 W	.	TM	EW22	18	NO
19	ZWEDRU	10	6.10 N	8.22 W	.	TM	EH32	19	YES

AVERAGE PRECIPITATION (MM)

OBS.	JAN.	FEB.	MAR.	APR.	MAY	JUN.	JUL.	AUG.	SEP.	OCT.	NOV.	DEC.	ANNUAL
13	19	59	152	175	199	197	204	159	361	220	98	36	1881
14	42	66	124	126	207	278	169	118	315	212	68	19	1745
15	19	49	155	205	249	360	443	389	436	288	248	133	2973
16	45	84	155	211	282	233	157	125	301	293	117	89	2094
17	30	46	133	202	130	201	208	437	455	251	98	44	2235
18	23	90	112	167	212	397	406	445	534	274	121	65	2847
19	27	70	118	199	218	231	162	118	344	308	155	83	2034

AVERAGE TEMPERATURE (DEGREES CENTIGRADE)

OBS.	JAN.	FEB.	MAR.	APR.	MAY	JUN.	JUL.	AUG.	SEP.	OCT.	NOV.	DEC.	ANNUAL
13	24.5	24.9	24.8	25.2	24.6	24.4	23.3	22.4	25.8	24.1	24.4	23.5	24.3
14	26.1	26.8	26.7	26.8	26.6	25.7	24.7	24.9	25.2	25.8	26.6	24.5	25.9
15	24.0	25.3	24.8	26.6	25.8	24.1	24.4	22.8	24.4	25.0	24.4	23.8	24.6
16	25.9	27.1	28.2	28.5	22.9	26.0	26.1	24.1	26.3	26.6	24.8	22.3	25.7
17	23.7	24.3	23.7	23.2	23.1	21.9	23.1	23.3	22.2	24.6	23.6	21.9	23.2
18	26.0	26.2	25.6	25.6	26.2	25.0	24.4	23.6	23.9	26.9	25.5	26.2	25.4
19	26.2	27.7	27.9	28.3	28.0	26.8	25.3	25.3	26.5	27.1	27.1	26.5	26.9

MALAGASY

Agriculture

Of 58,704,000 ha in total area, Malagasy was estimated in 1978 to have 2,929,000 ha of arable or permanent cropland, 34,000,000 ha in permanent pasture, 12,472,000 ha in forest and woodland, and 8,753,000 ha in other terrain categories. In 1978, approximately 450,000 ha were irrigated. Of the 1979 population, 84.0% or 3,500,000 individuals of the economically active population were engaged in agriculture. The total population in 1979 was 8,511,000. According to FAO statistics for 1979, there were 2000 horses, 1000 asses, 8,744,000 cattle, 580,000 pigs, 658,000 sheep, 1,583,000 goats, 14,082,000 chickens, 2,352,000 ducks, and 1,162,000 turkeys.[2] Agricultural production is summarized in Table 1. Conventional and energetic yield data on most of these will be found in Appendixes I and II.

Climate

Data from Wernstedt[3] are presented in the Climatological Data Table (see Introduction for explanation). Based on the means for data points cited in Wernstedt, the mean annual precipitation is 15.5 dm (mean of 107 sites) and the mean annual temperature 23.0°C (mean of 103 sites). The averages would suggest that much of the country could be classified as Subtropical Moist Forest Life Zone.[4]

Productivity

The vegetation map depicts the major vegetation types of Malagasy with estimated annual productivities of (in MT/ha/yr): broadleaf evergreen forest (with mangrove on west coast), 18 to 36; deciduous forest, 12 to 24; secondary growth (savoka), 6 to 18; savanna, 6 to 12; marsh, 6 to 18; and scrub, 6 to 12.

Economics

Food consumption per capita was 2486 cal/day, energy consumption 76 kg of coal equivalent per capita in 1977. In 1970, the caloric supply was 108.0% of requirements. The total population growth rate was 2.2% in 1970. Population density was 12 persons per square kilometer in 1970, with 14.9% of the total population urban. Life expectancy at birth of those born in 1970 was 42 years. In 1977, with GNP at $230 per capita, agriculture accounted for 38.4% of GDP, manufacturing for 13.9%, other industry for 5.5%, and services for 42.2%. In 1970, food and beverage constituted 78.9% of exports, nonfood agriculture for 5.0%, fuels, minerals, and metal for 8.9%, machines for 2.1%, and other manufacturing for 5.1% of total merchandise exports. Food and beverage constituted 12.4% of imports, nonfood agriculture 1.7%, fuel and lubricants 7.4%, nonfuel minerals and metals 1.0%, machinery and equipment 30.1%, and other manufactures constituted 47.5% of total merchandise imports.[10]

Energetics

Heavy oil deposits of Malagasy have not been fully assessed. Tar sands also exist.[10]

Table 1
AGRICULTURAL PRODUCTION IN MALAGASY

Crop	Area planted (ha)	Yield (kg/ha)	Total production (MT)
Wheat		1,667	
Rice paddy	1,216,000	1,914	2,327,000
Corn	100,000	1,000	100,000
Sorghum		441	
Potatoes	21,000	6,431	134,000
Sweet potatoes	63,000	5,378	340,000
Cassava	200,000	6,739	1,348,000
Dry beans	55,000	785	44,000
Peanuts in shell	35,000	860	30,000
Castor beans	5,000	260	1,000
Seed cotton	22,000	1,500	33,000
Cottonseed			19,000
Coconuts			19,000
Copra			1,000
Palm kernels			500
Palm oil			2,300
Cabbages	1,000	14,615	10,000
Tomatoes	3,000	10,889	29,000
Cauliflower		13,415	1,000
Cucumbers and gherkins		5,417	1,000
Eggplants		4,017	1,000
Dry onions	1,000	4,783	3,000
Garlic		2,366	
Green beans	1,000	2,615	2,000
Green peas		894	
Carrots	1,000	18,125	15,000
Grapes	2,000	8,813	14,000
Wine			10,000
Sugar cane	35,000	37,143	1,300,000
Sugar (centrifugal, raw)			113,000
Apples			4,000
Peaches and nectarines			6,000
Plums			7,000
Oranges			83,000
Avocados			9,000
Mangoes			143,000
Pineapples			53,000
Bananas			278,000
Cashew nuts			2,000
Green coffee	180,000	333	60,000
Cocoa beans	5,000	326	2,000
Tobacco leaves	6,000	833	5,000
Jute		325	
Sisal	22,000	1,126	25,000
Cotton lint			12,000
Fiber crops			7,000

From FAO 1979 Production Yearbook, Vol. 33, Food and Agriculture Organization, Rome, 1980.

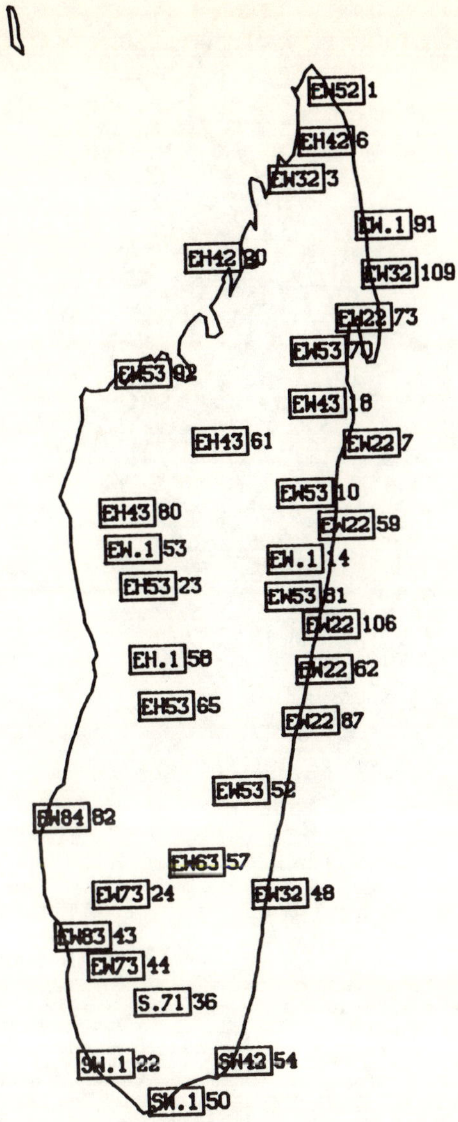

Selected weather stations for Malagasy.

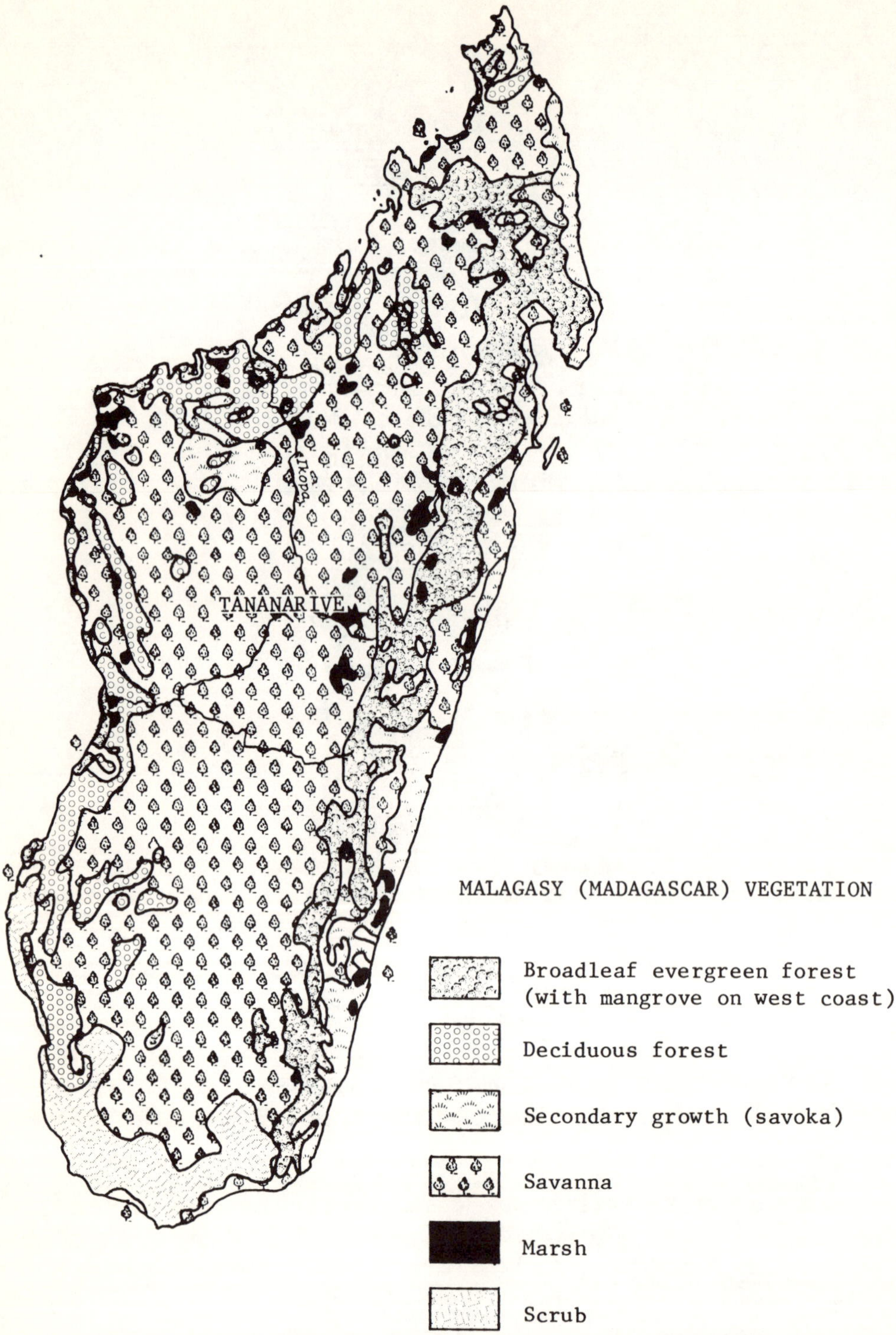
TANANARIVE
Ikopa
MALAGASY (MADAGASCAR) VEGETATION
Broadleaf evergreen forest (with mangrove on west coast)
Deciduous forest
Secondary growth (savoka)
Savanna
Marsh
Scrub

CLIMATOLOGICAL DATA FOR MALAGASY

GEOGRAPHY

OBS.	LOCATION	YEARS	LATITUDE	LONGITUDE	ELEVATION -M-	LIFEZONE	CCTX	ID	ONMAP
1	AMBAHIVAHIBE	17	12.58 S	49.27 E	370.0	TD	EW52	1	YES
2	AMBALAKONDRO	20	21.22 S	48.13 E	60.0	.	.	2	NO
3	AMBANJA	20	13.67 S	48.45 E	39.9	TM	EW32	3	YES
4	AMBATO-BOENI	20	16.45 S	46.72 E	39.9	.	.	4	NO
5	AMBATOMAINTY	15	17.70 S	45.67 E	349.9	TD	EW43	5	NO
6	AMBILOBE	20	13.20 S	49.05 E	29.9	TD	EH42	6	YES
7	AMBODIFOTOTRA	27	17.02 S	49.85 E	10.1	TM	EW22	7	YES
8	AMBODIRANO	10	18.77 S	48.45 E	922.6	.	.	8	NO
9	AMBOHIBARY	25	19.62 S	47.12 E	1649.6	WM	EC43	9	NO
10	AMBOHITSILAOZANA	21	17.63 S	48.50 E	783.6	SM	EW53	10	YES
11	AMBOSITRA	21	20.53 S	47.23 E	1167.7	SM	EC43	11	NO
12	AMBOVOMBE	20	25.18 S	46.10 E	135.0	SD	SW72	12	NO

AVERAGE PRECIPITATION (MM)

OBS.	JAN.	FEB.	MAR.	APR.	MAY	JUN.	JUL.	AUG.	SEP.	OCT.	NOV.	DEC.	ANNUAL
1	342	304	224	109	15	14	33	20	12	17	52	216	1358
2	.	.	.	.	.	.	.	.	.	.	.	.	.
3	538	462	287	138	50	34	24	35	42	65	167	328	2171
4	.	.	.	.	.	.	.	.	.	.	.	.	.
5	354	338	252	80	9	4	4	8	12	57	157	272	1546
6	498	474	297	107	17	16	10	12	11	32	103	283	1861
7	379	443	546	469	304	343	276	198	109	86	136	293	3582
8	.	.	.	.	.	.	.	.	.	.	.	.	.
9	285	269	212	93	41	25	20	17	27	75	178	278	1519
10	288	269	193	38	11	9	9	7	3	23	88	214	1152
11	295	243	230	82	37	27	22	23	33	59	196	282	1528
12	96	91	59	27	39	60	27	24	19	25	47	88	604

AVERAGE TEMPERATURE (DEGREES CENTIGRADE)

OBS.	JAN.	FEB.	MAR.	APR.	MAY	JUN.	JUL.	AUG.	SEP.	OCT.	NOV.	DEC.	ANNUAL
1	25.0	25.1	25.2	24.8	23.8	22.6	21.7	21.8	22.6	23.6	24.7	25.0	23.9
2	24.2	24.3	24.4	23.1	20.7	18.9	17.8	18.3	19.2	21.0	22.9	24.0	21.6
3	26.6	26.8	27.1	26.9	25.6	24.3	23.3	23.9	24.6	26.0	26.8	27.0	25.7
4	27.0	27.6	27.7	27.7	25.8	24.3	24.0	24.8	26.5	28.1	28.8	28.2	26.7
5	26.8	26.9	26.9	26.7	24.6	22.9	22.4	23.8	25.6	27.2	28.0	27.2	25.7
6	27.3	27.5	28.0	27.9	26.9	25.4	24.6	25.1	26.1	27.7	28.4	28.0	26.9
7	26.7	26.8	26.2	25.4	24.0	22.6	21.7	21.7	22.4	23.7	25.2	26.2	24.4
8	23.3	22.8	22.1	21.2	18.9	16.8	15.5	16.2	17.7	20.1	22.1	22.8	20.4
9	19.3	19.1	18.9	17.6	15.2	12.8	12.3	13.0	14.6	16.9	18.1	18.8	16.3
10	23.4	23.2	23.0	21.8	19.6	17.7	16.8	17.2	18.4	20.4	22.3	23.2	20.6
11	20.8	20.6	20.2	19.0	16.7	14.9	13.9	14.3	15.9	18.5	20.0	20.6	17.9
12	26.4	26.4	25.6	24.2	21.0	19.2	18.6	19.4	20.9	23.0	24.9	26.1	23.0

CLIMATOLOGICAL DATA FOR MALAGASY

GEOGRAPHY

OBS.	LOCATION	YEARS	LATITUDE	LONGITUDE	ELEVATION -M-	LIFEZONE	CCTX	ID	ONMAP
13	AMPANDRANDAVA	10	24.12 S	45.72 E	699.8	.	.	13	NO
14	AMPANGABE	10	18.47 S	48.28 E	904.6	.	.	14	YES
15	AMPANIHY-OUEST	22	24.70 S	44.75 E	274.9	TV	SW73	15	NO
16	AMPARIHY-EST	10	23.97 S	47.35 E	29.9	.	.	16	NO
17	AMPASIMPOLAKA	17	25.12 S	46.40 E	67.1	TV	SW72	17	NO
18	AMPATAKAMAROREN Y	14	16.50 S	48.78 E	849.8	SM	EW43	18	YES
19	AMPOTAKA	15	25.05 S	44.68 E	67.1	TT	SW83	19	NO
20	ANALALAVA	42	14.63 S	46.78 E	39.9	TD	EH42	20	YES
21	ANALAMAZAOTRA	20	18.93 S	48.40 E	927.8	SM	EW43	21	NO
22	ANDROKA	10	25.03 S	44.07 E	4.0	.	.	22	YES
23	ANKAVANDRA	20	18.78 S	45.30 E	29.9	TD	EH53	23	YES
24	ANKAZOABO-SUR	20	22.78 S	44.53 E	427.9	TV	EW73	24	YES

AVERAGE PRECIPITATION (MM)

OBS.	JAN.	FEB.	MAR.	APR.	MAY	JUN.	JUL.	AUG.	SEP.	OCT.	NOV.	DEC.	ANNUAL
13	.	.	.	.	.	.	.	.	.	.	.	.	.
14	.	.	.	.	.	.	.	.	.	.	.	.	.
15	141	90	64	32	15	18	8	5	12	15	55	126	582
16	.	.	.	.	.	.	.	.	.	.	.	.	.
17	64	82	52	20	40	57	33	23	16	26	43	80	537
18	339	289	236	71	28	41	46	55	34	39	98	238	1514
19	81	86	62	10	20	19	11	8	10	23	24	90	445
20	533	471	219	85	11	8	5	3	11	51	188	311	1896
21	296	315	231	97	64	82	81	68	54	47	113	233	1680
22	.	.	.	.	.	.	.	.	.	.	.	.	.
23	322	298	222	71	9	3	4	7	20	74	146	240	1415
24	209	143	83	19	7	9	2	2	7	26	61	150	719

AVERAGE TEMPERATURE (DEGREES CENTIGRADE)

OBS.	JAN.	FEB.	MAR.	APR.	MAY	JUN.	JUL.	AUG.	SEP.	OCT.	NOV.	DEC.	ANNUAL
13	26.8	26.4	25.5	24.7	21.7	20.1	20.2	21.3	23.3	25.3	25.9	26.6	24.0
14	22.9	22.5	21.5	21.0	18.6	16.9	15.7	16.0	17.1	19.5	21.9	22.5	19.7
15	28.4	27.8	27.1	25.8	21.7	20.3	19.8	21.1	23.3	25.4	27.3	28.1	24.7
16	26.1	25.6	25.1	24.0	21.5	19.8	19.1	19.9	21.5	23.6	25.1	25.9	23.1
17	27.8	27.9	27.3	25.2	22.1	20.2	20.0	20.4	22.0	24.1	25.9	27.4	24.2
18	21.9	22.5	22.4	21.0	18.7	17.2	15.3	16.1	17.0	18.5	20.6	21.8	19.4
19	28.1	28.0	27.5	25.5	22.0	19.6	19.5	20.7	22.5	24.9	26.5	27.7	24.4
20	26.8	26.7	27.3	27.7	26.9	25.6	25.4	25.7	26.3	26.8	27.1	27.1	26.6
21	22.0	22.3	21.7	20.7	18.3	16.5	15.4	15.7	16.7	18.7	20.7	21.7	19.2
22	27.4	27.4	26.6	24.9	22.1	20.2	19.7	20.5	21.6	23.0	24.9	26.6	23.7
23	28.0	28.1	27.9	27.6	25.4	23.5	23.5	25.0	26.9	28.5	28.8	28.2	26.8
24	27.1	26.9	26.2	25.2	22.0	20.1	20.1	21.5	23.7	25.8	27.4	27.1	24.4

CLIMATOLOGICAL DATA FOR MALAGASY

GEOGRAPHY

OBS.	LOCATION	YEARS	LATITUDE	LONGITUDE	ELEVATION -M-	LIFEZONE	CCTX	ID	ONMAP
25	ANKAZOBE	20	18.28 S	47.12 E	1189.6	SM	EW43	25	NO
26	ANOSIBE	10	19.43 S	48.20 E	749.8	SW	EW33	26	NO
27	ANTALAHA	34	14.88 S	50.25 E	89.9	TM	EW32	109	YES
28	ANTANIMORA	17	24.82 S	45.67 E	299.9	TV	SW73	27	NO
29	ANTOKAZO	10	17.60 S	48.55 E	1049.7	.	.	28	NO
30	ANTSAKABARY	10	15.03 S	48.95 E	549.9	.	.	29	NO
31	ANTSIRABE	13	19.87 S	47.00 E	1505.4	WM	EC53	30	NO
32	ANTSOHIHY	20	14.88 S	48.00 E	18.0	TD	EH43	31	NO
33	ARIVONIMAMO	10	19.03 S	47.17 E	1459.4	SM	EC53	32	NO
34	BEALANANA	16	14.55 S	48.72 E	1124.7	SM	EW53	33	NO
35	BEFANDRIANA-NORD	20	15.25 S	48.53 E	314.9	.	.	34	NO
36	BEHARA	17	24.95 S	46.38 E	54.9	.	.	35	NO

AVERAGE PRECIPITATION (MM)

OBS.	JAN.	FEB.	MAR.	APR.	MAY	JUN.	JUL.	AUG.	SEP.	OCT.	NOV.	DEC.	ANNUAL
25	376	292	223	54	13	4	3	8	10	48	168	301	1502
26	384	376	308	155	85	110	102	78	75	46	157	342	2232
27	267	239	260	301	147	149	153	136	108	80	121	217	2177
28	103	119	63	21	15	27	10	8	7	34	54	94	555
29	.	.	.	.	.	.	.	.	.	.	.	.	.
30	.	.	.	.	.	.	.	.	.	.	.	.	.
31	294	246	206	78	36	13	15	17	22	77	165	257	1426
32	441	363	248	58	4	2	3	5	7	27	109	232	1500
33	325	207	218	35	17	9	4	12	6	64	130	211	1238
34	314	296	245	75	9	13	5	13	7	24	80	250	1332
35	533	470	337	72	10	5	5	5	2	44	162	340	1985
36	84	85	65	24	25	40	20	11	11	26	54	92	537

AVERAGE TEMPERATURE (DEGREES CENTIGRADE)

OBS.	JAN.	FEB.	MAR.	APR.	MAY	JUN.	JUL.	AUG.	SEP.	OCT.	NOV.	DEC.	ANNUAL
25	21.6	21.6	21.6	20.7	18.5	16.8	15.8	16.7	18.5	20.8	21.6	21.8	19.7
26	23.7	24.1	23.4	22.3	19.8	18.2	17.5	17.2	18.2	20.3	22.3	23.4	20.9
27	26.0	26.1	25.9	25.4	24.0	22.6	21.8	21.6	22.2	23.2	24.5	25.4	24.1
28	27.3	26.9	26.1	24.7	21.2	19.5	19.5	20.7	22.4	24.6	26.4	26.7	23.9
29	22.3	22.1	22.0	20.9	19.0	17.0	16.1	16.3	17.8	19.8	21.4	22.3	19.7
30	25.5	25.7	25.8	24.9	23.0	21.1	19.8	20.4	21.2	23.8	25.2	26.0	23.5
31	19.9	19.6	19.2	17.8	15.0	13.1	12.4	13.5	15.4	17.7	19.2	19.6	16.9
32	27.7	27.6	28.1	27.8	26.6	24.8	24.3	25.0	26.4	28.2	28.7	28.3	27.0
33	20.6	20.1	19.9	18.8	16.5	14.6	14.0	14.7	16.3	18.5	19.8	20.1	17.8
34	21.6	21.8	21.9	21.1	19.9	17.8	16.6	16.9	17.9	20.1	21.5	21.5	19.9
35	.	.	.	.	.	.	.	.	.	.	.	.	.
36	.	.	.	.	.	.	.	.	.	.	.	.	.

CLIMATOLOGICAL DATA FOR MALAGASY

GEOGRAPHY

OBS.	LOCATION	YEARS	LATITUDE	LONGITUDE	ELEVATION -M-	LIFEZONE	CCTX	ID	ONMAP
37	BEKILY	17	24.22 S	45.32 E	399.9	.	.	36	YES
38	BEKODOKA	22	16.95 S	45.10 E	299.9	.	.	37	NO
39	BELOHA	18	25.17 S	45.05 E	100.0	.	.	38	NO
40	BENENITRA	15	23.45 S	45.05 E	220.1	.	.	39	NO
41	BERONONO	10	15.98 S	47.15 E	100.0	.	.	40	NO
42	BEROROHA	16	21.67 S	45.17 E	179.8	TV	EW63	41	NO
43	BESALAMPY	21	16.75 S	44.48 E	20.1	TD	EH53	42	NO
44	BETANIMENA	16	23.32 S	43.70 E	4.9	TT	EW83	43	YES
45	BETIOKY-SUD	20	23.72 S	44.37 E	263.0	TV	EW73	44	YES
46	BETOMBA	10	19.65 S	44.95 E	100.0	.	.	45	NO
47	BETROKA	33	23.28 S	46.08 E	794.6	SD	EW63	46	NO
48	DIEGO-SUAREZ	39	12.28 S	49.30 E	29.0	TV	EH62	47	NO

AVERAGE PRECIPITATION (MM)

OBS.	JAN.	FEB.	MAR.	APR.	MAY	JUN.	JUL.	AUG.	SEP.	OCT.	NOV.	DEC.	ANNUAL
37	170	138	75	31	11	18	8	6	6	34	70	180	746
38	379	304	234	95	13	4	5	9	20	35	142	221	1456
39	71	89	49	14	19	23	10	8	9	17	28	85	422
40	167	121	98	20	17	13	5	8	9	29	60	148	696
41	.	.	.	.	.	.	.	.	.	.	.	.	.
42	192	178	104	11	7	4	2	4	7	21	66	172	768
43	411	322	189	40	6	1	0	2	6	11	75	180	1243
44	64	97	36	10	20	12	4	5	7	17	26	44	342
45	167	115	69	15	11	11	4	4	9	17	72	139	634
46	.	.	.	.	.	.	.	.	.	.	.	.	.
47	219	136	90	29	13	13	8	6	12	39	96	196	857
48	265	205	182	59	9	7	6	6	4	9	24	125	901

AVERAGE TEMPERATURE (DEGREES CENTIGRADE)

OBS.	JAN.	FEB.	MAR.	APR.	MAY	JUN.	JUL.	AUG.	SEP.	OCT.	NOV.	DEC.	ANNUAL
37	.	.	.	.	.	.	.	.	.	.	.	.	.
38	.	.	.	.	.	.	.	.	.	.	.	.	.
39	.	.	.	.	.	.	.	.	.	.	.	.	.
40	.	.	.	.	.	.	.	.	.	.	.	.	.
41	27.4	27.2	27.4	26.5	24.5	22.7	21.6	22.7	24.0	26.5	28.0	27.6	25.5
42	28.5	28.5	27.9	27.0	23.5	21.3	21.0	22.7	25.3	28.0	29.3	29.1	26.0
43	27.6	27.6	27.8	28.0	26.0	24.4	24.1	24.9	26.1	27.3	28.1	28.1	26.6
44	27.9	27.7	26.9	25.3	22.7	20.7	20.0	20.8	22.1	23.7	25.9	27.1	24.2
45	27.8	27.7	28.0	25.8	22.0	19.9	19.9	21.3	23.6	25.6	27.3	27.7	24.7
46	29.1	28.9	28.6	28.4	25.4	23.4	23.3	25.1	27.5	29.3	29.9	29.5	27.3
47	24.9	24.7	24.2	22.8	19.0	17.4	17.3	18.5	20.8	23.0	24.7	24.8	21.9
48	27.5	27.8	28.3	28.3	27.5	26.2	25.3	25.3	25.5	26.5	27.7	28.1	27.0

CLIMATOLOGICAL DATA FOR MALAGASY

GEOGRAPHY

OBS.	LOCATION	YEARS	LATITUDE	LONGITUDE	ELEVATION -M-	LIFEZONE	CCTX	ID	ONMAP
49	FARAFANGANA	38	22.82 S	47.83 E	3.0	SW	EW32	48	YES
50	FARATSIHO	23	19.40 S	46.93 E	1749.6	WM	EC43	49	NO
51	FAUX-CAP	10	25.55 S	45.53 E	20.1	.	.	50	YES
52	FENERIVE-EST	20	17.38 S	49.42 E	4.9	TM	EW22	51	NO
53	FIANARANTSOA	41	21.45 S	47.08 E	1167.7	SM	EW53	52	YES
54	FOLAKARA-ANKISA	10	18.30 S	45.02 E	229.8	.	.	53	YES
55	FORT DAUPHIN	39	25.03 S	46.98 E	43.9	SM	SW42	54	YES
56	HELL VILLE	26	13.40 S	48.28 E	18.9	TM	EW32	55	NO
57	IFANADIANA	20	21.30 S	47.62 E	449.9	SW	EW23	56	NO
58	IHOSY	20	22.38 S	46.12 E	729.7	SD	EW63	57	YES
59	ISALO	10	19.72 S	45.45 E	39.9	.	.	58	YES
60	IVOLOINA	19	18.05 S	49.32 E	20.1	TM	EW22	59	YES

AVERAGE PRECIPITATION (MM)

OBS.	JAN.	FEB.	MAR.	APR.	MAY	JUN.	JUL.	AUG.	SEP.	OCT.	NOV.	DEC.	ANNUAL
49	320	308	382	240	188	205	159	116	69	55	172	219	2433
50	304	330	234	107	29	9	8	19	31	90	192	287	1639
51	.	.	.	.	.	.	.	.	.	.	.	.	.
52	338	393	374	273	186	172	156	151	119	111	161	294	2728
53	276	218	167	50	30	21	19	19	21	36	133	226	1215
54	.	.	.	.	.	.	.	.	.	.	.	.	.
55	195	201	229	[illegible]	124	154	99	87	47	71	83	130	1530
56	475	424	278	147	58	47	34	40	50	103	199	374	2230
57	482	467	407	177	100	120	100	85	80	53	167	331	2569
58	231	142	86	24	9	8	3	5	9	27	103	198	844
59	.	.	.	.	.	.	.	.	.	.	.	.	.
60	365	408	441	346	255	249	211	206	132	78	165	215	3071

AVERAGE TEMPERATURE (DEGREES CENTIGRADE)

OBS.	JAN.	FEB.	MAR.	APR.	MAY	JUN.	JUL.	AUG.	SEP.	OCT.	NOV.	DEC.	ANNUAL
49	25.7	25.7	25.1	24.1	21.9	20.5	19.6	19.8	20.9	22.2	23.8	25.1	22.9
50	18.7	18.6	18.5	17.6	15.7	14.1	13.3	14.3	15.9	17.6	18.3	18.4	16.7
51	26.4	26.3	25.7	24.4	22.3	20.6	20.2	20.2	21.3	22.6	24.3	25.9	23.4
52	26.4	26.4	26.1	25.1	23.3	21.7	20.8	20.9	21.4	22.9	24.5	25.6	23.7
53	21.3	20.9	20.6	19.5	16.9	15.2	14.2	15.2	16.5	19.1	20.9	21.1	18.5
54	27.4	27.2	27.1	25.7	24.8	22.9	22.6	23.8	25.7	27.3	27.8	27.2	25.9
55	25.7	25.6	24.9	24.0	21.8	20.4	19.8	20.0	20.8	22.4	23.9	25.1	22.9
56	27.0	27.1	27.2	27.1	26.0	24.6	23.7	24.1	25.0	26.2	26.9	27.2	26.0
57	24.2	24.1	23.5	22.4	19.9	18.2	17.2	17.9	19.1	21.4	23.2	24.0	21.3
58	24.5	24.2	23.6	22.7	19.6	17.9	17.4	18.5	20.4	23.0	25.3	24.5	21.7
59	28.8	28.6	28.4	28.0	25.5	23.3	22.8	24.6	26.7	28.8	29.5	28.8	27.0
60	26.6	26.5	26.2	25.2	23.4	21.7	20.9	20.9	21.5	22.9	24.5	25.7	23.8

CLIMATOLOGICAL DATA FOR MALAGASY

GEOGRAPHY

OBS.	LOCATION	YEARS	LATITUDE	LONGITUDE	ELEVATION -M-	LIFEZONE	CCTX	ID	ONMAP
61	KANDREHO	19	17.47 S	46.08 E	289.0	TD	EH43	60	NO
62	MAEVATANANA	27	16.95 S	46.83 E	77.1	TD	EH43	61	YES
63	MAHANORO	31	19.90 S	48.82 E	29.0	SW	EW22	62	YES
64	MAINTIRANO	38	18.05 S	44.03 E	13.1	TV	EW63	63	NO
65	MAJUNGA	42	15.73 S	46.32 E	36.0	TD	EH42	64	NO
66	MALAIMBANDY	15	20.33 S	45.60 E	179.8	TD	EH53	65	YES
67	MANAKARA	21	22.13 S	48.03 E	4.0	SW	EW22	66	NO
68	MANANARA-NORD	15	16.17 S	49.78 E	10.1	TM	EW32	67	NO
69	MANANDRIANA	10	14.87 S	49.38 E	929.6	.	.	68	NO
70	MANANJARY	44	21.25 S	48.33 E	13.1	SW	EW22	69	NO
71	MANDRITSARA	33	15.83 S	48.82 E	349.9	TD	EW53	70	YES
72	MANGINDRANO	16	14.28 S	48.93 E	1199.7	SM	EW53	71	NO

AVERAGE PRECIPITATION (MM)

OBS.	JAN.	FEB.	MAR.	APR.	MAY	JUN.	JUL.	AUG.	SEP.	OCT.	NOV.	DEC.	ANNUAL
61	391	338	232	90	9	4	5	8	10	49	133	282	1550
62	471	370	285	69	5	1	2	2	8	36	161	316	1725
63	360	383	455	313	251	205	216	166	132	86	138	254	2959
64	269	205	170	28	8	5	4	3	7	16	65	160	940
65	474	358	272	60	9	3	1	2	3	25	109	243	1559
66	369	290	228	36	6	7	5	2	6	55	100	228	1333
67	355	379	437	276	209	234	163	141	105	80	149	229	2757
68	336	318	349	254	137	170	147	133	99	102	106	223	2374
69	.	.	.	.	.	.	.	.	.	.	.	.	.
70	401	396	478	253	200	236	162	135	109	72	166	225	2834
71	315	291	197	35	7	3	3	4	4	9	80	200	1148
72	308	282	308	100	34	27	16	12	11	31	95	245	1469

AVERAGE TEMPERATURE (DEGREES CENTIGRADE)

OBS.	JAN.	FEB.	MAR.	APR.	MAY	JUN.	JUL.	AUG.	SEP.	OCT.	NOV.	DEC.	ANNUAL
61	27.3	27.5	27.6	27.6	25.9	24.2	23.8	24.9	26.4	27.9	28.5	27.9	26.6
62	27.8	27.8	28.0	28.4	27.0	25.4	25.0	26.0	27.4	28.7	29.1	28.6	27.4
63	26.0	25.9	25.5	24.7	22.9	21.3	20.5	20.6	21.4	22.8	24.3	25.3	23.4
64	27.4	27.6	27.4	27.1	25.0	22.9	22.6	23.4	24.7	26.1	27.3	27.6	25.7
65	27.4	27.6	27.8	28.1	26.8	25.4	25.0	25.6	26.6	27.5	28.1	27.8	27.0
66	28.2	28.2	28.1	27.6	24.7	22.9	22.8	24.0	26.4	28.2	28.8	28.4	26.5
67	26.0	25.8	25.3	24.2	22.0	20.6	19.8	19.9	20.7	22.2	23.7	25.0	22.9
68	27.0	26.9	26.6	25.8	24.1	22.6	21.8	21.8	22.4	23.6	25.1	26.2	24.5
69	22.0	22.2	22.1	20.9	18.9	17.1	16.2	16.3	16.5	18.7	20.1	21.4	19.4
70	26.2	26.1	25.6	24.5	22.5	20.8	19.9	20.1	20.9	22.5	24.4	25.4	23.2
71	26.5	26.6	26.7	25.8	23.9	22.3	21.2	21.5	22.4	24.4	26.1	26.7	24.5
72	22.4	22.0	21.9	21.3	19.6	17.4	16.4	17.0	17.9	20.1	21.7	21.9	19.9

CLIMATOLOGICAL DATA FOR MALAGASY

GEOGRAPHY

OBS.	LOCATION	YEARS	LATITUDE	LONGITUDE	ELEVATION -M-	LIFEZONE	CCTX	ID	ONMAP
73	MANTASOA	16	19.02 S	47.83 E	1399.6	WM	EC43	72	NO
74	MAROANTSETRA	20	15.43 S	49.73 E	2.1	TM	EW22	73	YES
75	MAROLAMBO	15	20.05 S	48.13 E	399.9	SW	EW33	74	NO
76	MAROTANDRANO	15	16.18 S	48.83 E	499.9	SM	EW53	75	NO
77	MAROVITSIKA	10	18.83 S	48.08 E	922.6	.	.	76	NO
78	MAROVOAY-MADIRO	20	16.12 S	46.62 E	20.1	TD	EH43	77	NO
79	MIANDRIVAZO	10	19.53 S	45.47 E	71.0	TD	EH53	78	NO
80	MIDONGY-DU-SUD	10	23.58 S	47.03 E	499.9	SM	EW43	79	NO
81	MORAFENOBE	17	17.83 S	44.93 E	146.9	TD	EH43	80	YES
82	MORAMANGA	21	18.95 S	48.22 E	911.7	SM	EW53	81	YES
83	MOROMBE	21	21.75 S	43.37 E	4.9	TT	EW84	82	YES
84	MORONDAVA	42	20.28 S	44.28 E	4.9	TV	EW63	83	NO

AVERAGE PRECIPITATION (MM)

OBS.	JAN.	FEB.	MAR.	APR.	MAY	JUN.	JUL.	AUG.	SEP.	OCT.	NOV.	DEC.	ANNUAL
73	282	308	199	63	32	28	42	37	22	57	137	312	1519
74	383	433	479	486	368	319	347	252	135	84	138	308	3730
75	318	400	381	163	95	86	106	64	62	44	125	263	2107
76	345	328	195	44	5	8	4	8	4	14	77	195	1227
77	.	.	.	.	.	.	.	.	.	.	.	.	.
78	466	352	260	77	4	3	1	5	9	22	96	227	1521
79	442	260	217	48	8	8	1	8	15	53	136	229	1426
80	359	329	281	105	51	81	44	60	44	62	146	299	1861
81	410	328	281	71	5	2	5	6	20	69	164	295	1656
82	289	256	195	72	42	43	43	35	25	31	120	263	1416
83	104	157	63	5	6	8	3	1	6	4	25	72	452
84	247	221	118	14	8	7	1	2	7	11	18	124	779

AVERAGE TEMPERATURE (DEGREES CENTIGRADE)

OBS.	JAN.	FEB.	MAR.	APR.	MAY	JUN.	JUL.	AUG.	SEP.	OCT.	NOV.	DEC.	ANNUAL
73	19.7	20.1	19.5	17.7	15.5	13.5	12.3	12.9	14.2	16.8	18.8	19.3	16.7
74	26.3	26.4	26.1	25.1	23.4	21.9	20.9	20.8	21.4	22.9	24.4	25.6	23.8
75	25.4	25.2	24.6	23.6	21.3	19.7	18.7	19.0	20.2	22.4	24.1	25.0	22.4
76	25.4	25.5	25.4	24.7	22.9	21.3	20.1	20.6	21.5	23.3	25.1	25.7	23.4
77	22.9	22.6	21.8	20.8	18.9	16.5	15.7	16.0	17.5	20.0	21.9	22.4	19.8
78	28.0	28.0	28.2	27.9	26.4	24.5	24.2	25.3	26.6	28.3	29.1	28.6	27.1
79	28.4	28.4	28.3	28.4	26.0	24.0	23.8	25.2	27.5	29.1	29.5	28.9	27.3
80	23.8	23.6	22.9	21.7	18.9	16.3	16.5	17.5	18.8	21.2	22.7	23.5	20.6
81	27.0	27.2	27.2	27.3	25.2	23.5	23.4	24.5	26.2	27.5	27.9	27.6	26.2
82	22.4	22.6	21.8	20.7	18.4	16.5	15.5	15.7	16.9	19.3	21.3	22.1	19.4
83	28.0	28.5	27.5	26.0	23.0	21.4	20.8	21.5	22.9	24.2	25.9	27.3	24.7
84	27.8	27.7	27.5	26.2	23.4	21.4	21.0	21.6	23.3	25.0	26.7	27.5	24.9

CLIMATOLOGICAL DATA FOR MALAGASY

GEOGRAPHY

OBS.	LOCATION	YEARS	LATITUDE	LONGITUDE	ELEVATION -M-	LIFEZONE	CCTX	ID	ONMAP
85	NAHAMPOANA	15	24.95 S	46.98 E	25.0	SW	SW32	84	NO
86	NANISANA	31	18.88 S	47.55 E	1258.5	SM	EW53	85	NO
87	NANOKELY	12	19.52 S	47.03 E	2019.3	WM	EC43	86	NO
88	NOSY-VARIKA	20	20.57 S	48.53 E	15.8	TM	EW22	87	YES
89	PORT BERGE	21	15.55 S	47.67 E	21.9	TD	EH43	88	NO
90	RENDRIRENDRY	10	17.95 S	49.22 E	549.9	.	.	89	NO
91	SAKARAHA	10	22.92 S	44.53 E	460.9	.	.	90	NO
92	SAMBAVA	10	14.27 S	50.15 E	4.0	.	.	91	YES
93	SOALALA	21	16.07 S	45.33 E	4.0	TD	EH53	92	YES
94	SOANIERANA-IVONGO	18	16.92 S	49.58 E	2.1	TM	EW22	93	NO
95	SOAVINANDRIANA	20	19.17 S	46.75 E	1574.6	SW	EC33	94	NO
96	TAMATAVE	39	18.15 S	49.40 E	4.9	TM	EW22	95	NO

AVERAGE PRECIPITATION (MM)

OBS.	JAN.	FEB.	MAR.	APR.	MAY	JUN.	JUL.	AUG.	SEP.	OCT.	NOV.	DEC.	ANNUAL
85	240	295	297	120	153	188	134	144	94	124	117	210	2117
86	304	278	193	44	11	7	5	8	9	46	184	256	1344
87	376	307	236	91	35	13	16	9	21	107	185	271	1667
88	370	386	431	270	225	248	156	147	113	74	152	243	2815
89	479	407	264	43	5	1	2	3	2	18	110	243	1576
90	.	.	.	.	.	.	.	.	.	.	.	.	.
91	.	.	.	.	.	.	.	.	.	.	.	.	.
92	.	.	.	.	.	.	.	.	.	.	.	.	.
93	409	316	221	40	9	1	1	3	2	5	54	157	1217
94	439	426	448	456	313	335	326	258	166	120	161	272	3721
95	455	375	488	98	30	12	13	10	23	69	190	390	2154
96	393	437	506	414	315	313	260	218	139	87	186	261	3529

AVERAGE TEMPERATURE (DEGREES CENTIGRADE)

OBS.	JAN.	FEB.	MAR.	APR.	MAY	JUN.	JUL.	AUG.	SEP.	OCT.	NOV.	DEC.	ANNUAL
85	25.9	25.7	25.0	23.7	21.2	19.8	19.2	19.5	21.2	22.4	24.0	25.1	22.7
86	21.4	21.2	20.8	19.5	17.2	15.1	14.2	14.7	16.2	18.6	20.3	20.9	18.3
87	16.1	16.1	16.0	14.7	12.4	10.9	10.3	10.5	12.9	14.9	15.9	16.1	13.9
88	26.1	26.0	25.6	24.7	22.8	21.2	20.5	20.7	21.5	22.9	24.5	25.7	23.5
89	27.8	28.0	28.2	27.5	25.7	23.9	23.1	24.0	25.8	27.8	29.0	28.7	26.6
90	25.5	26.0	24.8	23.7	21.7	19.8	19.1	19.1	19.7	21.2	23.3	24.5	22.4
91	26.5	26.3	25.5	24.0	20.4	18.2	18.0	19.3	21.8	24.4	26.2	26.4	23.1
92	27.0	27.5	27.2	26.6	25.2	23.5	22.5	22.7	23.1	24.3	25.6	26.6	25.1
93	27.5	27.7	27.9	28.2	26.3	24.6	24.2	24.7	25.8	26.8	27.8	28.5	26.6
94	26.2	26.1	25.8	25.0	23.3	21.7	21.1	21.0	21.6	22.8	24.3	25.5	23.7
95	19.4	19.2	19.1	18.4	16.3	14.9	14.2	15.2	16.7	18.3	19.1	19.3	17.5
96	26.7	26.7	26.2	25.3	23.6	22.1	21.3	21.3	22.1	23.4	24.9	26.2	24.1

CLIMATOLOGICAL DATA FOR MALAGASY

GEOGRAPHY

OBS.	LOCATION	YEARS	LATITUDE	LONGITUDE	ELEVATION -M-	LIFEZONE	CCTX	ID	ONMAP
97	TAMBOHORANO	23	17.50 S	43.95 E	20.1	TD	EW53	96	NO
98	TANANARIVE	11	18.92 S	47.55 E	1380.4	WM	EC53	97	NO
99	TRANOROA	10	24.72 S	45.05 E	191.4	.	.	98	NO
100	TSARATANANA	20	16.78 S	47.65 E	549.9	TD	EW43	99	NO
101	TSIHOMBE	24	25.30 S	45.48 E	64.0	TV	SW73	100	NO
102	TSIMBAZAZA	14	18.93 S	47.53 E	1260.7	SM	EW53	101	NO
103	TSINJOARIVO	20	19.62 S	47.70 E	1639.5	WM	EC53	102	NO
104	TSIROANOMANDIDY	16	18.75 S	46.05 E	899.8	SM	EW43	103	NO
105	TSIVORY	20	24.05 S	46.07 E	399.0	SD	SW63	104	NO
106	TULEAR	45	23.35 S	43.68 E	6.1	TT	EW83	105	NO
107	VATOMANDRY	46	19.33 S	48.97 E	3.0	TM	EW22	106	YES
108	VOHEMAR	45	13.37 S	50.03 E	2.1	TD	EW52	107	NO

AVERAGE PRECIPITATION (MM)

OBS.	JAN.	FEB.	MAR.	APR.	MAY	JUN.	JUL.	AUG.	SEP.	OCT.	NOV.	DEC.	ANNUAL
97	327	260	178	24	11	3	3	4	7	17	47	166	1047
98	309	241	204	51	18	9	7	9	12	52	156	287	1354
99	.	.	.	.	.	.	.	.	.	.	.	.	.
100	566	399	302	62	10	2	1	3	2	46	140	353	1887
101	82	102	59	21	24	33	13	9	15	17	34	97	504
102	294	290	199	56	11	9	7	9	8	52	146	250	1330
103	243	223	190	62	35	32	24	26	24	42	135	245	1280
104	393	364	258	89	14	7	6	7	13	59	167	304	1681
105	239	158	124	39	16	22	13	10	14	24	101	212	973
106	73	68	46	7	19	11	4	4	10	15	39	46	343
107	356	409	469	280	227	280	198	185	125	80	145	233	2988
108	202	175	179	169	72	78	73	74	61	64	97	200	1445

AVERAGE TEMPERATURE (DEGREES CENTIGRADE)

OBS.	JAN.	FEB.	MAR.	APR.	MAY	JUN.	JUL.	AUG.	SEP.	OCT.	NOV.	DEC.	ANNUAL
97	27.8	27.7	27.8	27.4	24.7	23.5	23.2	23.7	24.9	26.3	27.3	28.0	26.0
98	20.1	20.1	19.4	18.4	16.3	14.4	13.4	13.7	15.1	17.9	19.4	19.8	17.3
99	27.9	27.6	26.8	25.3	21.6	19.5	19.4	20.8	23.1	25.5	27.3	27.8	24.4
100	25.8	26.0	26.2	26.2	24.4	23.0	22.3	23.1	24.3	26.0	26.8	26.7	25.1
101	27.8	27.7	26.9	25.3	21.6	19.7	19.5	20.6	22.3	24.6	26.3	27.5	24.2
102	21.7	21.6	21.3	19.9	17.6	15.4	14.6	15.2	17.0	19.4	20.9	21.4	18.8
103	18.7	18.8	18.4	17.1	15.1	13.1	11.9	12.5	14.2	16.4	18.3	18.6	16.1
104	24.0	23.9	23.9	23.3	21.6	19.4	18.8	20.2	22.0	23.7	24.1	24.0	22.4
105	26.5	26.3	25.5	24.2	20.5	17.5	18.3	19.6	22.0	24.6	24.6	26.6	23.2
106	27.4	27.5	26.7	24.9	22.2	20.3	19.9	20.4	21.9	23.2	25.1	26.5	23.8
107	26.2	26.2	25.8	24.8	23.1	21.7	21.0	20.8	21.5	22.7	24.3	25.5	23.6
108	27.3	27.3	26.7	26.7	25.6	24.1	23.2	23.2	23.8	24.8	26.1	26.9	25.5

CLIMATOLOGICAL DATA FOR MALAGASY

GEOGRAPHY

OBS.	LOCATION	YEARS	LATITUDE	LONGITUDE	ELEVATION -M-	LIFEZONE	CCTX	ID	ONMAP
109	VOHIDIALA	10	17.85 S	47.25 E	772.7	.	.	108	NO

AVERAGE PRECIPITATION (MM)

OBS.	JAN.	FEB.	MAR.	APR.	MAY	JUN.	JUL.	AUG.	SEP.	OCT.	NOV.	DEC.	ANNUAL
109	.	.	.	.	.	.	.	.	.	.	.	.	.

AVERAGE TEMPERATURE (DEGREES CENTIGRADE)

OBS.	JAN.	FEB.	MAR.	APR.	MAY	JUN.	JUL.	AUG.	SEP.	OCT.	NOV.	DEC.	ANNUAL
109	24.0	23.6	23.6	22.6	20.6	18.9	17.8	18.0	19.4	21.8	23.4	24.0	21.5

MALAWI

Agriculture

Of 11,848,000 ha in total area, Malawi was estimated in 1978 to have 2,298,000 ha of arable or permanent cropland, 1,840,000 ha in permanent pasture, 2,314,000 ha in forest and woodland, and 2,956,000 ha in other terrain categories. In 1978, approximately 10,000 ha were irrigated. Of the 1979 population, 84.3% or 2,253,000 individuals of the economically active population were engaged in agriculture. The total population in 1979 was 5,963,000. According to FAO statistics for 1979, there were 790,000 cattle, 174,000 pigs, 140,000 sheep, 860,000 goats, and 8,000,000 chickens.[2] Agricultural production is summarized in Table 1. Conventional and energetic yield data on most of these will be found in Appendixes I and II.

Climate

Data from Wernstedt[3] are presented in the Climatological Data Table (see Introduction for explanation). Based on the means for data points cited in Wernstedt, the mean annual precipitation is 10.7 dm (mean of 21 sites) and the mean annual temperature 21.9°C (mean of 22 sites). The averages would suggest that much of the country could be classified as Subtropical Moist Forest Life Zone.[4]

Productivity

The vegetation map depicts very roughly the major land forms in Malawi, the savanna probably exhibiting productivity of ca. 3 to 6 MT/ha/yr, the thorn forest 6 to 12, and the tropical montane forest 12 to 24, perhaps even higher in the hotter wetter areas. Swamps around the lake might be capable of growing energy weeds like reed, cattail, water hyacinth, etc.

One Salvador-type *Leucaena* (Hawaiian Giant K8) is being grown as fuelwood for generating steam at a sugar factory.[28] *Gmelina arborea* can yield >30 m³/yr, the wood with sp gr of 0.42 to 0.62, the sapwood with 4800 kcal/kg. The wood is used for curing tobacco in Malawi. *Trema guineensis* is scarcely distinguishable from *T. orientalis.* Both have been known as "charcoal tree".[33]

Economics

Food consumption per capita was 2237 cal/day, energy consumption 51 kg coal equivalent per capita in 1977. In 1970, the caloric supply was 103% of requirements. The total population growth rate was 2.8% in 1970. Population density was 38 persons per square kilometer in 1970, with 10.0% of the total population urban. Life expectancy at birth of those born in 1970 was 42 years. In 1977, with GNP at $150 per capita, agriculture accounted for 44.1% of GDP, manufacturing for 14.9%, other industry for 8.1%, and services for 32.9%. In 1977, food and beverage constituted 91.0% of exports, nonfood agriculture for 1.7%, fuels, minerals, and metal for 0.1%, machines for 2.3%, and other manufacturing for 5.0% of total merchandise exports. Food and beverage constituted 8.1% of imports, nonfood agriculture 0.7%, fuel and lubricants 13.0%, nonfuel minerals and metals 1.7%, machinery and equipment 30.1%, and other manufactured goods constituted 46.4% of total merchandise imports.[10]

Energetics

Malawi has 14 million ton coal, and a hydroelectric potential of 100 MW. Importing more than 50% (51 to 75%) of its commercial energy demands as oil, Malawi is expected to be experiencing fuelwood problems by the year 2000.[10]

Table 1
AGRICULTURAL PRODUCTION IN MALAWI

Crop	Area planted (ha)	Yield (kg/ha)	Total production (MT)
Wheat		2,500	1,000
Rice paddy	52,000	962	50,000
Corn	1,000,000	1,200	1,200,000
Potatoes	30,000	3,667	110,000
Cassava	15,000	5,333	80,000
Dry beans	96,000	625	60,000
Chick peas	23,000	739	17,000
Peanuts in shell	250,000	680	170,000
Sunflower seed	4,000	800	3,000
Seed cotton	34,000	1,131	38,000
Cottonseed			23,000
Cabbages	2,000	11,765	20,000
Tomatoes	3,000	9,000	27,000
Dry onions	4,000	3,077	12,000
Sugar cane	11,000	100,000	1,100,000
Sugar (centrifugal, raw)			109,000
Citrus fruit nes			1,000
Mangoes	26,000		
Bananas			11,000
Plantains			15,000
Green coffee	2,000	667	1,000
Tea	18,000	1,863	33,000
Tobacco leaves	79,000	681	54,000
Cotton lint			10,000

From FAO 1979 Production Yearbook, Vol. 33, Food and Agriculture Organization, Rome, 1980.

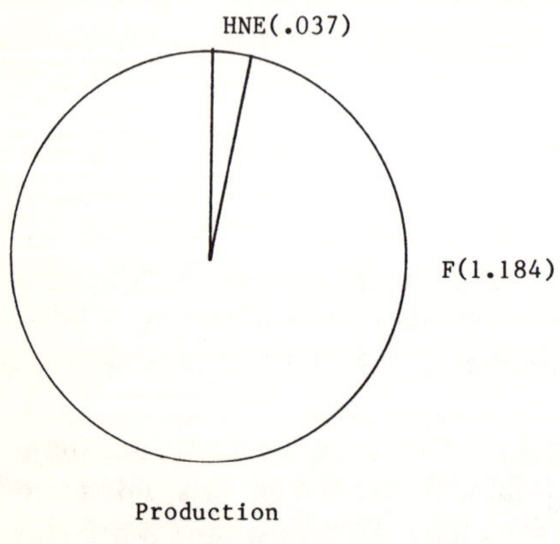

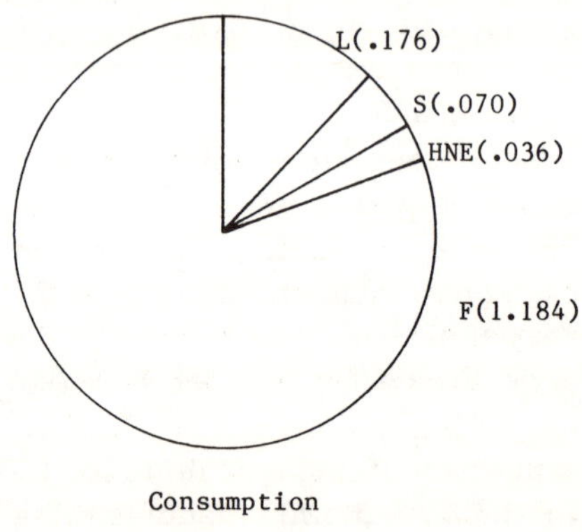

1978 Energy production and consumption in Malawi; fuel types listed in million metric tons coal equivalents (MMTCE). 1 MMTCE = 27.78×10^{12} Btu. L = liquid fuel, S = solid fuel, HNE = hydro, nuclear, and electric fuel, F = firewood, NG = natural gas. (Data from United Nations, World Energy Supplies, 1973—1978, Statistical Papers Series J, 22, Department of International Economic and Social Affairs, Statistical Office, New York, 1979.)

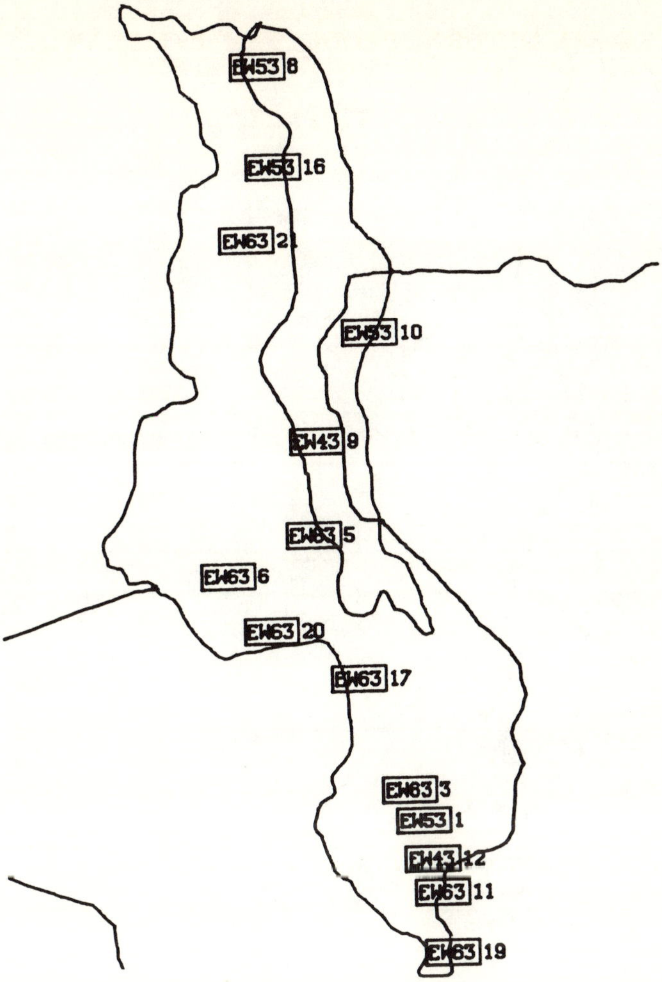

Selected weather stations for Malawi.

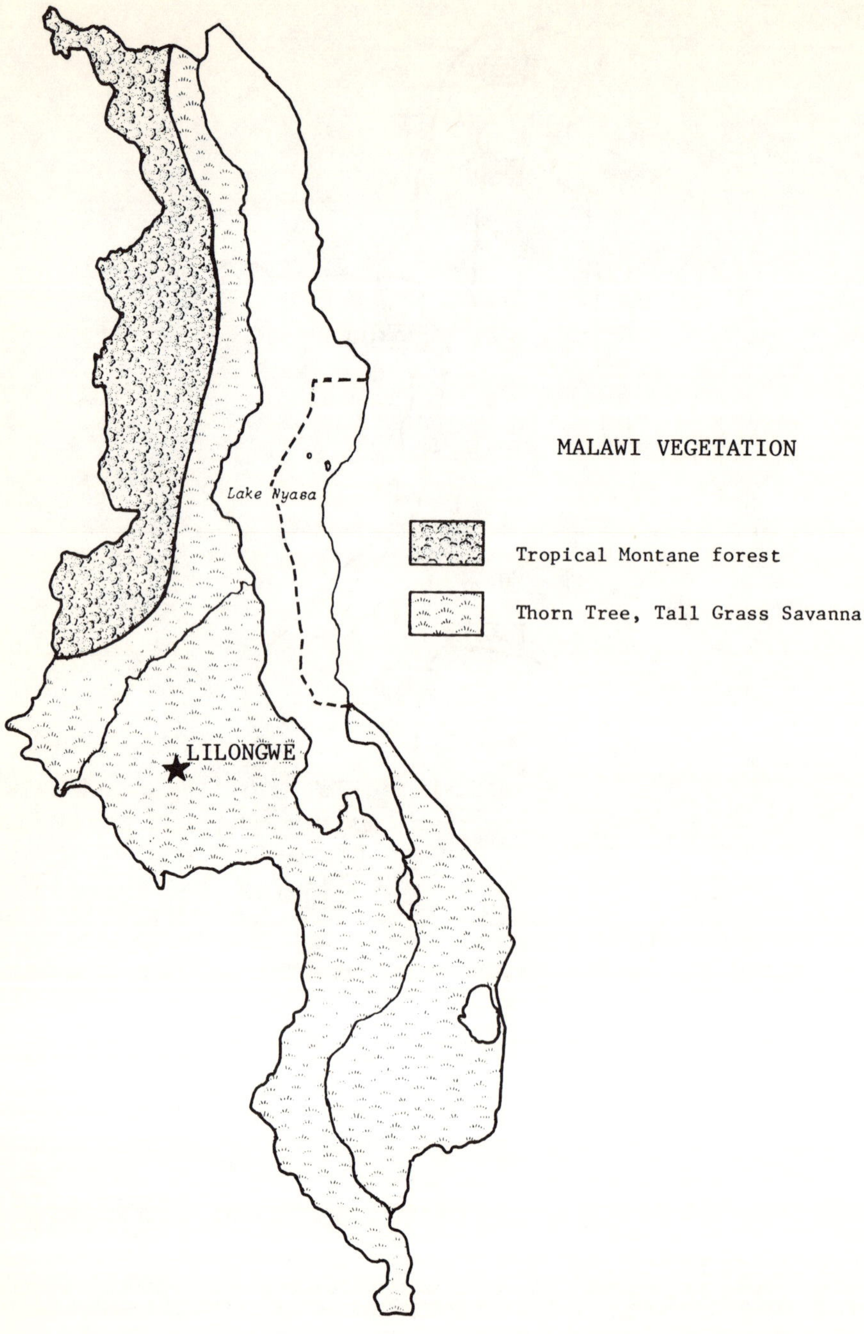
Lake Nyasa
LILONGWE
MALAWI VEGETATION
Tropical Montane forest
Thorn Tree, Tall Grass Savanna

CLIMATOLOGICAL DATA FOR MALAWI

GEOGRAPHY

OBS.	LOCATION	YEARS	LATITUDE	LONGITUDE	ELEVATION -M-	LIFEZONE	CCTX	ID	ONMAP
1	BVUMBWE	15	15.92 S	35.07 E	1146.0	SM	EW53	1	YES
2	CHIKWAWA	11	16.03 S	34.78 E	152.4	TV	EW63	2	NO
3	CHILEKA	23	15.68 S	34.97 E	768.1	SD	EW63	3	YES
4	CHIRIMBA	11	15.75 S	35.02 E	975.4	SM	EW53	4	NO
5	CHITALA	11	13.67 S	34.28 E	603.5	SD	EW63	5	YES
6	CHITEDZE RESEARCH	12	13.98 S	33.63 E	1149.1	SD	EW63	6	YES
7	FORT JOHNSTON	15	14.43 S	35.25 E	484.3	TV	EW63	7	NO
8	KARONGA	19	9.95 S	33.93 E	484.6	TD	EW53	8	YES
9	KOTA KOTA	15	12.93 S	34.32 E	500.2	TD	EW43	9	YES
10	LIKOMA ISLAND	11	12.08 S	34.73 E	512.7	TD	EW53	10	YES
11	LILONGWE	21	13.97 S	33.70 E	1134.5	SD	EW63	22	NO
12	MAKANGA	12	16.48 S	35.20 E	76.2	TV	EW63	11	YES

AVERAGE PRECIPITATION (MM)

OBS.	JAN.	FEB.	MAR.	APR.	MAY	JUN.	JUL.	AUG.	SEP.	OCT.	NOV.	DEC.	ANNUAL
1	242	240	190	94	20	26	12	13	11	34	114	230	1226
2	201	168	102	52	22	20	18	13	6	15	67	149	834
3	211	185	124	40	8	4	3	1	5	23	86	155	845
4	234	225	150	52	6	6	5	3	5	24	110	233	1055
5	248	240	142	43	1	0	0	1	3	3	47	149	877
6	230	211	128	36	7	1	0	1	1	2	82	192	891
7	240	205	120	39	5	2	2	3	5	14	48	144	826
8	172	151	301	210	35	4	1	2	0	6	34	145	1061
9	346	387	387	156	43	8	6	3	3	5	60	216	1620
10	312	397	171	57	7	1	0	1	1	0	20	218	1183
11	202	223	152	39	5	1	0	2	5	5	69	142	845
12	157	132	94	51	19	17	9	9	6	14	94	172	774

AVERAGE TEMPERATURE (DEGREES CENTIGRADE)

OBS.	JAN.	FEB.	MAR.	APR.	MAY	JUN.	JUL.	AUG.	SEP.	OCT.	NOV.	DEC.	ANNUAL
1	21.1	20.9	20.2	19.6	17.5	15.4	15.3	16.7	19.2	21.7	22.0	21.3	19.2
2	27.7	27.8	26.9	26.1	23.0	20.3	20.9	22.2	25.2	29.0	28.8	28.4	25.6
3	23.9	23.8	23.3	22.6	20.5	18.5	18.3	19.9	22.7	25.5	25.6	24.5	22.4
4	22.7	22.6	22.1	21.5	19.3	17.3	17.2	18.6	21.2	24.1	24.1	23.2	21.2
5	24.7	24.3	24.4	23.2	21.3	19.1	19.0	20.4	23.0	26.1	27.0	26.1	23.2
6	22.0	21.8	21.2	20.4	18.4	16.2	15.6	17.3	19.9	22.6	23.6	22.4	20.1
7	25.7	25.4	25.3	24.5	22.2	20.3	20.0	21.4	24.2	27.2	27.8	26.4	24.2
8	25.8	25.6	25.3	25.1	23.7	21.6	20.9	21.7	23.4	25.8	27.1	26.3	24.3
9	24.7	24.6	24.7	24.3	22.4	20.6	20.2	21.1	23.4	26.2	27.3	25.7	23.8
10	25.8	25.3	25.8	25.9	24.9	23.1	22.4	23.5	25.3	27.7	28.4	26.7	25.4
11	22.1	21.9	21.6	20.3	17.8	15.5	15.0	16.6	19.4	22.3	23.4	22.7	19.9
12	27.7	27.6	26.9	25.7	22.8	20.2	20.4	22.2	24.9	28.5	28.6	27.9	25.3

CLIMATOLOGICAL DATA FOR MALAWI

GEOGRAPHY

OBS.	LOCATION	YEARS	LATITUDE	LONGITUDE	ELEVATION -M-	LIFEZONE	CCTX	ID	ONMAP
13	MASAMBANJATI	11	16.22 S	35.13 E	993.6	SM	EW43	12	YES
14	MLANJE	11	16.08 S	35.67 E	628.2	SM	EW43	13	NO
15	MZIMBA	15	11.87 S	33.67 E	1351.2	SD	EW63	14	NO
16	MZUZU	12	11.45 S	33.92 E	1257.3	SM	EW53	15	NO
17	NCHENACHEMA	12	10.75 S	34.03 E	1289.3	SM	EW53	16	YES
18	NCHEU	14	14.80 S	34.60 E	1280.2	SD	EW63	17	YES
19	NKHOTAKOTA	10	12.92 S	34.28 E	499.9	.	.	18	NO
20	PORT HERALD	21	16.95 S	35.27 E	57.9	TV	EW63	19	YES
21	TIWI	12	14.42 S	33.95 E	1249.7	SD	EW63	20	YES
22	ZOMBWE	13	11.33 S	33.82 E	1310.6	SD	EW63	21	YES

AVERAGE PRECIPITATION (MM)

OBS.	JAN.	FEB.	MAR.	APR.	MAY	JUN.	JUL.	AUG.	SEP.	OCT.	NOV.	DEC.	ANNUAL
13	289	268	235	135	68	60	46	40	22	34	144	247	1588
14	318	316	340	172	71	60	34	36	18	67	188	288	1906
15	221	213	158	45	3	1	1	0	2	3	60	163	870
16	229	245	240	263	107	37	32	23	14	41	94	173	1498
17	224	215	222	160	58	13	7	3	7	9	53	158	1129
18	233	259	155	49	8	2	1	3	1	10	65	184	970
19	.	.	.	.	.	.	.	.	.	.	.	.	.
20	193	179	117	45	17	18	11	10	9	12	112	153	876
21	175	228	190	50	15	1	1	3	1	7	67	177	914
22	185	215	148	46	15	2	2	0	2	8	44	105	772

AVERAGE TEMPERATURE (DEGREES CENTIGRADE)

OBS.	JAN.	FEB.	MAR.	APR.	MAY	JUN.	JUL.	AUG.	SEP.	OCT.	NOV.	DEC.	ANNUAL
13	21.8	21.9	21.1	20.3	18.0	15.9	16.2	17.5	20.1	22.8	22.9	22.3	20.1
14	24.0	23.9	23.2	21.9	19.3	17.2	17.0	18.2	20.9	24.3	24.7	24.3	21.6
15	20.8	20.7	20.7	20.3	18.4	16.3	15.7	17.1	19.8	22.3	22.9	21.5	19.7
16	20.4	20.6	20.4	19.6	17.4	14.9	14.0	15.4	17.5	20.3	21.7	20.9	18.6
17	21.3	21.1	21.2	20.3	18.4	16.3	16.2	17.4	19.3	21.7	22.8	22.0	19.8
18	20.9	20.6	20.4	19.9	18.2	16.3	16.0	17.4	19.9	22.5	22.7	21.3	19.7
19	24.8	24.6	24.7	24.3	22.9	20.7	20.2	21.2	23.5	26.2	27.3	25.9	23.9
20	28.1	27.9	27.1	26.1	23.4	20.9	20.9	22.8	25.7	28.7	29.1	28.4	25.7
21	20.8	20.8	20.5	19.5	16.9	14.8	14.4	16.0	18.3	21.2	22.7	21.8	19.0
22	22.0	21.6	21.6	20.8	18.4	16.0	15.6	16.8	18.9	22.1	23.2	22.2	19.9

MALI

Agriculture

Of 124,000,000 ha in total area, Mali was estimated in 1978 to have 2,050,000 ha of arable or permanent cropland, 30,000,000 ha in permanent pasture, 4,457,000 ha in forest and woodland, and 85,493,000 ha in other terrain categories. In 1978, approximately 95,000 ha were irrigated. Of the 1979 population, 87.4% or 3,050,000 individuals of the economically active population were engaged in agriculture. The total population in 1979 was 6,465,000. According to FAO statistics for 1979, there were 180,000 horses, 489,000 asses, 4,459,000 cattle, 208,000 camels, 31,000 pigs, 6,067,000 sheep, 5,757,000 goats, and 10,884,000 chickens.[2] Agricultural production is summarized in Table 1. Conventional and energetic yield data on most of these will be found in Appendixes I and II.

Climate

Data from Wernstedt[3] are presented in the Climatological Data Table (see Introduction for explanation). Based on the means for data points cited in Wernstedt, the mean annual precipitation is 6.9 dm (mean of 75 sites) and the mean annual temperature 28.3°C (mean of 17 sites). The averages would suggest that much of the country could be classified as Tropical Very Dry Forest Life Zone.[4]

Productivity

The vegetation map shows the major vegetation types, whose annual productivity potentials are estimated at (in MT/ha/yr): deciduous forest, 6 to 24; acacia savanna, 6 to 12; tall grass and brush, 3 to 12; marsh grass and crops, 3 to 12; and desert vegetation, 0.

Economics

Food consumption per capita was 2117 cal/day, energy consumption 30 kg coal equivalent per capita in 1977. In 1970, the caloric supply was 75.0% of requirements. The total population growth rate was 2.4% in 1970. Population density was 4 persons per square kilometer in 1970, with 14.9% of the total population urban. Life expectancy at birth of those born in 1970 was 39 years. In 1977, with GNP at $120 per capita, agriculture accounted for 38.0% of GDP, manufacturing for 12.1%, other industry for 6.2%, and services for 43.7%. In 1976, food and beverage constituted 46.5% of exports, nonfood agriculture for 52.0%, fuels, minerals, and metal for 0.1%, machines for 0.5%, and other manufacturing for 0.9% of total merchandise exports. Food and beverage constituted 19.1% of imports, nonfood agriculture 0.7%, fuel and lubricants 14.3%, nonfuel minerals and metals 1.5%, machinery and equipment 29.9%, and other manufactured goods constituted 34.5% of total merchandise imports.[10]

Energetics

Mali has a hydroelectric potential of 3520 MW. Importing more than 75% of its commercial energy demands as oil, Mali is expected to be experiencing fuelwood problems by the year 2000.[10]

Table 1
AGRICULTURAL PRODUCTION IN MALI

Crop	Area planted (ha)	Yield (kg/ha)	Total production (MT)
Wheat	2,000	1,333	2,000
Rice paddy	160,000	1,107	177,000
Corn	90,000	667	60,000
Millet	1,400,000	531	744,000
Sweet potatoes	3,000	12,467	37,000
Cassava	5,000	8,000	40,000
Peanuts in shell	200,000	893	179,000
Seed cotton	109,000	1,239	135,000
Cottonseed			80,000
Tomatoes	2,000	3,750	6,000
Dry onions		12,963	4,000
Sugar cane	2,000	66,727	147,000
Sugar (centrifugal, raw)			16,000
Tobacco leaves	1,000	800	
Jute	3,000	500	1,000
Cotton lint			48,000

From FAO 1979 Production Yearbook, Vol. 33, Food and Agriculture Organization, Rome, 1980.

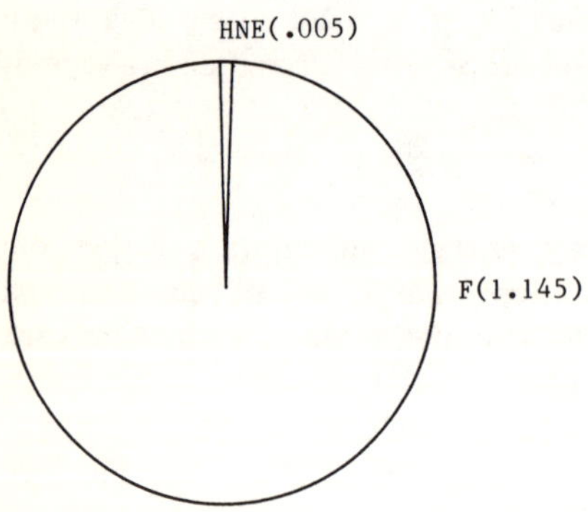

Production

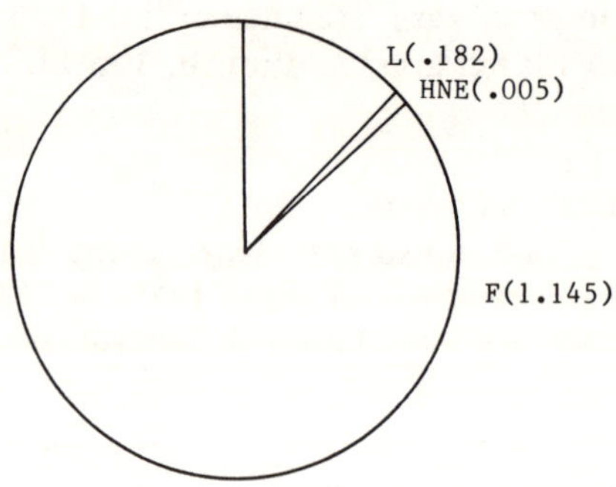

Consumption

1978 Energy production and consumption in Mali; fuel types listed in million metric tons coal equivalents (MMTCE). 1 MMTCE = 27.78×10^{12} Btu. L = liquid fuel, S = solid fuel, HNE = hydro, nuclear, and electric fuel, F = firewood, NG = natural gas. (Data from United Nations, World Energy Supplies, 1973—1978, Statistical Papers Series J, 22, Department of International Economic and Social Affairs, Statistical Office, New York, 1979.)

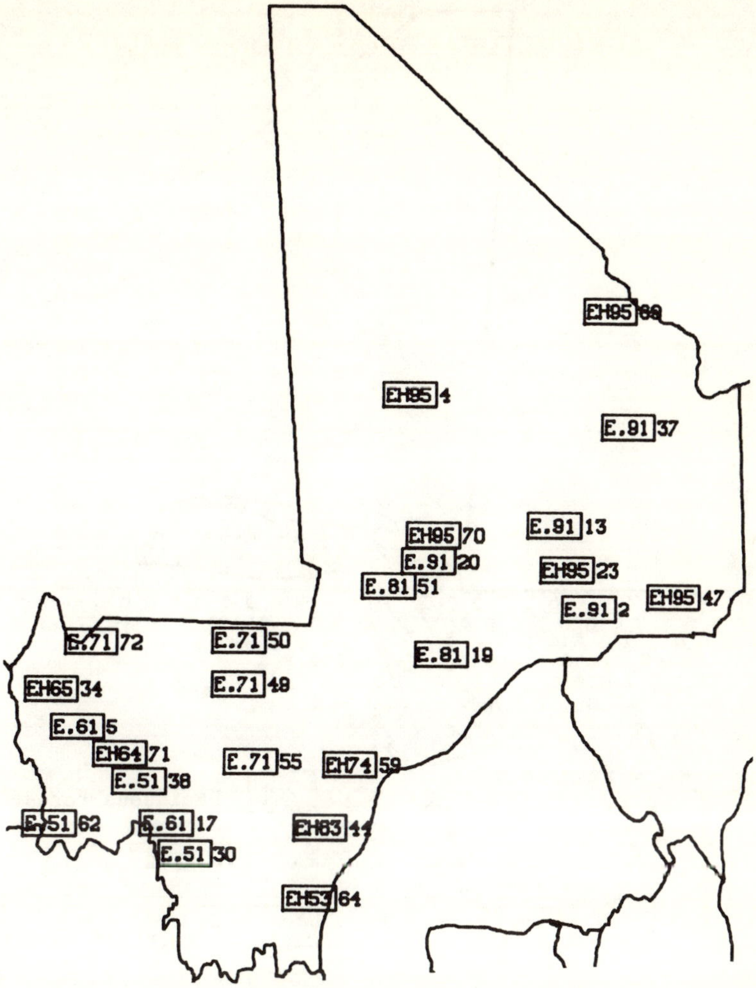

Selected weather stations for Mali.

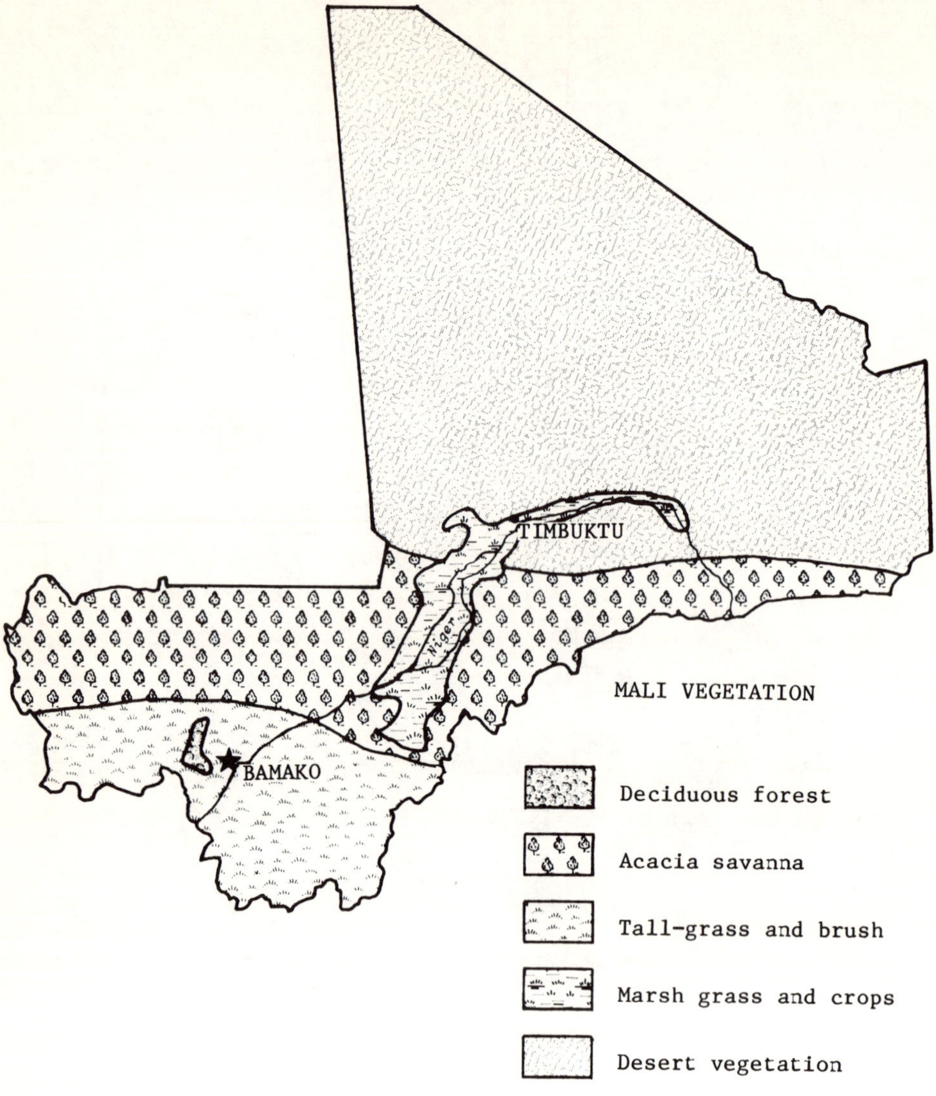
TIMBUKTU
Niger
BAMAKO
MALI VEGETATION
Deciduous forest
Acacia savanna
Tall-grass and brush
Marsh grass and crops
Desert vegetation

CLIMATOLOGICAL DATA FOR MALI

GEOGRAPHY

OBS.	LOCATION	YEARS	LATITUDE	LONGITUDE	ELEVATION -M-	LIFEZONE	CCTX	ID	ONMAP
1	AMBIDEDI	10	14.58 N	11.78 E	.	.	.	1	NO
2	ANSONGO	25	15.68 N	0.42 E	264.9	.	.	2	YES
3	AOUROU	10	14.95 N	11.58 E	.	.	.	3	NO
4	ARAOUAN	10	18.90 N	3.55 W	.	TO	EH95	4	YES
5	BAFOULABE	18	13.87 N	10.83 W	92.0	.	.	5	YES
6	BAMAKO	36	12.65 N	7.97 W	331.0	TD	EH53	6	NO
7	BANAMBA	17	13.55 N	7.33 W	527.9	.	.	7	NO
8	BANDIAGARA	29	14.33 N	3.60 W	425.8	.	.	8	NO
9	BANKASS	10	14.07 N	3.52 E	.	.	.	9	NO
10	BAROUELI	17	17.05 N	6.88 W	327.1	.	.	10	NO
11	BELEKO	10	13.47 N	7.12 E	.	.	.	11	NO
12	BOUGOUNI	34	11.42 N	7.50 W	349.9	TD	EH53	12	NO

AVERAGE PRECIPITATION (MM)

OBS.	JAN.	FEB.	MAR.	APR.	MAY	JUN.	JUL.	AUG.	SEP.	OCT.	NOV.	DEC.	ANNUAL
1	0	2	0	0	14	108	140	264	181	45	2	1	757
2	0	0	0	1	21	27	73	136	34	5	0	0	297
3	0	2	0	4	4	79	136	183	140	37	7	0	592
4	0	1	0	0	0	5	5	12	16	1	1	1	42
5	0	0	0	5	39	135	217	326	207	50	3	1	983
6	1	0	4	17	69	137	231	335	209	62	10	0	1075
7	1	0	5	12	22	105	197	289	134	28	4	1	797
8	0	0	2	12	22	65	149	193	107	28	3	0	580
9	0	0	2	3	29	87	205	200	99	26	6	0	657
10	0	0	3	23	24	101	218	280	130	22	4	0	805
11	0	0	5	19	51	158	249	305	187	43	8	2	1027
12	1	2	10	34	98	161	280	363	255	92	14	1	1311

AVERAGE TEMPERATURE (DEGREES CENTIGRADE)

OBS.	JAN.	FEB.	MAR.	APR.	MAY	JUN.	JUL.	AUG.	SEP.	OCT.	NOV.	DEC.	ANNUAL
1	.	.	.	.	.	.	.	.	.	.	.	.	.
2	.	.	.	.	.	.	.	.	.	.	.	.	.
3	.	.	.	.	.	.	.	.	.	.	.	.	.
4	18.2	20.9	26.0	31.2	34.2	37.1	35.0	33.3	33.9	30.7	24.0	18.6	28.6
5	.	.	.	.	.	.	.	.	.	.	.	.	.
6	25.2	27.8	30.5	32.1	31.7	28.5	26.6	26.0	26.4	27.4	26.8	25.0	27.8
7	.	.	.	.	.	.	.	.	.	.	.	.	.
8	.	.	.	.	.	.	.	.	.	.	.	.	.
9	.	.	.	.	.	.	.	.	.	.	.	.	.
10	.	.	.	.	.	.	.	.	.	.	.	.	.
11	.	.	.	.	.	.	.	.	.	.	.	.	.
12	25.0	27.5	29.8	30.6	29.5	27.2	25.6	25.2	25.8	26.6	26.5	24.8	27.0

CLIMATOLOGICAL DATA FOR MALI

GEOGRAPHY

OBS.	LOCATION	YEARS	LATITUDE	LONGITUDE	ELEVATION -M-	LIFEZONE	CCTX	ID	ONMAP
13	BOUREM	23	16.93 N	0.33 W	263.0	.	.	13	YES
14	DIA	10	14.32 N	4.98 W	.	.	.	14	NO
15	DIAMOU	10	14.08 N	11.27 E	.	.	.	15	NO
16	DIEMA	20	14.53 N	9.20 E	.	.	.	16	NO
17	DIOILA	10	12.45 N	8.83 W	324.9	.	.	17	YES
18	DJENNE	28	13.97 N	4.55 W	289.9	.	.	18	NO
19	DOUENTZA	23	14.98 N	2.83 W	304.8	.	.	19	YES
20	EL-OUALADJI	13	16.38 N	3.13 W	267.9	.	.	20	YES
21	FALADIE	18	13.18 N	8.33 W	332.8	.	.	21	NO
22	GALOUGO	10	13.83 N	11.07 E	.	.	.	22	NO
23	GAO	36	16.27 N	0.05 W	257.9	TT	EH95	23	YES
24	GOUALALA	10	11.22 N	8.23 E	.	.	.	24	NO

AVERAGE PRECIPITATION (MM)

OBS.	JAN.	FEB.	MAR.	APR.	MAY	JUN.	JUL.	AUG.	SEP.	OCT.	NOV.	DEC.	ANNUAL
13	0	0	0	1	3	20	46	66	20	1	0	0	157
14	0	0	0	0	16	46	126	250	118	8	0	0	564
15	0	1	0	2	16	124	180	283	194	56	4	0	860
16	0	0	22	16	38	140	163	225	135	39	4	1	783
17	0	0	2	19	56	114	186	288	139	52	2	0	856
18	1	0	2	5	22	59	163	217	103	16	1	0	588
19	0	1	4	6	15	77	127	159	71	17	1	0	477
20	1	0	1	0	2	21	81	120	54	10	0	0	290
21	1	1	4	15	41	125	239	329	187	47	8	1	998
22	0	1	0	4	27	137	178	268	220	72	6	0	912
23	1	0	1	1	6	24	70	126	38	4	0	0	271
24	0	4	12	46	103	184	268	418	279	119	22	1	1456

AVERAGE TEMPERATURE (DEGREES CENTIGRADE)

OBS.	JAN.	FEB.	MAR.	APR.	MAY	JUN.	JUL.	AUG.	SEP.	OCT.	NOV.	DEC.	ANNUAL
13	.	.	.	.	.	.	.	.	.	.	.	.	.
14	.	.	.	.	.	.	.	.	.	.	.	.	.
15	.	.	.	.	.	.	.	.	.	.	.	.	.
16	.	.	.	.	.	.	.	.	.	.	.	.	.
17	.	.	.	.	.	.	.	.	.	.	.	.	.
18	.	.	.	.	.	.	.	.	.	.	.	.	.
19	.	.	.	.	.	.	.	.	.	.	.	.	.
20	.	.	.	.	.	.	.	.	.	.	.	.	.
21	.	.	.	.	.	.	.	.	.	.	.	.	.
22	.	.	.	.	.	.	.	.	.	.	.	.	.
23	22.3	24.9	28.3	31.7	34.1	34.0	31.0	29.8	30.8	31.3	27.7	23.1	29.1
24	.	.	.	.	.	.	.	.	.	.	.	.	.

CLIMATOLOGICAL DATA FOR MALI

GEOGRAPHY

OBS.	LOCATION	YEARS	LATITUDE	LONGITUDE	ELEVATION -M-	LIFEZONE	CCTX	ID	ONMAP
25	GOUNDAM	29	16.45 N	3.67 W	274.9	.	.	25	NO
26	GOURMA-RHAROUS	23	16.93 N	1.93 W	270.1	.	.	26	NO
27	HOMBORI	35	15.33 N	1.68 W	286.8	TT	EH85	27	NO
28	KABARA	26	16.68 N	3.00 W	267.0	.	.	28	NO
29	KAMI	10	14.55 N	4.20 E	.	.	.	29	NO
30	KANGABA	10	12.00 N	8.42 W	370.0	.	.	30	YES
31	KARA	10	13.83 N	5.48 E	.	.	.	31	NO
32	KARANGASSO	10	12.47 N	5.02 E	.	.	.	32	NO
33	KATIBOUGOU	12	12.87 N	7.55 W	325.8	.	.	33	NO
34	KAYES	34	14.43 N	11.43 W	46.0	TV	EH65	34	YES
35	KAYO	10	13.25 N	7.60 E	.	.	.	35	NO
36	KENIEBA	14	12.83 N	11.25 W	135.9	TD	EH53	36	NO

AVERAGE PRECIPITATION (MM)

OBS.	JAN.	FEB.	MAR.	APR.	MAY	JUN.	JUL.	AUG.	SEP.	OCT.	NOV.	DEC.	ANNUAL
25	1	0	0	1	2	20	68	104	43	7	1	0	246
26	0	0	2	1	3	10	59	63	25	2	1	0	165
27	0	0	0	2	10	38	112	169	48	11	0	0	390
28	1	1	0	2	2	15	57	94	36	5	0	1	213
29	0	0	1	3	22	69	188	205	91	17	1	0	597
30	1	0	2	26	68	151	194	311	208	80	8	1	1048
31	0	0	0	3	17	77	176	210	101	10	2	0	596
32	0	0	2	19	164	143	304	207	128	46	8	1	1022
33	0	0	3	18	50	127	197	285	151	40	6	0	876
34	1	1	0	2	23	96	171	244	161	47	4	0	750
35	0	0	2	2	29	69	186	236	129	22	2	0	677
36	0	0	1	7	55	139	268	448	225	120	12	2	1353

AVERAGE TEMPERATURE (DEGREES CENTIGRADE)

OBS.	JAN.	FEB.	MAR.	APR.	MAY	JUN.	JUL.	AUG.	SEP.	OCT.	NOV.	DEC.	ANNUAL
25	.	.	.	.	.	.	.	.	.	.	.	.	.
26	.	.	.	.	.	.	.	.	.	.	.	.	.
27	24.4	26.5	30.3	33.2	34.9	33.3	30.2	28.3	29.8	31.4	29.2	25.2	29.7
28	.	.	.	.	.	.	.	.	.	.	.	.	.
29	.	.	.	.	.	.	.	.	.	.	.	.	.
30	.	.	.	.	.	.	.	.	.	.	.	.	.
31	.	.	.	.	.	.	.	.	.	.	.	.	.
32	.	.	.	.	.	.	.	.	.	.	.	.	.
33	.	.	.	.	.	.	.	.	.	.	.	.	.
34	25.3	27.8	31.1	33.7	34.9	31.6	28.4	27.2	27.5	28.7	28.2	20.5	28.7
35	.	.	.	.	.	.	.	.	.	.	.	.	.
36	26.0	28.3	30.8	32.2	32.4	29.1	26.8	26.4	26.8	27.2	26.4	25.3	28.1

CLIMATOLOGICAL DATA FOR MALI

GEOGRAPHY

OBS.	LOCATION	YEARS	LATITUDE	LONGITUDE	ELEVATION -M-	LIFEZONE	CCTX	ID	ONMAP
37	KIDAL	26	18.43 N	1.35 E	135.9	.	.	37	YES
38	KITA	18	13.07 N	9.45 W	338.9	.	.	38	YES
39	KOGONI	10	13.82 N	7.47 E	.	.	.	39	NO
40	KOLOKANI	24	13.53 N	7.98 W	299.9	.	.	40	NO
41	KONSEQUELA	10	12.40 N	5.88 E	.	.	.	41	NO
42	KOULIKORO	10	12.87 N	7.55 W	.	.	.	42	NO
43	KOULOUBA	10	13.48 N	9.15 E	.	.	.	43	NO
44	KOUTIALA	34	12.40 N	5.47 W	343.8	TV	EH63	44	YES
45	MACINA	10	14.07 N	5.35 W	.	.	.	45	NO
46	MARKALA	10	13.68 N	6.08 E	.	.	.	46	NO
47	MENAKA	22	15.87 N	2.32 E	379.8	TX	EH95	47	YES
48	MOPTI-SEVARE	34	14.50 N	4.20 W	267.9	TV	EH74	48	NO

AVERAGE PRECIPITATION (MM)

OBS.	JAN.	FEB.	MAR.	APR.	MAY	JUN.	JUL.	AUG.	SEP.	OCT.	NOV.	DEC.	ANNUAL
37	0	1	0	1	8	8	33	59	22	2	0	0	133
38	1	0	1	14	43	156	243	367	221	63	6	1	1116
39	0	0	2	0	10	63	177	206	115	19	3	0	595
40	1	1	1	14	30	117	201	288	150	39	6	0	847
41	0	1	9	19	54	126	174	272	205	57	9	0	926
42	0	0	2	11	33	107	191	289	176	33	14	0	856
43	0	0	4	13	54	142	228	317	207	57	7	0	1029
44	0	1	5	32	60	141	218	274	180	62	8	1	982
45	1	0	0	10	13	76	173	227	103	5	1	0	609
46	0	0	2	11	18	80	179	233	129	24	1	1	678
47	0	0	1	0	7	20	61	113	43	4	0	0	249
48	0	0	1	5	24	57	147	195	93	17	1	0	540

AVERAGE TEMPERATURE (DEGREES CENTIGRADE)

OBS.	JAN.	FEB.	MAR.	APR.	MAY	JUN.	JUL.	AUG.	SEP.	OCT.	NOV.	DEC.	ANNUAL
37	.	.	.	.	.	.	.	.	.	.	.	.	.
38	.	.	.	.	.	.	.	.	.	.	.	.	.
39	.	.	.	.	.	.	.	.	.	.	.	.	.
40	.	.	.	.	.	.	.	.	.	.	.	.	.
41	.	.	.	.	.	.	.	.	.	.	.	.	.
42	.	.	.	.	.	.	.	.	.	.	.	.	.
43	.	.	.	.	.	.	.	.	.	.	.	.	.
44	23.0	25.8	28.9	30.9	31.0	28.6	26.5	26.3	26.6	27.7	26.3	23.3	27.1
45	.	.	.	.	.	.	.	.	.	.	.	.	.
46	.	.	.	.	.	.	.	.	.	.	.	.	.
47	23.0	25.6	29.4	32.2	34.5	33.9	31.1	28.9	30.3	31.0	27.8	23.6	29.2
48	22.5	25.3	28.5	31.2	32.8	31.1	28.4	26.9	27.6	28.1	26.2	23.2	27.7

CLIMATOLOGICAL DATA FOR MALI

GEOGRAPHY

OBS.	LOCATION	YEARS	LATITUDE	LONGITUDE	ELEVATION -M-	LIFEZONE	CCTX	ID	ONMAP
49	MOURDIAH	18	14.52 N	7.30 W	270.1	.	.	49	YES
50	NARA	25	15.17 N	7.30 W	260.0	.	.	50	YES
51	NIAFUNKE	27	16.00 N	4.00 W	254.8	.	.	51	YES
52	NIENEBALE	22	12.92 N	7.50 W	310.0	.	.	52	NO
53	NIONO	10	14.25 N	6.00 E	.	.	.	53	NO
54	NIORO-DU-SAHEL	36	15.23 N	9.37 W	228.0	TV	EH75	54	NO
55	NYAMINA	13	13.37 N	7.00 W	310.0	.	.	55	YES
56	OULOUMA	10	14.20 N	11.58 E	.	.	.	57	NO
57	OUSSOUBIDIAGNA	10	14.25 N	10.45 E	.	.	.	58	NO
58	SAN	34	13.33 N	4.83 W	264.9	TV	EH74	59	YES
59	SANSANDING	10	13.77 N	6.00 W	.	.	.	60	NO
60	SARAFERE	13	15.77 N	3.75 W	278.9	.	.	61	NO

AVERAGE PRECIPITATION (MM)

OBS.	JAN.	FEB.	MAR.	APR.	MAY	JUN.	JUL.	AUG.	SEP.	OCT.	NOV.	DEC.	ANNUAL
49	0	1	2	4	13	82	153	214	101	21	0	0	591
50	2	0	0	4	16	39	143	195	87	21	1	1	508
51	1	0	0	3	3	20	86	124	77	9	0	0	322
52	0	2	2	14	60	132	212	294	151	37	5	1	908
53	0	0	2	0	18	75	186	228	132	16	1	0	658
54	0	1	0	7	22	75	185	248	140	28	2	1	709
55	0	0	2	9	21	100	200	234	123	22	5	0	715
56	1	5	0	0	13	98	177	264	264	77	20	0	919
57	0	1	0	9	30	139	191	260	198	74	4	1	907
58	0	0	2	15	44	107	183	247	119	27	5	0	749
59	0	0	0	14	18	67	195	230	101	15	2	0	642
60	1	0	1	0	2	18	66	128	61	15	2	0	294

AVERAGE TEMPERATURE (DEGREES CENTIGRADE)

OBS.	JAN.	FEB.	MAR.	APR.	MAY	JUN.	JUL.	AUG.	SEP.	OCT.	NOV.	DEC.	ANNUAL
49	.	.	.	.	.	.	.	.	.	.	.	.	.
50	.	.	.	.	.	.	.	.	.	.	.	.	.
51	.	.	.	.	.	.	.	.	.	.	.	.	.
52	.	.	.	.	.	.	.	.	.	.	.	.	.
53	.	.	.	.	.	.	.	.	.	.	.	.	.
54	23.0	25.8	29.3	32.5	34.5	33.0	29.2	27.6	28.1	29.8	27.6	23.5	28.6
55	.	.	.	.	.	.	.	.	.	.	.	.	.
56	.	.	.	.	.	.	.	.	.	.	.	.	.
57	.	.	.	.	.	.	.	.	.	.	.	.	.
58	22.9	26.8	29.8	32.2	32.9	30.2	27.8	26.6	27.3	28.8	27.7	24.4	28.6
59	.	.	.	.	.	.	.	.	.	.	.	.	.
60	.	.	.	.	.	.	.	.	.	.	.	.	.

CLIMATOLOGICAL DATA FOR MALI

GEOGRAPHY

OBS.	LOCATION	YEARS	LATITUDE	LONGITUDE	ELEVATION -M-	LIFEZONE	CCTX	ID	ONMAP
61	SATADOUGOU	10	12.43 N	11.40 W	.	.	.	62	YES
62	SEGOU	36	13.40 N	6.15 W	288.0	TV	EH74	63	NO
63	SIKASSO	36	11.35 N	5.68 W	374.0	TD	EH53	64	YES
64	SOFARA	13	14.02 N	4.13 W	295.0	.	.	65	NO
65	SOKOLO	13	14.73 N	6.12 W	292.9	.	.	66	NO
66	SONINKOURA	10	13.45 N	6.23 E	.	.	.	67	NO
67	SOTUBA	10	11.20 N	8.45 E	.	.	.	68	NO
68	TESSALIT	10	20.20 N	0.98 E	492.9	TO	EH95	69	YES
69	TOMBOUCTOU	24	16.77 N	3.02 W	268.8	TX	EH95	70	YES
70	TOUKOTO	35	13.47 N	9.88 W	178.9	TV	EH64	71	YES
71	YELIMANE	17	15.13 N	10.57 W	85.0	.	.	72	YES

AVERAGE PRECIPITATION (MM)

OBS.	JAN.	FEB.	MAR.	APR.	MAY	JUN.	JUL.	AUG.	SEP.	OCT.	NOV.	DEC.	ANNUAL
61	2	1	0	8	77	196	275	349	280	120	16	0	1324
62	0	0	2	14	25	85	190	238	123	25	0	0	702
63	0	6	17	51	125	175	273	341	262	106	16	1	1373
64	1	0	0	6	24	73	129	237	103	24	1	0	597
65	1	0	0	8	12	45	149	186	86	20	1	0	507
66	0	1	2	1	27	98	214	247	121	20	1	0	732
67	0	0	0	5	48	18	333	253	213	51	6	0	927
68	0	0	1	0	1	7	16	50	24	1	0	0	100
69	0	0	0	1	3	21	59	80	35	2	0	0	201
70	0	0	0	9	32	128	208	275	191	57	7	2	909
71	1	0	0	1	9	56	147	202	125	25	2	1	569

AVERAGE TEMPERATURE (DEGREES CENTIGRADE)

OBS.	JAN.	FEB.	MAR.	APR.	MAY	JUN.	JUL.	AUG.	SEP.	OCT.	NOV.	DEC.	ANNUAL
61	.	.	.	.	.	.	.	.	.	.	.	.	.
62	23.8	26.5	29.4	31.8	32.5	29.8	27.1	26.2	26.8	28.0	27.0	24.2	27.8
63	24.0	26.9	29.5	30.6	29.7	27.6	26.0	25.6	26.2	27.3	26.6	24.1	27.0
64	.	.	.	.	.	.	.	.	.	.	.	.	.
65	.	.	.	.	.	.	.	.	.	.	.	.	.
66	.	.	.	.	.	.	.	.	.	.	.	.	.
67	.	.	.	.	.	.	.	.	.	.	.	.	.
68	19.5	22.5	25.3	29.5	33.0	35.3	34.5	33.0	32.9	30.6	26.0	20.9	28.6
69	22.2	25.1	28.3	31.5	34.3	34.0	31.3	29.3	31.1	31.5	28.3	23.1	29.1
70	23.2	26.1	29.5	32.6	33.8	30.5	27.6	26.5	27.0	27.7	26.3	23.3	27.8
71	.	.	.	.	.	.	.	.	.	.	.	.	.

MAURITANIA

Agriculture

Of 103,070,000 ha in total area, Mauritania was estimated in 1978 to have 195,000 ha of arable or permanent cropland, 39,250,000 ha in permanent pasture, 15,134,000 ha in forest and woodland, and 48,461,000 ha in other terrain categories. In 1978, approximately 8000 ha were irrigated. Of the 1979 population, 83.3% or 405,000 individuals of the economically active population were engaged in agriculture. The total population in 1979 was 1,588,000. According to FAO statistics for 1979, there were 23,000 horses, 220,000 asses, 1,600,000 cattle, 720,000 camels, 5,200,000 sheep, 3,250,000 goats, and 3,000,000 chickens.[2] Agricultural production is summarized in Table 1. Conventional and energetic yield data on most of these will be found in Appendixes I and II.

Climate

Data from Wernstedt[3] are presented in the Climatological Data Table (see Introduction for explanation). Based on the means for data points cited in Wernstedt, the mean annual precipitation is 2.4 dm (mean of 24 sites) and the mean annual temperature 26.9°C (mean of 8 sites). The averages would suggest that much of the country could be classified as Tropical Desert Scrub Life Zone.[4]

Productivity

The vegetation map depicts the sparsity of vegetation in Mauritania. From zero productivity in the desert extremes, I expect productivity to grade into 10 MT, perhaps 12 MT, in the Acacia savannas, with higher productivity restricted to oases and marshes. Even here, salinity will probably limit productivity.

Economics

Food consumption per capita was 1976 cal/day, energy consumption 198 kg coal equivalent per capita in 1977. In 1970, the caloric supply was 87.0% of requirements. The total population growth rate was 2.5% in 1970. Population density was 1.3 persons per square kilometer in 1970, with 12.8% of the total population urban. Life expectancy at birth of those born in 1970 was 39 years. In 1977, with GNP at $270 per capita, agriculture accounted for 25.9% of GDP, manufacturing for 10.9%, other industry for 25.9%, and services for 37.3%. In 1970, food and beverage constituted 8.3% of exports, nonfood agriculture for 2.5%, fuels, minerals, and metal for 88.4%, machines for 0.4%, and other manufacturing for 0.5% of total merchandise exports. Food and beverage constituted 23.4% of imports, nonfood agriculture 0.8%, fuel and lubricants 7.7%, nonfuel minerals and metals 0.7%, machinery and equipment 37.9%, and other manufactured goods constituted 29.5% of total merchandise imports.[10]

Energetics

Mauritania has a hydroelectric potential of 2000 MW. Gum arabic is considered the best firewood in Mauritania. Importing more than 75% of its commercial energy demands as oil, Mauritania is expected to be experiencing fuelwood problems by the year 2000.[10]

Table 1
AGRICULTURAL PRODUCTION IN MAURITANIA

Crop	Area planted (ha)	Yield (kg/ha)	Total production (MT)
Wheat		5,000	
Rice paddy	2,000	2,000	4,000
Barley		667	
Corn	9,000	556	5,000
Millet	110,000	318	35,000
Potatoes		13,333	4,000
Sweet potatoes	5,000	400	2,000
Peanuts in shell	5,000	700	4,000
Watermelons	1,000	6,143	4,000
Dates			14,000

From FAO 1979 Production Yearbook, Vol. 33, Food and Agriculture Organization, Rome, 1980.

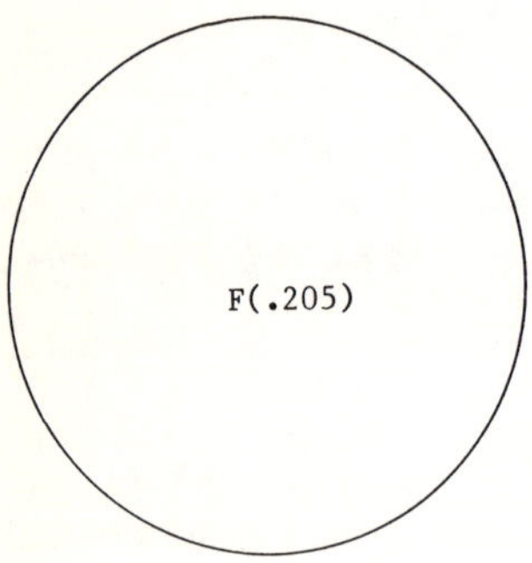

Production

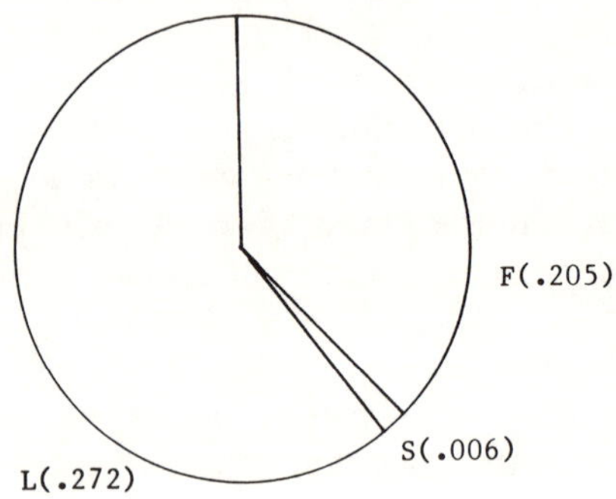

Consumption

1978 Energy production and consumption in Mauritiana; fuel types listed in million metric tons coal equivalents (MMTCE). 1 MMTCE = 27.78×10^{12} Btu. L = liquid fuel, S = solid fuel, HNE = hydro, nuclear, and electric fuel, F = firewood, NG = natural gas. (Data from United Nations, World Energy Supplies, 1973—1978, Statistical Papers Series J, 22, Department of International Economic and Social Affairs, Statistical Office, New York, 1979.)

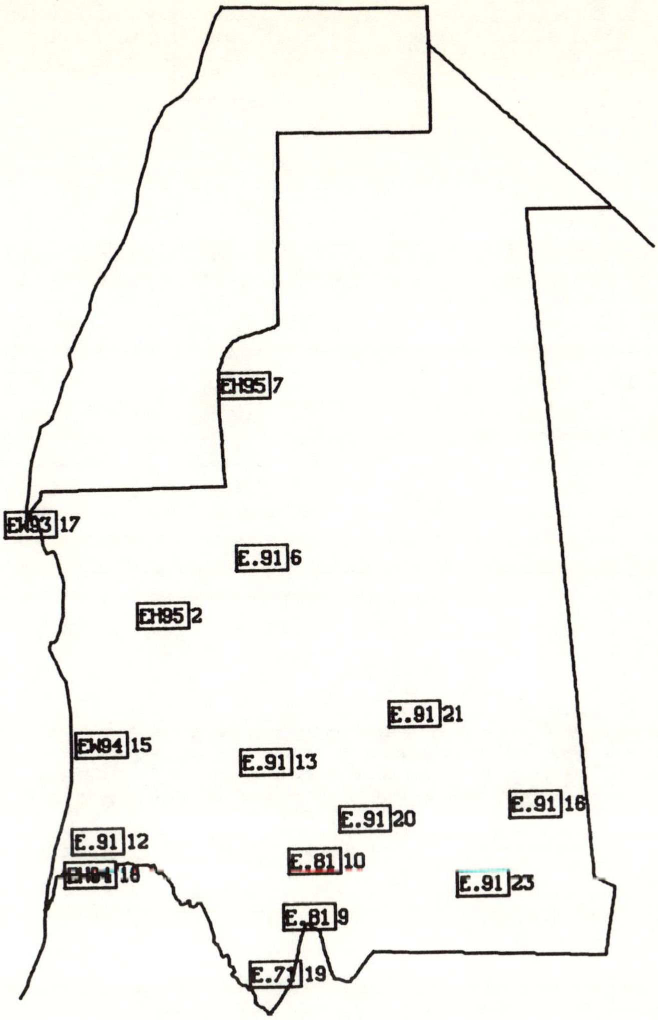

Selected weather stations for Mauritania.

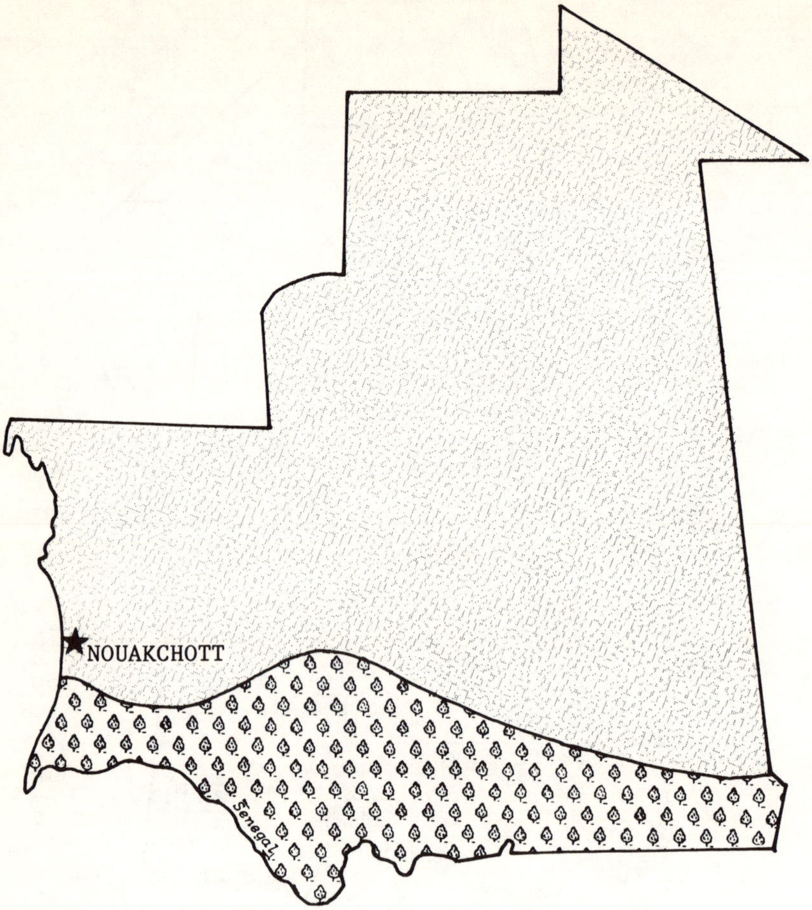

MAURITANIA VEGETATION

CLIMATOLOGICAL DATA FOR MAURITANIA

GEOGRAPHY

OBS.	LOCATION	YEARS	LATITUDE	LONGITUDE	ELEVATION -M-	LIFEZONE	CCTX	ID	ONMAP
1	AIOUN EL ATROU	10	16.67 N	9.62 W	.	.	.	1	NO
2	AKJOUJT	25	19.75 N	14.37 W	118.0	TO	EH95	2	YES
3	ALEG	28	17.05 N	13.92 W	.	.	.	3	NO
4	ATAR	35	20.52 N	13.07 W	225.9	TO	EH95	4	NO
5	BOGHE	28	16.58 N	14.28 W	17.1	.	.	5	NO
6	BOUTILIMIT	35	17.53 N	14.68 W	77.1	TX	EH94	19	NO
7	CHINGUETTI	18	20.47 N	12.37 W	499.9	.	.	6	YES
8	FORT GOURAUD	19	22.68 N	12.70 W	296.9	TO	EH95	7	YES
9	KAEDI	29	16.15 N	13.50 W	32.9	.	.	8	NO
10	KANKOSSA	10	15.93 N	11.52 W	.	.	.	9	YES
11	KIFFA	28	16.63 N	11.40 W	110.0	.	.	10	YES
12	M'BOUT	15	16.03 N	12.62 W	.	.	.	11	NO

AVERAGE PEECIPITATION (MM)

OBS.	JAN.	FEB.	MAR.	APR.	MAY	JUN.	JUL.	AUG.	SEP.	OCT.	NOV.	DEC.	ANNUAL
1	0	3	7	0	2	21	121	119	55	7	2	2	339
2	1	1	1	0	1	3	7	43	31	7	8	1	104
3	1	0	1	0	14	15	65	140	40	19	3	1	300
4	2	1	1	0	2	7	7	31	39	8	9	4	112
5	3	0	1	0	3	15	71	129	72	18	5	1	317
6	0	1	0	0	4	7	47	71	43	15	5	1	194
7	1	2	2	0	2	1	6	18	9	5	3	1	51
8	1	3	3	1	0	1	2	8	16	13	9	2	59
9	1	0	0	0	5	31	90	171	85	14	5	2	404
10	0	0	0	0	2	40	111	154	87	17	3	1	415
11	1	1	0	1	5	22	78	133	75	12	2	2	332
12	3	0	0	0	4	26	81	145	86	15	3	2	365

AVERAGE TEMPERATURE (DEGREES CENTIGRADE)

OBS.	JAN.	FEB.	MAR.	APR.	MAY	JUN.	JUL.	AUG.	SEP.	OCT.	NOV.	DEC.	ANNUAL
1	.	.	.	.	.	.	.	.	.	.	.	.	.
2	21.2	22.8	26.0	28.2	31.8	34.5	33.0	32.3	32.6	30.8	27.0	21.7	28.5
3	.	.	.	.	.	.	.	.	.	.	.	.	.
4	19.4	21.1	24.6	27.2	30.8	34.3	34.4	33.6	32.5	29.6	25.6	20.1	27.8
5	.	.	.	.	.	.	.	.	.	.	.	.	.
6	21.9	23.9	27.0	29.1	31.5	32.4	30.3	29.8	30.1	30.2	27.9	22.7	28.1
7	.	.	.	.	.	.	.	.	.	.	.	.	.
8	18.0	19.5	23.0	24.5	27.3	30.9	33.9	33.3	31.7	28.8	24.2	19.1	26.2
9	.	.	.	.	.	.	.	.	.	.	.	.	.
10	.	.	.	.	.	.	.	.	.	.	.	.	.
11	.	.	.	.	.	.	.	.	.	.	.	.	.
12	.	.	.	.	.	.	.	.	.	.	.	.	.

CLIMATOLOGICAL DATA FOR MAURITANIA

GEOGRAPHY

OBS.	LOCATION	YEARS	LATITUDE	LONGITUDE	ELEVATION -M-	LIFEZONE	CCTX	ID	ONMAP
13	MEDERDRA	18	16.92 N	15.67 W	25.0	.	.	12	YES
14	MOUDJERIA	28	17.88 N	12.33 W	.	.	.	13	YES
15	NEMA	35	16.60 N	7.27 W	268.8	TT	EH95	14	NO
16	NOUAKCHOTT	24	18.12 N	15.60 W	0.9	TX	EW94	15	YES
17	OUALATA	10	17.30 N	7.03 W	.	.	.	16	YES
18	PORT ETIENNE	35	20.93 N	17.05 W	7.9	SO	EW93	17	YES
19	ROSSO	21	16.50 N	15.82 W	4.9	TT	EH84	18	YES
20	SELIBABY	16	15.23 N	12.20 W	.	.	.	19	YES
21	TAMCHAKETT	17	17.15 N	10.40 W	.	.	.	20	YES
22	TICHITT	10	18.45 N	9.38 W	.	.	.	21	YES
23	TIDJIKJA	35	18.55 N	11.43 W	399.0	.	.	22	NO
24	TIMBEDRA	19	16.32 N	8.10 W	.	.	.	23	YES

AVERAGE PRECIPITATION (MM)

OBS.	JAN.	FEB.	MAR.	APR.	MAY	JUN.	JUL.	AUG.	SEP.	OCT.	NOV.	DEC.	ANNUAL
13	0	2	1	0	2	4	28	121	54	18	2	2	233
14	0	1	0	0	3	12	38	78	54	10	5	1	202
15	1	0	0	3	10	31	63	106	56	14	1	2	287
16	0	2	1	1	1	1	12	58	39	10	2	3	130
17	1	0	0	5	4	9	28	27	22	6	1	1	104
18	2	1	1	1	0	1	0	3	7	5	5	1	27
19	0	1	0	0	3	8	40	147	79	30	3	1	312
20	1	0	0	3	12	71	151	216	138	32	3	3	629
21	1	0	0	1	4	9	43	93	36	15	2	3	206
22	2	2	0	0	3	6	11	41	22	9	3	0	98
23	1	4	2	0	3	11	23	56	33	9	5	1	148
24	0	1	0	2	7	21	76	125	47	11	0	2	292

AVERAGE TEMPERATURE (DEGREES CENTIGRADE)

OBS.	JAN.	FEB.	MAR.	APR.	MAY	JUN.	JUL.	AUG.	SEP.	OCT.	NOV.	DEC.	ANNUAL
13	.	.	.	.	.	.	.	.	.	.	.	.	.
14	.	.	.	.	.	.	.	.	.	.	.	.	.
15	23.7	26.4	29.8	33.0	35.4	34.7	31.5	29.5	30.6	32.1	29.3	24.4	30.0
16	20.9	22.1	24.0	25.3	27.0	28.0	27.7	28.4	29.4	28.9	26.2	21.7	25.8
17	.	.	.	.	.	.	.	.	.	.	.	.	.
18	18.7	19.1	20.2	20.2	21.3	22.5	22.9	23.9	25.7	24.4	22.2	19.6	21.7
19	22.4	24.0	26.7	28.3	29.7	30.3	29.5	29.0	29.1	29.0	26.8	22.8	27.3
20	.	.	.	.	.	.	.	.	.	.	.	.	.
21	.	.	.	.	.	.	.	.	.	.	.	.	.
22	.	.	.	.	.	.	.	.	.	.	.	.	.
23	.	.	.	.	.	.	.	.	.	.	.	.	.
24	.	.	.	.	.	.	.	.	.	.	.	.	.

MEXICO*

Agriculture

Of 197,255,000 ha in total area, Mexico was estimated in 1978 to have 23,220,000 ha of arable or permanent cropland, 74,499,000 ha in permanent pasture, 70,400,000 ha in forest and woodland, and 24,185,000 ha in other terrain categories. In 1978, approximately 5,000,000 ha were irrigated. Of the 1979 population, 36.9% or 7,206,000 individuals of the economically active population were engaged in agriculture. The total population in 1979 was 67,676,000. According to FAO statistics for 1979, there were 6,447,000 horses, 3,207,000 mules, 3,233,000 asses, 29,920,000 cattle, 12,578,000 pigs, 7,850,000 sheep, 8,103,000 goats, 150,000,000 chickens, 2,700,000 ducks, and 1,300,000 turkeys.[2] Agricultural production is summarized in Table 1. Conventional and energetic yield data on most of these will be found in Appendixes I and II.

Climate

Data from Wernstedt[3] are presented in the Climatological Data Table (see Introduction for explanation). Based on the means for data points cited in Wernstedt, the mean annual precipitation is 9.0 dm (mean of 333 sites) and the mean annual temperature 20.8°C (mean of 294 sites). The averages would suggest that much of the country could be classified as Subtropical Dry Forest Life Zone.[4]

Productivity

Major vegetation types of Mexico are shown in the vegetation map. After extensive field work in Mexico, I am comfortable with these annual productivity potentials of these units, assuming reasonable management:

Vegetation type	MT/ha/yr
Broadleaf evergreen forest	18—36
Broadleaf deciduous and needleleaf evergreen forest predominant; some tall grass	12—18
Broadleaf evergreen and deciduous forest predominant; some grass and shrubs	12—24
Semiarid grassland, with scattered shrubs and cacti	0—3
Desert scrub; some low grass	0—6
Marsh and swamp, including mangrove	6—36

Economics

Food consumption per capita was 2654 cal/day, energy consumption 1330 kg coal equivalent per capita in 1977. In 1970, the caloric supply was 105.0% of requirements. The total population growth rate was 3.3% in 1970. Population density was 26 persons per square kilometer in 1970, with 59.3% of the total population urban. Life expectancy at birth of those born in 1970 was 62.4 years. In 1977, with GNP at $1160 per capita, agriculture accounted for 9.8% of GDP, manufacturing for 27.7%, other industry for 8.6%, and services for 53.8%. In 1976, food and beverage constituted 32.7% of exports, nonfood agriculture for 9.4%, fuels, minerals, and metal for 28.7%, machines for 7.2%, and other manufacturing for 22.0% of total merchandise exports. Food and beverage constituted 7.5% of imports, nonfood agriculture 3.5%, fuel and lubricants 5.7%, nonfuel minerals and metals 3.7%, machinery and equipment 50.2%, and other manufactured goods constituted 29.4% of total merchandise imports.[10]

* Maps and tables follow text.

Table 1
AGRICULTURAL PRODUCTION IN MEXICO

Crop	Area planted (ha)	Yield (kg/ha)	Total production (MT)
Wheat	628,000	3,618	2,272,000
Rice	151,000	3,238	489,000
Barley	326,000	1,549	505,000
Corn	7,148,000	1,295	9,255,000
Oats	66,000	955	63,000
Sorghum	1,456,000	2,680	3,902,000
Potatoes	56,000	12,982	727,000
Sweet potatoes	9,000	10,795	95,000
Cassava	6,000	15,088	86,000
Dry beans	1,585,000	666	1,056,000
Dry broad beans	60,000	850	51,000
Dry peas	5,000	755	4,000
Chick peas	143,000	1,357	194,000
Lentils	6,000	1,000	6,000
Soybeans	386,000	1,816	701,000
Peanuts in shell	58,000	1,534	89,000
Castor beans	8,000	750	6,000
Sunflower seed	12,000	833	10,000
Rapeseed	3,000	1,000	3,000
Sesame seed	250,000	600	150,000
Linseed	7,000	1,000	7,000
Safflower seed	496,000	1,341	665,000
Seed cotton	381,000	2,323	885,000
Cottonseed			540,000
Olives			12,000
Olive oil			1,000
Coconuts			788,000
Copra			110,000
Palm kernels			14,000
Palm oil			7,000
Cabbages	3,000	21,462	56,000
Tomatoes	62,000	17,500	1,082,000
Cauliflower	3,000	9,688	31,000
Pumpkins, squash, gourds	12,000	11,174	129,000
Cucumbers and gerkins	7,000	14,815	100,000
Eggplants	1,000	22,000	22,000
Chilies and peppers, green	53,000	8,943	474,000
Garlic	6,000	5,000	30,000
Green beans	10,000	3,053	29,000
Green peas	9,000	4,222	38,000
Carrots	4,000	18,625	75,000
Watermelons	30,000	8,667	260,000
Cantaloupes and other melons	25,000	10,000	250,000
Grapes	50,000	9,900	495,000
Wine			52,000
Raisins			4,000
Dates			3,000
Sugar cane	502,000	70,548	35,415,000
Sugar beets	4,000	25,000	100,000
Sugar (centrifugal, raw)			3,060,000
Sugar (noncentrifugal)			50,000
Apples			390,000
Pears			51,000
Peaches and nectarines			200,000

Table 1 (continued)
AGRICULTURAL PRODUCTION IN MEXICO

Crop	Area planted (ha)	Yield (kg/ha)	Total production (MT)
Plums			50,000
Oranges			3,240,000
Tang., mand., clement., satsuma			72,000
Lemons and limes			465,000
Grapefruit and pomelo			65,000
Apricots			8,000
Avocados			373,000
Mangoes			570,000
Pineapples			455,000
Bananas			1,929,000
Papayas			200,000
Strawberries			107,000
Walnuts			7,000
Green coffee	345,000	661	228,000
Cocoa beans	61,000	656	40,000
Tobacco leaves	48,000	1,667	80,000
Cotton lint			336,000

From FAO 1979 Production Yearbook, Vol. 33, Food and Agriculture Organization, Rome, 1980.

Energetics

Mexico has 31,250,000 million b.o.e. in oil, 3720 in associated gas, 10,026 in non-associated gas, 3.5 billion b.o.e. in heavy oil reserves, 5448 million ton coal (875 technically and economically recoverable), and a hydroelectric potential of 20,344 MW. Mexico has nuclear facilities under construction. Mexico is expected to be producing geothermal energy in 1990. A net oil-exporting developing country, Mexico is not expected to experience major fuelwood problems by the year 2000.[10]

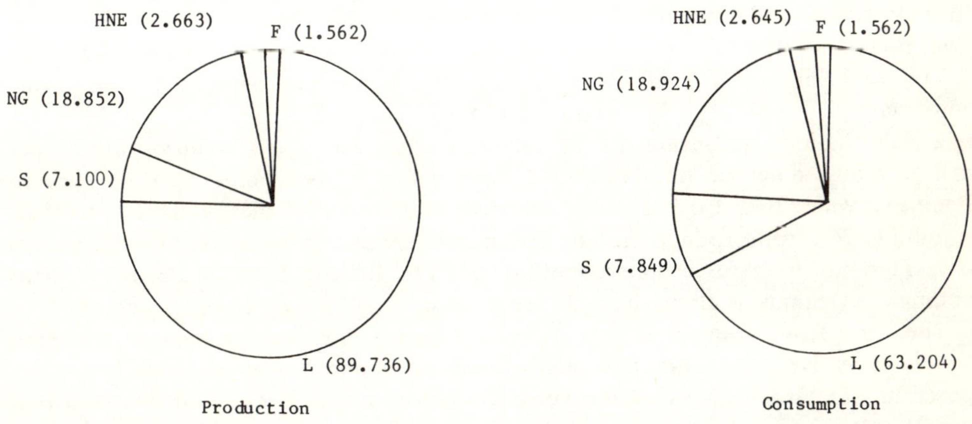

1978 Energy production and consumption in Mexico; fuel types listed in million metric tons coal equivalents (MMTCE). 1 MMTCE = 27.78×10^{12} Btu. L = liquid fuel, S = solid fuel, HNE = hydro, nuclear, and electric fuel, F = firewood, NG = natural gas. (Data from United Nations, World Energy Supplies, 1973—1978, Statistical Papers Series J, 22, Department of International Economic and Social Affairs, Statistical Office, New York, 1979.)

Soils

The soils of Mexico are shown on the Soils Map, in nine groups: (1) Sierozems, Reddish Brown and Reddish Chestnut soils, (2) Chernozemic soils, (3) Red Desert soils, (4) Lithosols, (5) Lithosols and Noncalcic Brown soils, (6) Rendzinas and Grumusols, (7) Terra Rossa soils, (8) Reddish Brown Latosols, and (9) Low-Humic Gley and associated soils.

The Sierozems, the Reddish Brown, and the Reddish Chestnut soils are extensive on plains and hills in the northern half of Mexico. The Sierozems are grayish soils with calcareous subsoils; they are low in organic matter. The Reddish Brown and Reddish Chestnut soils have somewhat darker surface layers and contain more organic matter; subsoils are generally reddish; these soils receive slightly more rainfall than the Sierozems and are used to some extent for the production of some dryland crops. Principal use for all these soils, however, is grazing. Under irrigation and with the addition of fertilizers, especially those containing nitrogen, the smoother areas of these soils could be made productive. However, local low areas with high salt accumulations, mainly Solonchak soils, would not be suitable for most crops.

The Chernozemic soils occur on plains and low hills in southern and western Mexico. Chernozemic soils are fertile soils that have thick, dark, loamy surfaces rich in organic matter overlying brownish subsoils. They are used extensively for crop production, except on the steeper slopes where the soils are shallow.

The driest soils of Mexico are the Red Desert soils, which occur on plains in the northwest. They are light colored loamy and sandy soils with reddish calcareous subsoils; they are low in nitrogen. Without irrigation, moisture is limiting for crop production.

The Lithosols occur on mountains and steep hills dominantly in arid and semiarid regions; rock outcrops are common among these soils. Lithosols in some places, and especially among the Sierozems, Reddish Brown and Reddish Chestnut soils, cannot be shown at the small scale of the soil map. The Lithosols support only scrub vegetation and have little potential for crop production; they do provide some browse.

Lithosols occur also in association with the Noncalcic Brown soils in the mountains and steep hills of the subhumid regions. The Noncalcic Brown soils have loamy surface layers over finer grained, reddish dense subsoils. Though generally too steep and stony for cultivation in Mexico, the soils of this association do support considerable vegetation, mainly forest.

The Rendzinas and Grumusols are dark gray to black clays developed from calcareous materials, mostly on low plains. The Rendzinas, which occur also on hills, are relatively shallow, well-drained soils overlying limestone. The Grumusols are deeper and generally somewhat poorly drained. Rendzina soils are dominant on the Yucatan Peninsula where they are extensively forested; shallowness of these soils restricts their suitability for crop production. The Grumusols are sticky when wet and hard when dry. They are used for crop production, but are difficult to cultivate. Some areas among the Grumusols along the Gulf are quite salty and poorly suited for crops.

The Terra Rossa soils, which occur on the plains of the Yucatan Peninsula, are red friable clays that are moderately shallow over limestone. Limestone rock is also exposed in many places. Most of the Terra Rossa soils are in forest, probably their best use. Where soils are thicker, crops could be grown.

The Reddish Brown Latosols are deep, permeable soils, mainly on low plains in the more humid areas. The infertile soils can be made productive with fertilizers.

The Low-Humic Gley and associated soils, poorly drained, occur on low plains in the humid areas of southeastern Mexico. They have thin dark surface layers moderately high in organic matter and gray clayey subsoils. Associated with them are Humic Gley soils, which have thick highly organic surfaces, and Planosols, which have a dense claypan layer 30 to 60 cm below the surface. The claypan restricts root penetration and downward movement of water. Soils of this association, if adequately drained, could be used for crops.

Too small to show on the map, yet of major agricultural importance, are small widespread areas of Alluvial soils. These soils occupy narrow strips along most drainageways, where they are in a favorable position for irrigation. Near larger rivers, they are generally fine textured, fertile, and highly productive under irrigation; where not irrigated, they are moderately productive. In areas where flash floods occur, the soils are sandy and gravelly and too droughty for crop production. Some areas are too salty for many crops.

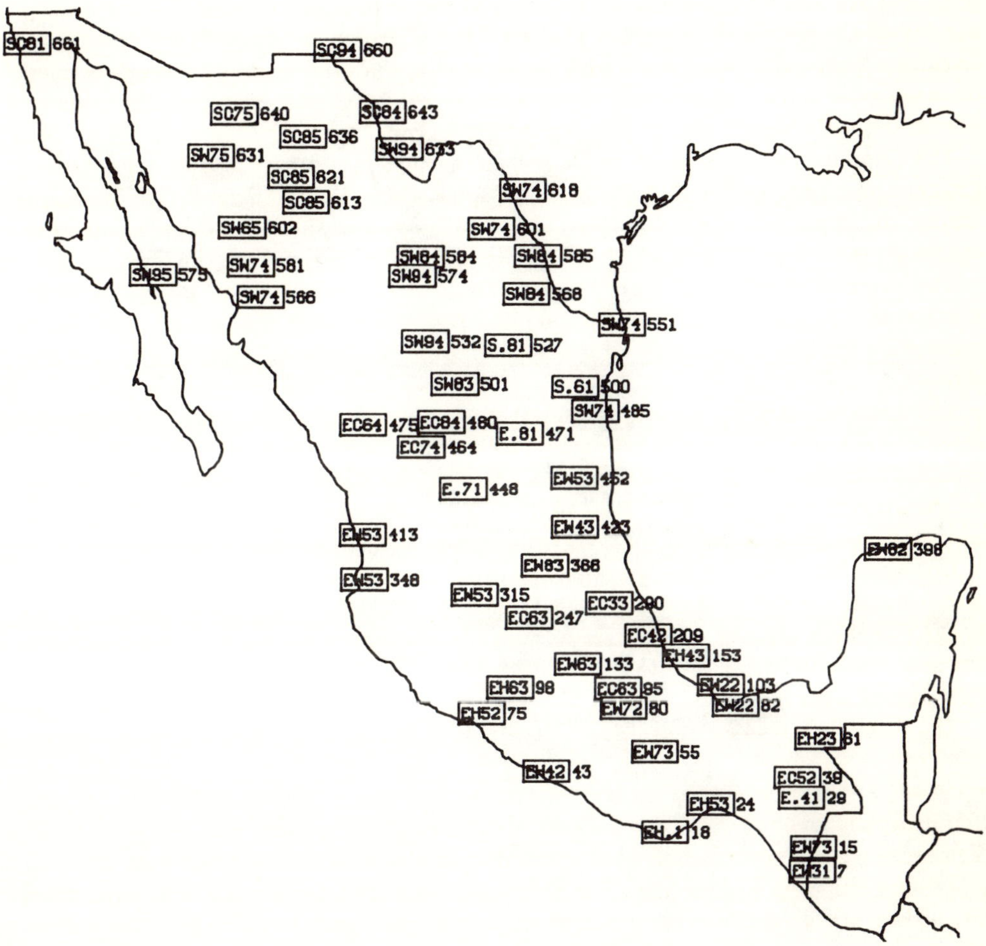

Selected weather stations for Mexico.

MEXICO VEGETATION

- Broadleaf evergreen forest
- Broadleaf deciduous and needleleaf evergreen forest predominant; some tall grass
- Broadleaf evergreen and deciduous forest predominant; some grass and shrubs
- Semiarid grassland, with scattered shrubs and cacti
- Desert scrub; some low grass
- Marsh and swamp, including mangrove

MEXICO SOILS

Red Desert soils

Reddish-Chestnut, Reddish-Brown, and Sierozem (gray) soils

Grumusols (dark clays)

Planosols (soils with clay pan)

Rendzina (black) and Terra Rossa (red) soils (shallow over limestone)

Low-Humic Gley soils (wet)

Lithosols (shallow, stony soils) of mountains, mainly scrub-covered

Lithosolic soils of mountains, mainly forested

CLIMATOLOGICAL DATA FOR MEXICO

GEOGRAPHY

OBS.	LOCATION	YEARS	LATITUDE	LONGITUDE	ELEVATION -M-	LIFEZONE	CCTX	ID	ONMAP
1	ABASOLO	15	20.45 N	101.53 W	1759.6	SD	EW63	326	NO
2	ABASOLO	15	24.07 N	98.38 W	84.1	TV	SW74	485	YES
3	ACAPONETA	15	22.50 N	105.38 W	25.0	TD	EW53	449	NO
4	ACAPULCO	15	16.83 N	99.93 W	3.0	TD	EH42	43	YES
5	ACTOPAN	15	20.28 N	98.97 W	1989.4	WT	EC83	309	NO
6	AGUA BUENA	15	21.98 N	99.40 W	349.9	TM	EW33	437	NO
7	AGUACALIENTES	15	21.88 N	102.30 W	1878.5	SD	EC73	431	NO
8	AHOME	15	25.92 N	109.18 W	39.0	TT	SW84	553	NO
9	AHUACATLAN	15	21.08 N	104.47 W	983.9	SD	EW63	380	NO
10	ALCOZAUCA	15	17.30 N	98.45 W	1199.7	SD	EW63	57	NO
11	ALDAMA	15	22.90 N	98.08 W	.	.	.	455	NO
12	ALTAR	15	30.73 N	111.77 W	396.8	ST	SW95	650	NO

AVERAGE PRECIPITATION (MM)

OBS.	JAN.	FEB.	MAR.	APR.	MAY	JUN.	JUL.	AUG.	SEP.	OCT.	NOV.	DEC.	ANNUAL
1	16	9	6	2	21	112	201	192	182	30	19	20	811
2	27	14	4	44	58	108	59	92	165	57	30	15	674
3	18	2	1	0	1	92	346	322	287	57	24	69	1218
4	9	1	0	0	29	430	216	246	360	170	30	11	1504
5	10	6	13	16	39	66	67	65	84	50	20	5	439
6	58	42	23	22	124	240	471	258	445	217	63	37	2027
7	11	6	4	1	18	111	136	106	89	34	19	18	551
8	3	5	5	1	1	7	49	80	77	27	12	58	324
9	23	15	2	2	14	120	231	185	147	46	14	22	821
10	2	3	5	18	55	167	161	185	166	64	7	5	836
11	10	33	4	14	135	176	199	108	202	58	38	21	998
12	15	10	9	4	1	10	74	58	43	13	21	18	274

AVERAGE TEMPERATURE (DEGREES CENTIGRADE)

OBS.	JAN.	FEB.	MAR.	APR.	MAY	JUN.	JUL.	AUG.	SEP.	OCT.	NOV.	DEC.	ANNUAL
1	15.9	18.0	20.1	22.2	24.3	23.9	22.4	22.0	21.6	20.6	18.2	16.8	20.5
2	18.4	20.3	21.1	26.2	28.6	29.3	29.2	29.8	28.1	25.6	21.0	18.6	24.7
3	20.8	21.9	22.6	24.1	25.8	27.8	27.4	27.6	27.4	27.1	24.7	21.6	24.9
4	25.7	25.7	26.2	26.8	28.2	28.2	28.4	28.3	27.6	27.6	27.0	26.1	27.1
5	12.7	14.4	16.2	18.0	19.4	18.5	17.8	17.6	17.4	15.6	13.9	13.7	16.3
6	17.3	20.0	21.8	26.0	27.9	28.0	26.7	27.1	26.5	23.9	19.6	17.6	23.5
7	13.0	14.7	17.4	20.0	22.0	21.3	19.8	19.8	19.2	18.0	15.6	13.8	17.9
8	17.5	18.4	20.1	22.8	25.4	29.2	31.7	31.2	30.4	26.7	21.4	18.0	24.4
9	17.8	18.5	20.2	22.0	24.1	25.5	24.4	24.3	24.4	24.1	20.8	18.2	22.1
10	18.5	20.7	22.3	23.2	23.8	23.6	22.0	21.6	21.5	20.5	19.0	18.8	21.3
11	.	.	.	.	.	.	.	.	.	.	.	.	.
12	12.8	15.3	16.5	20.0	24.3	28.7	31.4	30.4	28.9	23.4	17.8	14.2	22.0

CLIMATOLOGICAL DATA FOR MEXICO

GEOGRAPHY

OBS.	LOCATION	YEARS	LATITUDE	LONGITUDE	ELEVATION -M-	LIFEZONE	CCTX	ID	ONMAP
13	ALTOTONGA	15	19.72 N	97.25 W	1186.6	WM	EC43	222	NO
14	ALVARADO	15	18.78 N	95.75 W	9.1	SW	EW33	121	NO
15	ALVARO OBREGON	15	21.00 N	100.40 W	1869.6	WD	EC73	374	NO
16	ALVARO OBREGON	15	18.53 N	92.65 W	1.8	TD	EW52	101	NO
17	AMEALCO	15	20.18 N	100.15 W	2599.3	.	.	291	NO
18	AMECA	15	20.57 N	104.07 W	1234.7	SD	EW63	342	NO
19	ANGAMACUTIRO	15	20.13 N	101.70 W	2499.4	.	.	282	NO
20	ANGEL PUERTO	15	15.65 N	96.52 W	43.0	.	.	18	YES
21	APATZINGAN	15	19.08 N	102.25 W	681.8	TV	EH73	150	NO
22	APIZACO	15	19.42 N	98.13 W	2407.3	WD	EC63	191	NO
23	ARAMBARRI	15	24.10 N	99.82 W	1075.6	.	.	487	NO
24	ARANCAS	15	20.75 N	102.33 W	1789.5	.	.	359	NO

AVERAGE PRECIPITATION (MM)

OBS.	JAN.	FEB.	MAR.	APR.	MAY	JUN.	JUL.	AUG.	SEP.	OCT.	NOV.	DEC.	ANNUAL
13	91	33	33	47	103	215	181	190	342	295	226	160	1917
14	47	46	41	83	169	300	387	306	430	152	61	56	2078
15	8	5	9	10	36	83	101	81	109	38	13	13	506
16	59	70	42	28	62	176	140	131	189	308	133	127	1312
17	14	11	12	12	104	119	249	167	224	53	13	8	986
18	26	5	3	7	35	169	208	195	119	57	20	37	875
19	12	8	3	6	42	142	230	168	170	59	19	18	877
20	.	.	.	.	.	.	.	.	.	.	.	.	.
21	4	7	2	1	14	89	187	173	168	46	8	16	716
22	8	2	7	26	120	150	229	161	184	51	13	10	960
23	26	15	8	15	53	83	73	37	87	50	15	5	467
24	8	7	11	1	50	198	205	202	179	54	18	40	974

AVERAGE TEMPERATURE (DEGREES CENTIGRADE)

OBS.	JAN.	FEB.	MAR.	APR.	MAY	JUN.	JUL.	AUG.	SEP.	OCT.	NOV.	DEC.	ANNUAL
13	12.0	14.8	16.5	19.1	20.0	19.1	17.6	18.1	17.3	15.8	13.5	12.3	16.3
14	15.5	17.2	17.2	20.0	20.9	19.8	19.6	19.6	19.7	19.1	17.1	15.8	18.5
15	12.7	14.5	16.9	19.1	20.2	19.1	18.1	17.8	17.4	15.5	14.1	13.6	16.6
16	22.9	23.8	24.9	26.8	27.5	27.4	27.2	27.5	27.1	26.0	24.3	23.5	25.7
17	.	.	.	.	.	.	.	.	.	.	.	.	.
18	16.9	17.5	19.2	20.5	23.2	24.1	23.9	23.7	23.5	22.4	20.3	18.3	21.1
19	.	.	.	.	.	.	.	.	.	.	.	.	.
20	27.6	28.0	28.0	28.1	29.0	28.2	28.6	28.4	27.9	28.4	28.4	27.8	28.2
21	25.3	26.6	28.6	30.2	31.6	30.9	28.6	28.0	28.2	28.4	27.2	25.4	28.2
22	10.7	12.0	14.2	14.8	16.5	15.6	14.6	14.9	14.9	14.1	11.8	11.7	13.8
23	.	.	.	.	.	.	.	.	.	.	.	.	.
24	.	.	.	.	.	.	.	.	.	.	.	.	.

CLIMATOLOGICAL DATA FOR MEXICO

GEOGRAPHY

OBS.	LOCATION	YEARS	LATITUDE	LONGITUDE	ELEVATION -M-	LIFEZONE	CCTX	ID	ONMAP
25	ATOYAC	15	17.20 N	100.43 W	199.9	TD	EH52	53	NO
26	ATZALAN	15	19.80 N	97.22 W	1319.8	WW	EC32	236	NO
27	AURORA	15	15.20 N	92.42 W	254.8	TW	EH12	13	NO
28	BACOACHI	15	30.67 N	109.95 W	.	.	.	649	NO
29	BADIRAGUATO	15	25.37 N	107.48 W	170.1	TV	SW64	526	NO
30	BAMOA	15	25.72 N	108.32 W	42.1	.	.	545	NO
31	BASIS	15	24.88 N	105.75 W	1564.5	.	.	512	NO
32	BELLAVISTA	15	32.02 N	116.62 W	292.0	WT	SC84	662	NO
33	BOLEO EL	15	27.32 N	112.33 W	4.9	TO	SW95	587	NO
34	BRAVOS	15	17.55 N	99.50 W	1249.7	.	.	65	NO
35	BURGOS	15	24.95 N	98.78 W	.	.	.	514	NO
36	BURRO EL	15	29.22 N	101.92 W	.	SX	SW94	625	NO

AVERAGE PRECIPITATION (MM)

OBS.	JAN.	FEB.	MAR.	APR.	MAY	JUN.	JUL.	AUG.	SEP.	OCT.	NOV.	DEC.	ANNUAL
25	12	0	0	0	38	263	163	194	356	153	56	14	1249
26	87	57	67	73	99	258	277	232	337	348	166	64	2067
27	34	21	50	126	457	906	599	727	797	531	134	51	4433
28	30	33	8	8	1	29	107	100	34	35	23	35	444
29	14	20	7	0	5	49	310	263	167	61	12	76	983
30	.	.	.	.	.	.	.	.	.	.	.	.	.
31	52	22	7	7	27	65	170	180	86	70	11	86	782
32	52	75	40	41	11	2	1	1	4	21	21	81	349
33	2	3	0	1	2	0	5	16	50	10	7	28	123
34	2	3	2	4	72	152	198	167	156	86	24	21	886
35	48	22	16	39	111	66	69	94	186	74	38	22	785
36	8	3	1	6	14	16	39	16	24	9	4	10	151

AVERAGE TEMPERATURE (DEGREES CENTIGRADE)

OBS.	JAN.	FEB.	MAR.	APR.	MAY	JUN.	JUL.	AUG.	SEP.	OCT.	NOV.	DEC.	ANNUAL
25	28.5	28.5	29.0	29.7	31.2	30.2	30.6	30.6	29.4	29.9	29.6	29.3	29.7
26	12.1	13.4	15.1	16.9	18.3	17.7	16.9	17.0	16.4	15.1	12.7	11.8	15.3
27	23.9	27.0	28.2	28.1	28.3	27.1	26.7	27.2	26.9	27.0	26.9	26.1	26.9
28	.	.	.	.	.	.	.	.	.	.	.	.	.
29	18.0	19.1	21.0	23.9	25.8	30.0	28.3	28.1	28.3	26.8	22.6	19.0	24.3
30	18.4	20.1	22.4	24.3	27.0	30.4	31.2	30.1	28.7	27.7	22.9	20.2	25.4
31	.	.	.	.	.	.	.	.	.	.	.	.	.
32	11.6	12.9	13.8	15.1	17.5	19.5	22.5	23.2	20.7	17.9	15.9	12.2	16.9
33	16.1	18.0	20.0	22.7	26.1	29.2	31.4	31.5	30.2	26.5	21.3	17.6	24.2
34	.	.	.	.	.	.	.	.	.	.	.	.	.
35	.	.	.	.	.	.	.	.	.	.	.	.	.
36	12.9	16.0	18.8	22.4	24.9	27.4	26.5	26.7	25.9	23.0	17.5	14.1	21.3

CLIMATOLOGICAL DATA FOR MEXICO

GEOGRAPHY

OBS.	LOCATION	YEARS	LATITUDE	LONGITUDE	ELEVATION -M-	LIFEZONE	CCTX	ID	ONMAP
37	BUSTILLOS	15	28.47 N	106.65 W	2042.5	WT	.	615	NO
38	CABO CORRIENTES	15	20.40 N	105.72 W	86.0	TV	EW63	322	NO
39	CALAYA	15	20.53 N	100.82 W	1753.5	SD	EW73	335	NO
40	CALLES PRESA	15	22.10 N	102.42 W	2024.5	WD	EC73	440	NO
41	CALNALI	15	20.90 N	98.57 W	948.8	.	.	365	NO
42	CAMARGO	15	27.70 N	105.17 W	1652.6	ST	SW85	599	NO
43	CAMARGO	15	26.32 N	98.83 W	68.0	TV	SW74	564	NO
44	CAMARON	15	27.30 N	100.07 W	186.8	ST	SW84	585	YES
45	CAMPECHE	15	19.95 N	90.53 W	25.0	TV	EW62	255	NO
46	CAMPOS PUNTA	15	19.02 N	104.35 W	96.9	.	.	139	NO
47	CANANEA	15	30.98 N	110.30 W	1605.7	WD	SC74	652	NO
48	CARBO	15	29.72 N	110.97 W	463.9	ST	SW85	634	NO

AVERAGE PRECIPITATION (MM)

OBS.	JAN.	FEB.	MAR.	APR.	MAY	JUN.	JUL.	AUG.	SEP.	OCT.	NOV.	DEC.	ANNUAL
37	14	.	5	.	16	15	89	148	35	34	18	9	383
38	20	29	1	0	2	57	145	272	307	91	15	13	954
39	8	9	6	10	31	101	150	135	150	36	14	12	662
40	20	8	3	2	26	74	152	109	104	29	24	7	558
41	65	58	32	81	130	342	463	324	538	253	110	64	2462
42	13	4	4	5	8	25	80	59	55	31	13	21	317
43	31	31	34	20	61	85	41	23	115	58	43	57	600
44	36	27	19	33	121	42	22	32	87	31	26	20	495
45	18	10	13	6	44	154	175	169	144	87	34	35	890
46	25	15	0	0	6	72	174	195	291	68	19	7	872
47	27	50	21	16	17	27	147	113	80	33	38	56	623
48	1	1	8	4	6	3	122	86	44	8	6	71	361

AVERAGE TEMPERATURE (DEGREES CENTIGRADE)

OBS.	JAN.	FEB.	MAR.	APR.	MAY	JUN.	JUL.	AUG.	SEP.	OCT.	NOV.	DEC.	ANNUAL
37	4.6	5.7	9.6	16.4	17.9	19.5	19.5	18.4	17.3	14.7	9.1	4.4	13.1
38	23.2	22.9	22.7	23.9	25.4	26.7	28.1	27.9	27.5	27.8	26.1	23.8	25.5
39	14.8	16.8	20.0	23.3	25.4	24.5	23.6	22.8	21.8	19.5	16.9	15.1	20.4
40	12.1	13.5	15.4	18.1	20.4	20.9	19.3	19.3	18.9	17.1	14.2	13.3	16.9
41	.	.	.	.	.	.	.	.	.	.	.	.	.
42	11.8	14.4	16.8	20.4	24.8	29.1	28.4	26.8	25.6	21.7	15.9	11.6	20.6
43	15.8	18.8	22.1	26.0	28.4	30.0	31.0	31.7	29.3	25.8	20.7	15.6	24.6
44	12.8	15.7	18.4	23.4	26.4	30.0	30.3	30.9	28.0	24.1	16.6	12.6	22.4
45	22.6	23.6	25.0	26.8	27.7	27.5	26.9	27.1	27.2	26.3	24.4	23.4	25.7
46	.	.	.	.	.	.	.	.	.	.	.	.	.
47	6.6	7.4	9.4	12.9	17.8	24.5	23.8	22.6	20.5	16.8	11.4	6.2	15.0
48	15.8	16.2	17.2	20.5	23.9	27.4	29.0	29.2	28.4	24.2	20.3	14.6	22.2

CLIMATOLOGICAL DATA FOR MEXICO

GEOGRAPHY

OBS.	LOCATION	YEARS	LATITUDE	LONGITUDE	ELEVATION -M-	LIFEZONE	CCTX	ID	ONMAP
49	CARDENAS	15	17.98 N	93.38 W	17.1	.	.	77	NO
50	CARMEN	15	24.08 N	99.17 W	.	SD	SW74	486	NO
51	CARMEN CIUDAD	15	18.63 N	91.75 W	3.0	TD	EW42	112	NO
52	CARRILLO	15	26.87 N	103.95 W	1101.9	ST	SW94	574	YES
53	CASAS GRANDES	15	30.37 N	104.98 W	1477.7	WT	SC84	643	YES
54	CEDRAL	15	23.82 N	100.80 W	.	ST	EC83	479	NO
55	CERRITOS	15	22.42 N	100.28 W	1152.8	TV	EW73	447	NO
56	CHALCHICOMULA	15	18.98 N	97.45 W	2539.3	.	.	137	NO
57	CHAMPOTON	15	19.35 N	90.72 W	2.1	TD	EH53	183	NO
58	CHARCAS	15	23.12 N	101.12 W	2019.6	WT	EC83	460	NO
59	CHETUMAL	15	18.50 N	88.33 W	4.0	.	.	100	NO
60	CHIAPILLA	15	16.57 N	92.75 W	549.9	.	.	35	NO

AVERAGE PRECIPITATION (MM)

OBS.	JAN.	FEB.	MAR.	APR.	MAY	JUN.	JUL.	AUG.	SEP.	OCT.	NOV.	DEC.	ANNUAL
49	109	110	37	20	122	237	259	270	307	471	78	160	2180
50	9	15	16	66	102	117	61	29	153	62	19	10	653
51	102	42	51	38	113	172	168	212	230	278	124	152	1683
52	5	5	10	4	7	40	52	56	73	11	5	21	288
53	9	16	7	10	13	21	66	86	53	34	19	22	383
54	25	9	5	4	57	40	77	38	35	28	28	5	350
55	21	8	18	17	55	97	78	65	97	53	25	14	549
56	.	.	.	.	.	.	.	.	.	.	.	.	.
57	45	25	14	21	72	127	217	193	192	181	34	53	1198
58	12	11	11	6	39	64	60	38	72	40	23	11	387
59	77	22	29	30	138	176	131	106	141	214	87	96	1247
60	1	2	4	5	114	272	167	139	254	98	7	7	1070

AVERAGE TEMPERATURE (DEGREES CENTIGRADE)

OBS.	JAN.	FEB.	MAR.	APR.	MAY	JUN.	JUL.	AUG.	SEP.	OCT.	NOV.	DEC.	ANNUAL
49	.	.	.	.	.	.	.	.	.	.	.	.	.
50	15.7	18.9	21.9	24.6	26.5	28.3	28.2	28.3	26.5	23.8	18.9	15.6	23.1
51	23.0	24.3	25.3	27.0	28.0	27.4	27.3	27.4	27.2	25.7	23.8	23.4	25.8
52	11.5	13.2	16.2	19.6	22.5	25.2	24.1	24.4	22.8	19.2	15.0	11.7	18.8
53	7.6	10.1	13.2	14.9	20.2	25.4	25.7	24.3	21.8	16.9	11.0	7.1	16.5
54	11.5	13.8	16.5	19.5	21.6	21.6	20.6	20.8	20.4	18.0	14.2	12.7	17.6
55	17.4	20.9	23.3	26.6	28.2	27.4	26.3	27.2	26.2	24.3	20.7	17.8	23.9
56	10.4	12.1	15.1	15.8	16.2	15.7	15.3	14.7	14.7	13.8	11.3	11.7	13.9
57	23.1	24.0	26.4	28.0	29.1	28.5	27.7	27.8	27.6	26.6	24.4	23.8	26.4
58	11.6	13.5	15.1	17.7	19.6	19.6	19.6	19.0	18.4	16.8	13.8	12.8	16.3
59	.	.	.	.	.	.	.	.	.	.	.	.	.
60	.	.	.	.	.	.	.	.	.	.	.	.	.

CLIMATOLOGICAL DATA FOR MEXICO

GEOGRAPHY

OBS.	LOCATION	YEARS	LATITUDE	LONGITUDE	ELEVATION -M-	LIFEZONE	CCTX	ID	ONMAP
61	CHIAUTLA	15	15.45 N	98.60 W	909.8	TV	EW63	17	NO
62	CHICALOTE	15	22.00 N	102.27 W	1701.7	.	.	438	NO
63	CHICO NOCHIXTLAN	15	18.08 N	97.67 W	.	.	.	79	NO
64	CHICONTEPEC	15	20.98 N	98.17 W	595.0	SM	EW43	372	NO
65	CHIHUAHUA	15	28.63 N	106.07 W	1422.5	ST	SW85	617	NO
66	CHOIX	15	26.72 N	105.27 W	310.0	TV	SW64	571	NO
67	CHURUMUCO	15	18.63 N	101.73 W	.	.	.	111	NO
68	CINTALAPA	15	16.70 N	93.75 W	554.7	SD	EW63	37	NO
69	COLIMA	15	19.23 N	103.75 W	493.8	TV	EW62	166	NO
70	COMALCALCO	15	18.25 N	93.23 W	10.1	.	.	87	NO
71	COMITAN	15	16.25 N	92.12 W	1634.6	SM	EC52	26	NO
72	COMONFORT	15	20.73 N	100.75 W	1793.4	SD	EW73	358	NO

AVERAGE PRECIPITATION (MM)

OBS.	JAN.	FEB.	MAR.	APR.	MAY	JUN.	JUL.	AUG.	SEP.	OCT.	NOV.	DEC.	ANNUAL
61	2	3	10	2	71	152	150	199	188	65	4	1	847
62	29	14	0	0	31	95	167	129	114	35	25	7	646
63	.	.	.	.	.	.	.	.	.	.	.	.	.
64	66	70	77	48	145	291	246	178	360	212	104	67	1865
65	4	5	8	8	10	24	80	96	94	37	8	21	395
66	22	24	9	6	4	45	205	182	129	47	22	71	765
67	6	11	1	2	19	156	183	147	166	65	10	1	766
68	5	5	4	3	69	207	154	140	161	84	9	4	845
69	13	8	0	0	8	144	192	181	194	79	24	35	877
70	170	135	51	26	81	200	132	186	334	465	215	263	2258
71	13	10	8	34	141	214	121	114	202	117	19	13	1004
72	15	7	4	14	40	113	172	171	129	43	11	15	734

AVERAGE TEMPERATURE (DEGREES CENTIGRADE)

OBS.	JAN.	FEB.	MAR.	APR.	MAY	JUN.	JUL.	AUG.	SEP.	OCT.	NOV.	DEC.	ANNUAL
61	20.2	22.2	24.6	25.8	26.7	25.6	24.7	24.4	24.3	23.2	21.7	20.6	23.7
62	.	.	.	.	.	.	.	.	.	.	.	.	.
63	15.6	18.1	19.7	20.8	21.5	19.7	18.4	18.7	18.0	18.0	17.4	16.1	18.5
64	15.1	17.9	19.8	23.1	25.1	25.5	24.3	24.9	24.1	23.1	19.2	16.1	21.5
65	9.6	11.8	14.9	18.7	23.1	26.7	25.1	24.2	22.2	18.2	13.3	9.2	18.1
66	17.7	19.1	21.0	24.1	27.6	31.2	29.8	28.5	28.2	26.3	22.1	18.1	24.5
67	.	.	.	.	.	.	.	.	.	.	.	.	.
68	19.9	21.0	23.1	24.9	26.4	24.9	24.8	24.9	24.3	22.9	21.2	20.6	23.2
69	22.4	22.6	23.4	25.0	26.2	26.4	25.9	25.6	25.1	25.4	24.3	22.9	24.6
70	.	.	.	.	.	.	.	.	.	.	.	.	.
71	16.0	17.0	18.3	19.5	19.6	18.9	18.6	18.7	18.6	17.8	16.6	16.6	18.0
72	15.4	16.8	19.3	20.8	22.8	23.1	22.0	21.9	21.0	19.9	17.6	16.6	19.8

CLIMATOLOGICAL DATA FOR MEXICO

GEOGRAPHY

OBS.	LOCATION	YEARS	LATITUDE	LONGITUDE	ELEVATION -M-	LIFEZONE	CCTX	ID	ONMAP
73	COMPOSTELA	15	21.25 N	104.88 W	874.8	.	.	393	NO
74	CONCORDIA	15	25.78 N	103.12 W	1104.6	SX	SW94	548	NO
75	COPAINALA	15	17.08 N	93.17 W	449.9	SM	EW53	49	NO
76	COQUIMATLAN	15	19.20 N	103.82 W	318.5	TV	EW63	162	NO
77	COSALA	15	24.38 N	106.68 W	449.9	TV	.	494	NO
78	COTIJA	15	19.75 N	102.65 W	1519.7	SD	EW72	226	NO
79	COZUMEL	15	20.52 N	86.95 W	3.0	TD	EW42	333	NO
80	CREEL	15	27.77 N	107.65 W	1478.0	.	.	600	NO
81	CRUCES	15	29.43 N	107.37 W	1724.6	.	.	632	NO
82	CRUZ (LA)	15	23.92 N	106.92 W	18.9	.	.	482	NO
83	CUATROCIENEGAS	15	26.98 N	102.07 W	741.9	SX	SW93	578	NO
84	CUAUTLA	15	18.80 N	90.95 W	1290.5	SD	.	122	NO

AVERAGE PRECIPITATION (MM)

OBS.	JAN.	FEB.	MAR.	APR.	MAY	JUN.	JUL.	AUG.	SEP.	OCT.	NOV.	DEC.	ANNUAL
73	24	5	0	1	2	208	360	412	190	76	9	8	1296
74	11	5	2	3	7	16	33	19	41	20	13	21	190
75	19	14	19	17	76	199	133	131	190	183	26	40	1047
76	11	6	0	0	9	130	162	165	212	92	22	50	859
77	15	20	3	.	9	80	292	223	190	47	8	52	940
78	28	15	3	2	22	115	136	155	126	41	20	30	694
79	83	31	57	54	177	177	76	127	213	301	121	85	1502
80	.	.	.	.	.	.	.	.	.	.	.	.	.
81	1	9	3	3	4	15	91	137	59	23	5	27	378
82	.	.	.	.	.	.	.	.	.	.	.	.	.
83	7	6	1	3	17	15	13	13	18	16	8	17	134
84	0	3	1	7	43	87	140	119	180	53	7	.	639

AVERAGE TEMPERATURE (DEGREES CENTIGRADE)

OBS.	JAN.	FEB.	MAR.	APR.	MAY	JUN.	JUL.	AUG.	SEP.	OCT.	NOV.	DEC.	ANNUAL
73	.	.	.	.	.	.	.	.	.	.	.	.	.
74	12.8	15.9	18.0	21.4	24.7	26.1	26.1	26.4	24.7	22.0	16.2	12.4	20.6
75	19.4	20.4	21.9	24.1	25.3	24.1	23.5	23.7	23.8	22.3	20.2	19.5	22.3
76	22.6	22.6	23.8	25.1	26.0	26.7	25.9	26.0	25.1	25.1	24.2	23.2	24.7
77	18.0	19.3	20.5	23.4	25.9	28.4	27.7	27.2	28.4	25.9	22.0	18.7	23.7
78	16.9	18.0	18.5	19.5	20.9	20.8	20.4	19.6	19.4	18.6	17.5	16.4	18.9
79	23.0	23.2	24.3	25.9	26.8	26.8	27.1	27.2	26.7	26.1	24.3	23.2	25.4
80	3.1	4.0	6.1	9.2	12.9	17.3	17.6	16.4	14.9	11.6	6.6	3.1	10.2
81	.	.	.	.	.	.	.	.	.	.	.	.	.
82	17.5	18.1	18.7	19.4	22.5	24.6	28.0	28.0	28.3	27.0	23.4	18.8	22.9
83	13.6	17.4	20.0	24.2	27.0	29.2	29.3	29.4	27.9	24.1	19.0	14.6	23.0
84	20.9	21.1	22.9	23.8	24.8	22.9	22.6	23.3	21.5	21.7	21.0	20.5	22.2

CLIMATOLOGICAL DATA FOR MEXICO

GEOGRAPHY

OBS.	LOCATION	YEARS	LATITUDE	LONGITUDE	ELEVATION -M-	LIFEZONE	CCTX	ID	ONMAP
85	CUERNAVACA	15	18.92 N	99.23 W	1537.7	SM	EW53	129	NO
86	CUICATLAN	15	17.80 N	96.97 W	595.0	TT	EW96	71	NO
87	CULIACAN	15	24.47 N	107.40 W	53.0	TV	SW74	498	NO
88	DOCTOR	15	20.85 N	99.60 W	2714.2	.	.	363	NO
89	DOCTOR ARROYO	15	23.67 N	100.17 W	1755.6	.	.	472	NO
90	DOLORES HIDALGO	15	21.15 N	100.93 W	1894.6	SD	EW73	388	NO
91	DURANGO	15	24.02 N	104.67 W	1897.4	WT	SC84	484	NO
92	EJUTLA	15	16.55 N	96.72 W	1439.6	SD	.	34	NO
93	ENRAMADAS LAS	15	25.48 N	99.52 W	222.8	SD	SW74	531	NO
94	ENSENADA	15	31.87 N	116.63 W	13.1	WT	.	661	YES
95	ESCUINAPA	15	22.67 N	105.75 W	14.0	.	.	451	NO
96	ESPERANZA LA	15	19.38 N	103.50 W	999.7	SD	EW73	187	NO

AVERAGE PRECIPITATION (MM)

OBS.	JAN.	FEB.	MAR.	APR.	MAY	JUN.	JUL.	AUG.	SEP.	OCT.	NOV.	DEC.	ANNUAL
85	3	5	7	9	53	195	217	217	245	78	8	3	1039
86	0	0	2	17	10	117	42	33	62	8	0	10	299
87	11	11	4	1	2	31	149	171	119	41	10	54	603
88	14	3	8	8	45	113	161	120	230	123	23	5	852
89	13	21	19	24	50	82	94	76	80	37	22	17	534
90	11	6	9	7	30	82	129	94	129	35	10	14	504
91	13	9	1	3	13	61	125	91	101	31	14	19	483
92	0	0	1	9	88	146	81	76	152	48	2	.	602
93	33	16	14	36	100	82	85	80	166	45	23	18	697
94	78	67	32	31	11	5	2	.	3	17	20	65	331
95	.	.	.	.	.	.	.	.	.	.	.	.	.
96	13	9	0	1	8	107	139	171	150	62	17	13	690

AVERAGE TEMPERATURE (DEGREES CENTIGRADE)

OBS.	JAN.	FEB.	MAR.	APR.	MAY	JUN.	JUL.	AUG.	SEP.	OCT.	NOV.	DEC.	ANNUAL
85	18.4	19.4	21.0	22.5	23.2	21.3	20.2	20.0	19.8	19.8	19.4	19.1	20.3
86	23.1	22.7	24.3	26.4	26.5	25.5	25.0	25.7	24.2	23.4	23.4	23.8	24.5
87	19.2	20.4	21.3	23.6	26.0	28.7	28.6	28.0	27.9	26.8	23.0	19.6	24.4
88	.	.	.	.	.	.	.	.	.	.	.	.	.
89	.	.	.	.	.	.	.	.	.	.	.	.	.
90	13.6	15.5	17.7	19.9	21.4	20.9	20.6	20.5	20.0	18.1	15.9	14.3	18.2
91	11.6	13.1	15.6	18.4	20.8	22.2	20.2	20.1	19.3	17.7	14.4	12.1	17.1
92	18.7	21.3	22.9	23.6	24.0	22.9	22.1	22.9	22.4	21.0	19.5	19.3	21.7
93	15.7	17.4	20.5	24.7	28.0	28.5	29.7	30.0	27.8	24.8	19.0	15.0	23.4
94	12.7	13.4	14.3	15.3	17.1	18.2	20.2	21.0	19.3	17.8	16.1	13.9	16.6
95	18.8	19.8	20.0	21.9	24.2	27.1	27.5	27.1	26.8	26.1	23.0	18.9	23.4
96	21.1	21.9	22.7	24.2	25.4	24.7	23.8	23.3	23.2	23.0	22.2	21.5	23.1

CLIMATOLOGICAL DATA FOR MEXICO

GEOGRAPHY

OBS.	LOCATION	YEARS	LATITUDE	LONGITUDE	ELEVATION -M-	LIFEZONE	CCTX	ID	ONMAP
97	ETLA	15	17.22 N	96.78 W	.	SD	EW73	55	YES
98	FLOR DE JIMULCO	15	25.12 N	103.30 W	1919.6	SX	SW94	517	NO
99	FRESNILLO	15	23.17 N	102.88 W	2249.4	WT	EC84	462	NO
100	FUERTE EL	15	26.42 N	108.63 W	114.9	TV	SW74	566	YES
101	GALEANA	15	24.83 N	100.07 W	1653.5	ST	SC83	509	NO
102	GARCIA	15	29.98 N	108.33 W	.	.	.	637	NO
103	GOGORRON	15	21.83 N	100.85 W	.	ST	EW83	427	NO
104	GONZALEZ	15	21.48 N	101.22 W	2139.4	WD	EC73	409	NO
105	GONZALEZ VILLA	15	30.63 N	106.52 W	1204.6	WX	SC95	647	NO
106	GUADALAJARA	15	20.68 N	103.33 W	1588.6	SD	EW63	354	NO
107	GUADALUPE DE CAL.	15	26.10 N	107.10 W	1112.8	WM	SC54	561	NO
108	GUADALUPE VICTORIA	15	24.42 N	104.13 W	1981.5	WT	SC83	496	NO

AVERAGE PRECIPITATION (MM)

OBS.	JAN.	FEB.	MAR.	APR.	MAY	JUN.	JUL.	AUG.	SEP.	OCT.	NOV.	DEC.	ANNUAL
97	1	3	3	23	73	121	112	93	83	51	12	12	662
98	8	8	2	3	10	24	52	42	42	17	16	11	236
99	10	1	0	5	16	57	104	69	80	30	15	19	406
100	21	19	4	1	2	22	165	134	93	42	13	54	568
101	23	20	9	11	58	73	53	38	101	44	26	16	473
102	9	4	0	2	9	31	77	53	53	15	16	6	275
103	4	3	15	8	27	75	54	42	53	26	23	15	343
104	12	7	10	6	31	94	99	73	98	33	24	18	503
105	3	7	6	9	10	8	42	43	28	20	8	9	191
106	17	5	3	2	19	191	250	198	177	52	20	19	953
107	60	56	35	6	24	84	257	221	156	64	29	84	1076
108	8	5	3	2	5	47	101	97	86	27	8	14	401

AVERAGE TEMPERATURE (DEGREES CENTIGRADE)

OBS.	JAN.	FEB.	MAR.	APR.	MAY	JUN.	JUL.	AUG.	SEP.	OCT.	NOV.	DEC.	ANNUAL
97	16.0	17.7	19.9	21.4	21.7	21.2	20.1	20.2	20.0	19.0	17.3	16.6	19.3
98	14.2	16.8	18.3	22.6	25.4	27.2	26.3	26.0	24.8	22.5	17.9	13.9	21.3
99	11.4	13.4	15.9	19.0	21.5	21.7	19.8	20.0	19.7	18.0	14.5	12.8	17.3
100	18.0	19.5	21.5	24.4	27.4	31.8	31.8	30.7	30.4	28.3	23.0	18.7	25.5
101	13.4	15.3	16.1	19.0	20.5	21.2	21.0	20.8	20.3	18.8	15.5	13.8	18.0
102	.	.	.	.	.	.	.	.	.	.	.	.	.
103	13.9	15.8	18.0	20.0	20.7	20.5	20.0	20.2	19.4	17.9	16.5	15.2	18.2
104	12.5	14.5	16.7	18.7	20.3	19.8	18.7	18.8	18.4	16.6	14.6	13.5	16.9
105	4.9	7.4	11.5	16.6	22.3	25.6	25.5	24.4	22.5	17.4	9.4	4.0	16.0
106	14.6	16.0	18.4	21.2	22.8	22.2	20.3	20.2	19.9	18.9	16.8	15.2	18.9
107	6.0	6.9	8.0	11.3	14.4	17.6	17.1	16.5	16.7	14.1	8.9	7.2	12.1
108	14.2	13.8	15.2	17.0	18.5	19.3	19.4	18.0	17.4	16.5	15.4	14.3	16.6

CLIMATOLOGICAL DATA FOR MEXICO

GEOGRAPHY

OBS.	LOCATION	YEARS	LATITUDE	LONGITUDE	ELEVATION -M-	LIFEZONE	CCTX	ID	ONMAP
109	GUANACEVI	15	25.92 N	105.95 W	.	WD	SC74	554	NO
110	GUANAJUATO	15	21.02 N	101.25 W	2036.4	SD	EC73	376	NO
111	GUAYAMAS	15	27.92 N	110.88 W	4.0	TT	SW94	604	NO
112	GUERRERO	15	28.55 N	107.50 W	2009.5	WT	SC84	616	NO
113	GUERRERO VILLA	15	22.00 N	98.77 W	68.0	TD	EW53	439	NO
114	HERMOSILLO	15	29.08 N	110.97 W	210.9	TT	SW85	623	NO
115	HIDALGO	15	19.67 N	100.53 W	1999.5	WD	EC63	217	NO
116	HIDALGO	15	24.25 N	99.43 W	.	.	.	493	NO
117	HIDALGO DEL PARRAL	15	26.93 N	105.65 W	1651.7	WD	SC74	577	NO
118	HORMIGUERO EL	15	27.03 N	105.70 W	1753.5	WD	SC74	579	NO
119	HOSTOTIPAQUILLO	15	21.22 N	104.08 W	.	.	.	392	NO
120	HUAJUAPAN	15	17.80 N	97.78 W	1596.5	SD	EW73	70	NO

AVERAGE PRECIPITATION (MM)

OBS.	JAN.	FEB.	MAR.	APR.	MAY	JUN.	JUL.	AUG.	SEP.	OCT.	NOV.	DEC.	ANNUAL
109	17	13	13	8	14	65	145	155	117	38	13	50	648
110	12	8	6	6	27	136	166	140	153	51	19	14	740
111	8	6	5	3	3	1	47	75	55	9	11	29	251
112	10	11	5	6	13	33	118	119	81	24	10	29	457
113	35	32	9	22	70	168	222	177	232	94	32	22	1115
114	2	15	5	2	2	4	71	84	63	41	6	26	320
115	20	15	8	13	48	146	171	179	166	63	41	21	891
116	32	7	13	44	115	135	131	136	257	109	57	11	1046
117	5	4	5	2	8	44	124	117	117	44	19	19	509
118	15	10	4	2	11	57	139	133	126	41	17	21	576
119	11	17	5	5	22	127	159	203	98	26	14	13	699
120	1	3	10	17	95	140	120	117	159	57	9	12	740

AVERAGE TEMPERATURE (DEGREES CENTIGRADE)

OBS.	JAN.	FEB.	MAR.	APR.	MAY	JUN.	JUL.	AUG.	SEP.	OCT.	NOV.	DEC.	ANNUAL
109	8.9	10.4	12.4	15.5	18.2	21.7	19.1	18.6	18.0	15.7	12.2	9.3	15.0
110	13.9	15.4	17.9	20.1	21.4	20.1	18.8	18.8	18.4	17.5	15.8	14.8	17.7
111	17.7	18.9	20.7	22.8	25.5	28.3	30.7	30.5	30.2	27.3	22.6	18.6	24.5
112	4.1	6.0	8.2	11.7	14.9	18.5	19.3	18.2	16.8	8.3	8.4	4.8	12.2
113	19.8	20.8	23.8	27.8	28.8	28.4	28.3	28.3	27.8	26.3	22.6	20.0	25.2
114	15.7	17.3	20.2	22.9	26.4	31.0	32.0	31.4	30.3	26.0	21.1	15.6	24.2
115	13.6	14.6	16.0	17.7	19.4	19.5	18.8	18.6	18.4	17.5	15.7	14.3	17.0
116	.	.	.	.	.	.	.	.	.	.	.	.	.
117	10.7	11.9	13.4	17.3	20.5	23.3	22.9	22.1	20.6	18.3	14.2	10.6	17.2
118	10.4	12.3	13.8	18.6	22.0	23.9	22.3	21.1	20.7	18.4	13.9	11.2	17.4
119	.	.	.	.	.	.	.	.	.	.	.	.	.
120	17.2	18.9	21.2	22.9	24.0	23.0	22.1	21.9	21.3	20.0	18.0	17.6	20.7

CLIMATOLOGICAL DATA FOR MEXICO

GEOGRAPHY

OBS.	LOCATION	YEARS	LATITUDE	LONGITUDE	ELEVATION -M-	LIFEZONE	CCTX	ID	ONMAP
121	HUATUSCO	15	19.15 N	96.95 W	1343.6	SW	EW33	155	NO
122	HUAUCHINANGO	15	20.17 N	98.05 W	1599.6	WW	EC33	290	YES
123	HUEJOTZINGO	15	19.15 N	98.40 W	2200.4	WD	EC63	154	NO
124	HUEJUTLA	15	21.13 N	98.40 W	171.9	.	.	385	NO
125	HUETAMO	15	18.67 N	100.90 W	356.0	TV	EH63	116	NO
126	HUICHAPAN	15	20.38 N	99.65 W	2101.6	WT	EC83	320	NO
127	HUIZOQUILUCAN	15	19.37 N	99.37 W	2699.3	WM	EC52	186	NO
128	IGUALA	15	18.37 N	99.55 W	734.9	TD	EH53	91	NO
129	IRAPUATO	15	20.67 N	101.35 W	1733.7	SD	EW73	351	NO
130	IXMIQUILPAN	15	20.48 N	99.22 W	1744.7	ST	EW93	331	NO
131	IXTLAN DE JUAREZ	15	17.33 N	96.50 W	1699.6	WM	EC52	58	NO
132	JACALA	15	21.00 N	99.18 W	1391.7	.	.	375	NO

AVERAGE PRECIPITATION (MM)

OBS.	JAN.	FEB.	MAR.	APR.	MAY	JUN.	JUL.	AUG.	SEP.	OCT.	NOV.	DEC.	ANNUAL
121	47	46	41	83	169	300	387	306	430	152	61	56	2078
122	59	49	50	56	91	366	399	355	402	257	103	53	2239
123	3	5	4	18	72	160	197	143	152	67	15	3	838
124	54	58	43	83	121	246	308	244	459	224	94	47	1980
125	2	3	1	2	21	162	204	204	199	51	14	10	874
126	9	4	4	14	33	51	64	57	72	32	8	7	354
127	10	10	13	30	82	220	236	213	218	88	27	10	1159
128	1	1	4	6	54	226	232	259	237	73	19	4	1115
129	10	12	12	5	24	124	183	161	162	35	14	10	740
130	4	4	8	15	25	42	49	38	62	31	4	5	287
131	9	9	6	15	83	163	168	119	211	154	50	17	1003
132	13	5	5	6	30	98	104	78	129	83	15	27	592

AVERAGE TEMPERATURE (DEGREES CENTIGRADE)

OBS.	JAN.	FEB.	MAR.	APR.	MAY	JUN.	JUL.	AUG.	SEP.	OCT.	NOV.	DEC.	ANNUAL
121	15.5	17.2	17.2	20.0	20.9	19.8	19.6	19.6	19.7	19.1	17.1	15.8	18.5
122	11.2	12.9	14.5	17.7	19.3	18.6	17.8	17.7	17.2	15.2	12.9	11.8	15.6
123	13.2	15.2	17.6	17.9	19.4	18.9	18.4	17.0	16.8	16.9	14.6	13.3	16.6
124	.	.	.	.	.	.	.	.	.	.	.	.	.
125	24.1	25.8	28.4	31.2	33.2	31.2	29.6	28.6	28.3	28.0	26.7	25.7	28.4
126	12.4	13.9	16.2	18.2	19.2	18.1	17.1	17.0	16.8	15.2	14.0	13.2	15.9
127	11.2	12.6	14.9	15.7	16.1	15.2	14.2	14.3	13.1	12.9	11.9	11.8	13.7
128	24.4	26.1	28.3	30.0	30.7	28.7	27.4	27.0	26.7	26.5	25.4	24.4	27.1
129	14.6	16.5	18.9	21.3	23.1	22.6	21.2	21.0	20.5	19.6	16.9	15.3	19.3
130	14.1	15.4	18.2	20.2	22.0	21.6	21.2	20.8	20.4	18.2	16.0	14.6	18.6
131	13.4	14.9	16.9	16.9	17.8	16.7	15.9	15.8	15.8	14.9	13.7	14.2	15.6
132	.	.	.	.	.	.	.	.	.	.	.	.	.

CLIMATOLOGICAL DATA FOR MEXICO

GEOGRAPHY

OBS.	LOCATION	YEARS	LATITUDE	LONGITUDE	ELEVATION -M-	LIFEZONE	CCTX	ID	ONMAP
133	JALAPA	15	19.53 N	96.92 W	1398.7	SM	EC42	209	YES
134	JAMILTEPEC	15	16.28 N	97.82 W	239.9	.	.	27	NO
135	JERECUARO	15	20.15 N	100.42 W	1786.4	WD	EC63	288	NO
136	JONACATEPEC	15	18.68 N	98.80 W	1299.7	SD	EW63	117	NO
137	JONUTA	15	18.08 N	92.13 W	14.0	.	.	81	NO
138	JUAREZ	15	31.73 N	106.48 W	1136.6	WX	SC94	660	YES
139	JUCHIPILA	15	21.42 N	103.12 W	.	.	.	405	NO
140	JUNTA LA	15	28.42 N	107.33 W	2040.3	WT	SC85	613	YES
141	LAGOS	15	21.37 N	101.93 W	1879.4	SD	EW73	402	NO
142	LAMPAZOS	15	27.03 N	100.52 W	339.9	ST	SW85	580	NO
143	LEON	15	21.12 N	101.68 W	1808.7	SD	EW73	383	NO
144	LERDO	15	25.50 N	103.53 W	1139.6	ST	SW94	532	YES

AVERAGE PRECIPITATION (MM)

OBS.	JAN.	FEB.	MAR.	APR.	MAY	JUN.	JUL.	AUG.	SEP.	OCT.	NOV.	DEC.	ANNUAL
133	52	54	52	58	120	298	214	201	290	130	71	48	1589
134	5	0	1	0	164	309	312	396	544	285	34	5	2054
135	11	12	10	13	41	119	195	153	139	54	18	18	785
136	4	4	5	7	52	147	174	195	204	53	7	3	854
137	96	55	35	26	120	241	193	222	229	221	124	141	1703
138	6	10	8	8	15	12	29	33	24	24	9	9	187
139	33	16	4	2	24	116	180	174	128	40	26	4	747
140	3	2	2	1	9	58	129	120	34	35	4	16	414
141	17	9	7	2	21	112	149	122	118	45	17	17	634
142	18	11	10	25	47	31	41	13	138	58	19	24	435
143	12	6	4	2	23	108	165	138	133	38	18	13	660
144	6	4	3	3	15	27	44	31	68	19	10	11	252

AVERAGE TEMPERATURE (DEGREES CENTIGRADE)

OBS.	JAN.	FEB.	MAR.	APR.	MAY	JUN.	JUL.	AUG.	SEP.	OCT.	NOV.	DEC.	ANNUAL
133	14.5	15.5	17.0	19.5	20.2	19.4	19.1	19.1	18.8	18.1	15.9	15.2	17.8
134	.	.	.	.	.	.	.	.	.	.	.	.	.
135	13.5	15.5	17.0	19.3	20.7	20.1	18.5	18.4	18.3	17.2	15.6	14.9	17.4
136	19.3	20.9	22.5	24.2	25.5	24.0	22.8	22.5	22.0	21.4	20.2	19.7	22.1
137	.	.	.	.	.	.	.	.	.	.	.	.	.
138	5.4	8.9	12.2	17.8	21.2	25.7	27.0	26.0	22.8	17.2	9.8	6.1	16.7
139	.	.	.	.	.	.	.	.	.	.	.	.	.
140	4.7	6.3	10.0	12.5	16.5	19.7	20.0	19.4	18.4	16.0	8.1	6.0	13.1
141	14.3	16.2	18.8	21.8	24.4	24.6	23.0	22.9	22.0	20.1	17.3	15.2	20.0
142	12.3	15.2	20.2	24.1	26.7	29.2	29.5	29.7	27.0	22.9	17.3	12.1	22.2
143	14.4	16.1	18.7	21.2	22.9	22.0	20.1	20.1	19.8	18.7	16.3	15.4	18.8
144	13.7	16.3	19.2	23.0	25.9	27.5	26.5	26.4	24.5	21.6	17.3	13.6	21.3

CLIMATOLOGICAL DATA FOR MEXICO

GEOGRAPHY

OBS.	LOCATION	YEARS	LATITUDE	LONGITUDE	ELEVATION -M-	LIFEZONE	CCTX	ID	ONMAP
145	LINARES	15	24.87 N	99.57 W	684.0	SD	SW64	511	NO
146	LLANOS VERDE	15	19.73 N	98.05 W	.	WD	EC62	225	NO
147	LOBOS ISLA	15	21.47 N	97.22 W	0.9	TD	EW43	408	NO
148	LOMAS DEL MIRADOR	15	21.70 N	99.02 W	.	TD	EW43	423	YES
149	MACLOVIO HERRERA	15	23.15 N	105.13 W	983.9	SD	EW65	461	NO
150	MACUSPANA	15	17.75 N	92.58 W	60.0	.	.	69	NO
151	MADERA	15	29.28 N	107.87 W	2078.4	WT	SC84	627	NO
152	MANUAL DOBLADO	15	20.73 N	101.95 W	1777.6	SD	EW73	357	NO
153	MANZANILLO	15	19.07 N	104.33 W	3.0	TD	EH53	146	NO
154	MARAVATIO	15	19.88 N	100.40 W	2014.4	WD	EC63	247	YES
155	MARIA MADRE ISLA	15	21.58 N	106.50 W	4.0	TV	EW73	416	NO
156	MARTIN DON	15	27.50 N	100.73 W	239.9	ST	SW84	594	NO

AVERAGE PRECIPITATION (MM)

OBS.	JAN.	FEB.	MAR.	APR.	MAY	JUN.	JUL.	AUG.	SEP.	OCT.	NOV.	DEC.	ANNUAL
145	30	15	20	52	92	100	99	61	188	81	31	31	801
146	8	19	15	27	49	134	127	108	139	85	33	11	756
147	68	57	28	30	97	172	192	229	417	293	114	77	1849
148	37	32	20	44	89	176	251	205	406	188	43	20	1512
149	7	6	14	9	25	63	109	251	255	7	21	72	841
150	180	105	90	84	189	309	231	322	390	478	202	291	2870
151	17	17	5	8	5	15	80	97	43	22	17	40	366
152	7	1	3	5	33	148	183	149	128	38	16	23	735
153	23	12	0	0	5	102	137	184	386	127	20	55	1051
154	22	7	4	12	58	105	215	167	113	52	17	5	777
155	11	21	1	0	0	38	89	118	210	75	28	77	667
156	28	20	21	43	50	49	32	39	86	50	19	25	461

AVERAGE TEMPERATURE (DEGREES CENTIGRADE)

OBS.	JAN.	FEB.	MAR.	APR.	MAY	JUN.	JUL.	AUG.	SEP.	OCT.	NOV.	DEC.	ANNUAL
145	14.8	17.3	19.3	23.1	24.9	26.4	26.8	27.5	25.4	22.7	18.0	14.3	21.7
146	11.4	12.2	14.0	15.3	15.9	15.0	14.2	14.8	13.8	13.2	12.0	12.0	13.6
147	19.4	20.8	21.7	24.3	26.3	27.3	27.4	27.5	27.2	26.3	23.1	20.1	24.3
148	19.1	20.6	22.6	26.4	28.7	28.8	27.7	27.5	26.9	24.9	21.4	19.2	24.5
149	9.6	11.6	16.5	20.8	25.5	29.5	28.0	26.8	23.8	19.7	14.2	8.2	19.5
150	.	.	.	.	.	.	.	.	.	.	.	.	.
151	6.3	7.4	10.4	12.8	14.6	17.2	17.8	16.4	16.6	13.6	8.3	6.8	12.3
152	15.9	17.5	20.6	22.9	24.8	23.6	22.1	21.8	21.7	20.8	18.2	16.9	20.6
153	24.2	23.7	23.6	24.5	26.2	27.8	28.2	28.1	27.6	27.5	26.5	25.2	26.1
154	13.5	14.7	16.8	18.1	19.8	19.5	18.3	18.2	18.4	17.6	15.6	14.2	17.1
155	20.3	20.3	20.8	22.5	25.2	27.0	28.1	28.1	27.6	26.3	24.5	22.1	24.4
156	12.9	15.3	19.1	23.8	27.1	29.7	30.7	30.7	28.1	23.9	17.8	12.8	22.7

CLIMATOLOGICAL DATA FOR MEXICO

GEOGRAPHY

OBS.	LOCATION	YEARS	LATITUDE	LONGITUDE	ELEVATION -M-	LIFEZONE	CCTX	ID	ONMAP
157	MASCOTA	15	20.55 N	104.75 W	1199.7	SD	EW63	340	NO
158	MATAMOROS	15	25.53 N	103.25 W	1119.8	SX	SW94	535	NO
159	MATAMOROS	15	25.87 N	97.50 W	11.9	SD	SW74	551	YES
160	MATANZAS	15	21.63 N	101.55 W	.	.	.	418	NO
161	MATEHUALA	15	23.60 N	100.65 W	1596.5	.	.	471	YES
162	MAXCANU	15	20.58 N	98.98 W	11.9	.	.	344	NO
163	MAZAPIL	15	24.63 N	101.57 W	.	.	.	503	NO
164	MAZATLAN	15	23.18 N	106.42 W	78.0	TV	EW64	463	NO
165	MENDEZ	15	25.12 N	98.57 W	124.1	TV	SW73	518	NO
166	MERIDA	15	20.97 N	89.63 W	21.9	TV	EW62	370	NO
167	METZQUITITLAN	15	20.53 N	98.63 W	1352.7	.	.	336	NO
168	MEXICALI	15	32.65 N	115.50 W	0.9	SO	SW94	669	NO

AVERAGE PRECIPITATION (MM)

OBS.	JAN.	FEB.	MAR.	APR.	MAY	JUN.	JUL.	AUG.	SEP.	OCT.	NOV.	DEC.	ANNUAL
157	21	13	2	2	26	130	209	161	139	63	17	13	794
158	14	4	1	5	12	38	54	24	37	23	14	12	239
159	42	31	28	46	73	80	58	44	171	78	47	47	747
160	.	.	.	.	.	.	.	.	.	.	.	.	.
161	21	11	5	15	52	59	72	61	74	23	21	7	421
162	.	.	.	.	.	.	.	.	.	.	.	.	.
163	25	9	4	8	35	115	120	84	70	88	22	19	599
164	11	10	4	0	2	29	167	242	268	61	12	44	850
165	36	19	18	18	108	88	53	46	103	69	40	11	609
166	31	16	19	26	81	151	141	129	154	103	31	30	913
167	13	9	10	15	49	65	75	71	129	78	9	6	528
168	6	10	5	6	1	1	3	8	7	8	9	14	76

AVERAGE TEMPERATURE (DEGREES CENTIGRADE)

OBS.	JAN.	FEB.	MAR.	APR.	MAY	JUN.	JUL.	AUG.	SEP.	OCT.	NOV.	DEC.	ANNUAL
157	18.4	18.8	20.9	23.2	23.8	24.7	23.1	22.8	23.2	22.5	20.7	19.2	21.8
158	12.4	15.8	18.1	23.3	27.0	28.5	26.9	27.6	26.8	23.5	16.9	13.4	21.7
159	16.2	17.9	20.1	23.8	26.5	28.5	29.0	29.3	27.6	24.6	20.1	16.3	23.3
160	11.8	14.0	16.0	17.6	20.4	18.8	18.7	18.4	18.1	16.4	13.8	13.8	16.5
161	.	.	.	.	.	.	.	.	.	.	.	.	.
162	22.6	23.5	25.9	27.9	28.3	27.4	27.2	27.0	26.3	25.3	23.6	23.1	25.7
163	.	.	.	.	.	.	.	.	.	.	.	.	.
164	19.3	19.4	19.7	21.3	23.8	26.4	27.5	27.6	27.5	26.7	23.5	20.6	23.6
165	15.1	18.6	21.4	25.2	27.8	29.2	29.5	29.9	28.3	25.7	20.0	15.8	23.9
166	22.9	23.8	25.4	27.1	28.0	27.7	27.4	27.4	27.2	26.1	24.0	23.4	25.9
167	.	.	.	.	.	.	.	.	.	.	.	.	.
168	11.8	15.1	17.9	21.2	25.2	29.6	33.0	32.7	29.5	23.2	16.6	12.3	22.3

CLIMATOLOGICAL DATA FOR MEXICO

GEOGRAPHY

OBS.	LOCATION	YEARS	LATITUDE	LONGITUDE	ELEVATION -M-	LIFEZONE	CCTX	ID	ONMAP
169	MEXICO CITY	15	19.43 N	99.13 W	2302.8	WD	EC73	194	NO
170	MEZCALA ISLA	15	20.33 N	103.02 W	1077.8	SM	EW53	314	NO
171	MIAHUATLAN	15	18.08 N	97.67 W	.	SD	EW72	80	YES
172	MILAGRO EL	15	25.75 N	100.82 W	71.9	.	.	546	NO
173	MINAS NUEVAS	15	27.08 N	108.98 W	.	SD	SW74	581	YES
174	MINATITLAN	15	17.98 N	94.50 W	64.9	TM	EH23	76	NO
175	MISANTLA	15	19.93 N	96.85 W	410.0	SM	EW43	254	NO
176	MIXQUIAHUALA	15	20.22 N	99.22 W	2049.5	WD	EC73	298	NO
177	MOCORITO	15	25.48 N	107.92 W	837.9	TV	SW74	530	NO
178	MOLINOSDECABALLERO	15	20.05 N	100.17 W	.	WD	EC63	275	NO
179	MONCLOVA	15	26.92 N	101.43 W	585.8	ST	SW85	576	NO
180	MONTE MORELOS	15	25.20 N	99.83 W	431.9	TT	SW84	522	NO

AVERAGE PRECIPITATION (MM)

OBS.	JAN.	FEB.	MAR.	APR.	MAY	JUN.	JUL.	AUG.	SEP.	OCT.	NOV.	DEC.	ANNUAL
169	5	6	9	13	52	104	125	103	117	35	14	16	589
170	10	5	4	1	17	186	191	227	236	61	14	47	1001
171	1	0	2	17	56	127	101	66	119	28	2	7	526
172	.	.	.	.	.	.	.	.	.	.	.	.	.
173	30	26	3	2	5	33	189	172	116	39	14	80	707
174	162	79	39	36	78	307	296	358	602	401	367	152	2876
175	53	120	96	88	89	136	173	136	229	246	188	121	1675
176	11	9	11	17	62	74	106	98	104	46	10	6	554
177	10	20	11	2	1	38	215	168	119	42	20	87	734
178	18	11	12	29	31	183	185	156	183	53	36	26	924
179	15	15	8	18	35	36	53	30	108	35	17	15	386
180	11	10	19	23	84	83	7	18	71	39	24	11	398

AVERAGE TEMPERATURE (DEGREES CENTIGRADE)

OBS.	JAN.	FEB.	MAR.	APR.	MAY	JUN.	JUL.	AUG.	SEP.	OCT.	NOV.	DEC.	ANNUAL
169	12.3	13.9	16.0	17.7	18.3	17.6	16.7	16.7	16.2	15.0	13.7	12.4	15.5
170	18.3	19.7	21.7	23.7	24.7	23.2	22.3	21.9	21.5	21.6	20.2	18.6	21.4
171	18.3	19.0	20.4	21.7	22.3	21.9	21.3	21.3	21.1	20.3	18.9	18.5	20.4
172	13.6	16.0	19.7	24.3	27.1	26.9	27.1	27.9	25.8	22.9	17.4	13.6	21.9
173	16.2	17.2	19.2	22.0	24.7	28.2	27.7	26.4	26.1	24.5	20.6	17.0	22.5
174	23.2	23.9	25.1	27.1	28.9	28.2	27.7	27.5	27.5	26.7	24.6	23.9	26.2
175	16.4	18.9	20.0	22.4	25.0	25.6	24.8	24.7	24.3	23.5	20.9	17.1	22.0
176	12.6	14.8	16.0	18.2	19.9	18.5	18.4	18.0	18.3	16.3	15.0	14.1	16.7
177	18.4	19.3	20.7	23.3	26.2	29.6	29.8	28.8	28.2	25.6	22.6	19.1	24.3
178	11.1	12.2	14.4	15.8	17.3	16.9	15.9	16.1	15.9	14.4	12.9	11.9	14.6
179	12.6	15.2	18.2	22.3	25.5	27.7	27.7	28.1	25.1	21.8	16.7	12.8	21.1
180	15.5	17.2	24.1	25.7	28.3	28.7	30.7	29.9	27.8	25.2	20.2	15.3	24.0

CLIMATOLOGICAL DATA FOR MEXICO

GEOGRAPHY

OBS.	LOCATION	YEARS	LATITUDE	LONGITUDE	ELEVATION -M-	LIFEZONE	CCTX	ID	ONMAP
181	MONTE PUERTA	15	20.23 N	101.02 W	1699.6	SD	EC73	302	NO
182	MONTERREY	15	25.67 N	100.30 W	534.0	.	.	543	NO
183	MORELIA	15	19.70 N	101.12 W	1922.4	WD	EC63	219	NO
184	MOROLEON	15	20.13 N	101.20 W	1771.5	SD	EW63	283	NO
185	MOTOZINTLA	15	15.37 N	92.23 W	1454.5	SD	EW73	15	YES
186	MULEGE	15	26.88 N	112.00 W	35.1	SO	SW95	575	YES
187	MUZQUIZ	15	27.87 N	101.52 W	503.8	SD	SW74	601	YES
188	NACO	15	31.35 N	109.93 W	1403.6	WT	SC85	656	NO
189	NAUTLA	15	20.25 N	96.78 W	7.9	TD	EH43	306	NO
190	NAVOJOA	15	27.12 N	109.47 W	38.1	TT	SW85	582	NO
191	NAZAS	15	25.22 N	104.13 W	1274.7	ST	SW84	523	NO
192	NOGALES	15	31.35 N	110.97 W	1176.8	.	.	655	NO

AVERAGE PRECIPITATION (MM)

OBS.	JAN.	FEB.	MAR.	APR.	MAY	JUN.	JUL.	AUG.	SEP.	OCT.	NOV.	DEC.	ANNUAL
181	8	4	5	8	32	52	153	137	135	52	25	8	695
182	22	23	15	29	42	85	73	62	204	109	25	24	714
183	13	8	7	8	43	132	171	160	155	58	19	6	781
184	3	3	8	8	41	149	167	152	172	44	24	19	791
185	0	1	3	9	63	189	121	116	140	86	9	3	740
186	3	4	0	0	0	0	6	17	40	5	7	19	101
187	24	11	26	27	88	78	63	78	177	72	28	27	702
188	12	33	14	7	4	15	98	83	52	19	21	30	388
189	75	51	47	35	81	152	258	114	430	364	89	121	1816
190	3	1	4	0	10	6	73	100	56	57	14	72	396
191	8	5	3	5	20	36	78	61	56	22	9	10	314
192	.	.	.	.	.	.	.	.	.	.	.	.	.

AVERAGE TEMPERATURE (DEGREES CENTIGRADE)

OBS.	JAN.	FEB.	MAR.	APR.	MAY	JUN.	JUL.	AUG.	SEP.	OCT.	NOV.	DEC.	ANNUAL
181	13.9	14.9	17.9	18.8	20.9	21.2	19.8	20.0	19.8	18.1	16.1	14.6	18.0
182	.	.	.	.	.	.	.	.	.	.	.	.	.
183	13.9	15.4	17.3	19.2	20.5	19.5	18.2	18.0	17.9	17.1	15.4	14.4	17.2
184	16.1	18.0	21.1	23.0	24.1	22.9	21.1	21.1	20.9	19.8	17.9	16.6	20.2
185	19.4	20.4	21.4	23.1	23.4	22.3	22.2	22.1	21.8	21.3	19.8	20.0	21.5
186	14.0	16.0	17.8	20.1	22.9	27.2	30.5	30.4	29.0	24.6	19.5	14.9	22.2
187	15.1	16.5	18.7	23.5	25.7	28.4	28.8	29.3	27.4	24.0	17.9	13.4	22.4
188	7.6	9.6	11.6	14.7	19.3	24.4	24.6	23.1	21.6	17.5	12.2	7.2	16.1
189	22.7	23.8	24.8	27.7	28.8	29.0	29.1	29.2	28.8	27.8	25.7	23.9	26.9
190	17.9	19.4	20.8	23.8	27.0	31.2	33.8	32.4	31.9	29.0	23.3	19.3	25.8
191	13.6	16.4	19.0	22.6	25.9	27.5	26.2	26.0	24.7	21.8	17.0	14.0	21.2
192	8.3	10.2	12.9	16.3	20.1	25.0	27.4	25.4	23.5	18.7	12.8	9.1	17.5

CLIMATOLOGICAL DATA FOR MEXICO

GEOGRAPHY

OBS.	LOCATION	YEARS	LATITUDE	LONGITUDE	ELEVATION -M-	LIFEZONE	CCTX	ID	ONMAP
193	NOMBRE DE DIOS	15	23.82 N	104.23 W	1922.4	.	.	478	NO
194	NUEVO LAREDO	15	27.48 N	99.50 W	139.9	TT	SW84	593	NO
195	NURI	15	28.05 N	109.37 W	264.0	.	.	607	NO
196	OAXACA	15	17.07 N	96.70 W	1562.7	SD	EW73	48	NO
197	OBREGON	15	27.48 N	109.92 W	39.9	TX	.	592	NO
198	OCAMPO	15	21.63 N	101.40 W	1237.8	WT	EC83	419	NO
199	OCOSINGO	15	17.03 N	92.18 W	769.9	TD	EW42	47	NO
200	OJINAGA	15	29.57 N	104.42 W	840.6	SX	SW94	633	YES
201	OJO EL	15	24.15 N	104.07 W	1999.5	WT	SC84	490	NO
202	OMETEPEC	15	16.55 N	98.58 W	399.9	TT	EH82	33	NO
203	ORIZABA	15	18.85 N	97.08 W	1247.5	SW	EW33	124	NO
204	ORO EL	15	25.95 N	105.33 W	1829.4	SD	SC74	555	NO

AVERAGE PRECIPITATION (MM)

OBS.	JAN.	FEB.	MAR.	APR.	MAY	JUN.	JUL.	AUG.	SEP.	OCT.	NOV.	DEC.	ANNUAL
193	11	13	0	2	15	78	119	87	75	30	9	7	446
194	24	24	21	26	64	41	29	20	78	28	27	26	407
195	12	9	13	2	18	24	164	129	106	37	15	121	651
196	2	2	11	24	62	125	93	103	169	40	8	9	650
197	7	5	2	4	.	2	70	46	44	14	4	13	212
198	21	8	10	6	29	81	104	88	91	23	20	18	498
199	41	42	37	78	180	252	209	184	286	83	55	68	1668
200	3	6	4	3	9	12	27	44	41	32	14	12	208
201	8	10	0	2	12	62	126	104	83	42	15	20	484
202	0	0	0	3	19	92	58	71	107	35	5	1	392
203	46	40	26	49	136	351	371	329	430	184	100	55	2115
204	3	2	2	8	7	50	134	198	200	47	16	9	676

AVERAGE TEMPERATURE (DEGREES CENTIGRADE)

OBS.	JAN.	FEB.	MAR.	APR.	MAY	JUN.	JUL.	AUG.	SEP.	OCT.	NOV.	DEC.	ANNUAL
193	.	.	.	.	.	.	.	.	.	.	.	.	.
194	14.1	17.8	20.8	25.2	28.1	30.4	32.5	32.3	29.9	26.5	21.0	16.1	24.6
195	.	.	.	.	.	.	.	.	.	.	.	.	.
196	17.5	19.0	21.0	22.4	22.7	21.4	20.9	20.8	20.3	19.7	18.4	18.0	20.2
197	18.2	19.8	22.2	25.2	27.4	32.1	33.6	33.7	32.7	29.2	23.8	19.6	26.5
198	12.6	14.2	16.4	18.3	21.4	20.0	18.9	19.2	18.5	16.9	14.6	13.9	16.9
199	21.8	22.7	23.2	25.2	26.3	25.9	25.6	25.8	25.8	24.6	22.2	21.8	24.2
200	12.2	15.7	15.8	21.1	25.7	30.8	30.2	30.6	27.8	23.0	14.9	10.8	21.6
201	10.6	11.8	13.8	17.0	19.8	21.8	20.1	19.3	18.7	16.5	12.5	11.6	16.1
202	25.7	26.2	26.8	27.2	28.4	27.4	27.1	27.3	26.8	27.4	27.3	27.5	27.1
203	15.1	16.2	17.9	20.3	21.0	20.1	19.5	19.6	19.5	18.9	16.6	15.7	18.4
204	10.6	14.6	16.6	19.0	21.7	22.6	21.3	21.2	20.1	17.6	15.3	10.4	17.6

CLIMATOLOGICAL DATA FOR MEXICO

GEOGRAPHY

OBS.	LOCATION	YEARS	LATITUDE	LONGITUDE	ELEVATION -M-	LIFEZONE	CCTX	ID	ONMAP
205	ORO EL	15	19.82 N	100.08 W	1889.5	WD	EC62	237	NO
206	OZULUAMA	15	21.67 N	97.85 W	228.9	TD	EW43	422	NO
207	PACHUCA	15	20.13 N	98.75 W	2434.4	WT	EC82	284	NO
208	PALENQUE	15	17.50 N	92.00 W	210.0	TM	EH23	61	YES
209	PANUCO	15	23.40 N	105.97 W	602.0	TD	EW53	465	NO
210	PATZCUARO	15	19.50 N	101.52 W	2137.6	WM	EC53	201	NO
211	PAZ LA	15	24.17 N	110.30 W	57.9	.	.	491	NO
212	PENJAMO	15	20.43 N	101.72 W	1759.6	SD	EW63	325	NO
213	PENON-BLANCO	15	24.78 N	104.03 W	1609.6	WT	SC84	508	NO
214	PEROTE	15	10.57 N	97.23 W	2464.3	WT	EC83	2	NO
215	PETO	15	20.13 N	88.88 W	.	.	.	285	NO
216	PIAXTLA	15	18.20 N	98.27 W	1119.8	TV	EW73	84	NO

AVERAGE PRECIPITATION (MM)

OBS.	JAN.	FEB.	MAR.	APR.	MAY	JUN.	JUL.	AUG.	SEP.	OCT.	NOV.	DEC.	ANNUAL
205	9	11	7	12	38	90	170	191	136	56	27	7	755
206	46	20	14	23	82	246	218	191	388	234	53	42	1558
207	5	13	13	18	33	72	59	53	78	49	21	6	419
208	276	121	84	25	132	305	316	351	485	614	262	215	3187
209	7	12	9	0	0	132	412	231	244	153	10	68	1279
210	21	14	7	4	38	199	246	239	214	78	26	23	1110
211	3	11	1	0	0	0	6	42	52	9	13	34	173
212	17	5	4	1	41	124	202	181	140	39	20	14	789
213	5	6	0	3	23	40	99	107	74	21	11	7	397
214	13	16	12	19	24	90	39	30	109	88	26	13	480
215	51	44	30	74	89	200	188	137	200	148	41	77	1281
216	1	2	12	5	52	159	135	121	176	48	8	1	720

AVERAGE TEMPERATURE (DEGREES CENTIGRADE)

OBS.	JAN.	FEB.	MAR.	APR.	MAY	JUN.	JUL.	AUG.	SEP.	OCT.	NOV.	DEC.	ANNUAL
205	10.5	11.0	13.2	14.2	14.2	13.4	13.4	13.3	13.3	12.5	11.3	10.5	12.6
206	18.4	20.2	22.1	25.2	27.3	27.8	27.3	27.5	26.5	25.2	21.8	18.8	24.0
207	11.8	12.8	14.5	15.9	16.7	15.8	15.2	15.1	14.9	13.5	12.2	12.4	14.2
208	21.7	25.6	26.2	27.1	29.6	28.8	27.7	27.7	27.2	26.1	25.0	23.7	26.4
209	20.4	21.1	23.1	25.1	27.5	27.2	26.0	25.9	25.2	24.2	22.1	20.2	24.0
210	13.9	13.3	16.1	17.9	19.9	19.2	17.3	17.0	16.8	16.0	14.3	12.5	16.0
211	.	.	.	.	.	.	.	.	.	.	.	.	.
212	16.2	17.1	19.4	22.1	22.4	22.4	21.8	21.2	20.6	20.2	18.2	17.0	19.9
213	11.7	13.2	15.1	17.2	20.2	21.8	20.3	20.2	19.3	17.3	13.9	12.3	16.9
214	10.2	11.3	13.9	15.1	15.7	15.0	14.5	14.7	13.9	12.7	11.2	10.8	13.2
215	.	.	.	.	.	.	.	.	.	.	.	.	.
216	21.4	23.2	25.3	26.7	27.4	25.6	24.8	25.1	24.2	23.5	22.2	21.5	24.2

CLIMATOLOGICAL DATA FOR MEXICO

GEOGRAPHY

OBS.	LOCATION	YEARS	LATITUDE	LONGITUDE	ELEVATION -M-	LIFEZONE	CCTX	ID	ONMAP
217	PIEDAD LA	15	20.33 N	102.02 W	1774.5	SM	EW53	315	YES
218	PIEDRAS NEGRAS	15	28.70 N	100.52 W	220.1	SD	SW74	618	YES
219	PILARESDENACOZARI	15	30.32 N	109.72 W	1408.8	SD	SC75	640	YES
220	POLOTITLAN	15	20.23 N	99.82 W	2290.3	WD	EC73	303	NO
221	POTRERO	15	18.87 N	96.83 W	605.9	.	.	127	NO
222	PROGRESO	15	21.28 N	89.67 W	14.0	TT	EW82	398	YES
223	PROVIDENCIA LA	15	16.53 N	93.98 W	784.9	SM	EW52	32	NO
224	PUEBLA	15	19.03 N	98.18 W	2149.4	WD	EC63	143	NO
225	PUENTE DE IXTLA	15	18.62 N	99.32 W	899.8	TV	EW63	108	NO
226	PUERTO ANGEL	15	15.65 N	96.50 W	43.0	.	.	19	NO
227	PUERTO MEXICO	15	18.15 N	94.40 W	14.0	TM	EW22	82	YES
228	PUNTA JEREZ	15	22.90 N	97.77 W	0.9	SM	EW53	456	NO

AVERAGE PRECIPITATION (MM)

OBS.	JAN.	FEB.	MAR.	APR.	MAY	JUN.	JUL.	AUG.	SEP.	OCT.	NOV.	DEC.	ANNUAL
217	22	17	5	6	39	167	233	214	215	59	27	38	1034
218	23	26	24	47	94	51	35	25	95	42	25	31	518
219	19	28	18	9	8	34	178	144	59	41	20	47	579
220	10	5	15	29	51	178	140	81	134	42	23	8	637
221	.	.	.	.	.	.	.	.	.	.	.	.	.
222	33	17	15	18	55	73	45	46	55	69	20	26	473
223	0	5	4	9	112	289	199	188	271	132	11	2	1222
224	6	6	12	13	72	157	137	121	187	57	21	9	799
225	2	2	4	8	56	193	193	203	219	43	9	6	939
226	1	1	1	1	83	203	139	166	283	123	11	4	1016
227	131	65	65	36	96	263	217	376	490	559	368	218	2884
228	47	13	4	11	53	183	198	159	306	166	42	38	1221

AVERAGE TEMPERATURE (DEGREES CENTIGRADE)

OBS.	JAN.	FEB.	MAR.	APR.	MAY	JUN.	JUL.	AUG.	SEP.	OCT.	NOV.	DEC.	ANNUAL
217	14.2	15.8	18.1	20.8	22.5	22.7	21.2	21.0	20.6	19.2	16.8	15.1	19.0
218	10.9	14.6	17.8	22.2	25.4	28.5	29.9	30.2	27.7	26.6	16.0	10.8	21.7
219	9.5	11.4	13.9	17.0	20.8	25.3	25.1	24.2	24.1	20.2	14.8	10.3	18.0
220	11.5	14.1	16.6	17.9	19.6	18.5	18.1	17.8	17.1	15.6	13.9	13.2	16.2
221	18.9	19.5	21.1	23.7	25.0	24.7	23.7	23.7	23.4	23.0	20.6	19.5	22.2
222	22.7	23.1	24.1	25.5	26.3	26.7	26.7	26.8	26.8	26.3	24.4	23.2	25.2
223	20.9	21.5	22.5	24.1	24.9	23.9	24.2	24.3	24.0	22.8	21.4	21.2	23.0
224	12.8	14.5	16.8	18.3	18.9	17.9	17.2	17.4	16.8	16.0	14.7	13.6	16.2
225	20.4	22.6	25.0	26.2	27.8	26.4	25.6	25.2	24.7	24.4	22.6	21.0	24.3
226	.	.	.	.	.	.	.	.	.	.	.	.	.
227	21.8	22.6	23.9	25.9	27.3	27.0	26.9	26.8	26.5	25.5	23.7	22.4	25.0
228	17.4	19.1	20.7	23.8	26.0	26.6	27.2	27.3	26.8	25.1	21.3	17.8	23.3

CLIMATOLOGICAL DATA FOR MEXICO

GEOGRAPHY

OBS.	LOCATION	YEARS	LATITUDE	LONGITUDE	ELEVATION -M-	LIFEZONE	CCTX	ID	ONMAP
229	QUERETARO	15	20.60 N	100.38 W	1841.6	SD	EC73	345	NO
230	QUIRIEGO	15	27.52 N	109.25 W	.	TV	SW75	595	NO
231	RINCON DE ROMOS	15	22.23 N	102.32 W	1949.5	WT	EC84	446	NO
232	RINCONADA	15	19.35 N	96.55 W	313.0	.	.	182	NO
233	RIO GRANDE	15	23.83 N	103.02 W	.	ST	EC84	480	YES
234	RIO MANTE	15	22.70 N	99.03 W	79.9	TD	EW53	452	YES
235	RIOVERDE	15	21.93 N	99.98 W	986.6	SD	EW73	432	NO
236	RODEO EL	15	25.18 N	104.58 W	1809.6	ST	SW84	521	NO
237	ROSA MORADA	15	22.12 N	105.20 W	65.8	.	.	443	NO
238	ROSARIO	15	22.97 N	105.90 W	17.1	.	.	457	NO
239	ROSARIO	15	27.88 N	109.30 W	279.8	SD	SW65	602	YES
240	SABINAS	15	27.92 N	101.30 W	339.9	SD	SW75	605	NO

AVERAGE PRECIPITATION (MM)

OBS.	JAN.	FEB.	MAR.	APR.	MAY	JUN.	JUL.	AUG.	SEP.	OCT.	NOV.	DEC.	ANNUAL
229	10	4	4	11	27	93	104	87	122	34	10	12	518
230	20	24	5	1	7	10	214	180	94	45	17	46	663
231	17	3	5	2	36	64	138	70	88	24	21	7	473
232	.	.	.	.	.	.	.	.	.	.	.	.	.
233	10	9	6	4	17	57	95	63	111	31	22	13	439
234	42	11	12	18	119	165	231	198	251	108	28	9	1192
235	18	7	6	9	54	89	83	73	121	48	20	11	537
236	13	7	1	7	11	48	119	76	83	30	3	9	407
237	.	.	.	.	.	.	.	.	.	.	.	.	.
238	6	0	2	0	0	62	157	114	152	37	7	63	601
239	20	26	8	3	21	45	287	174	105	44	26	60	819
240	23	56	46	42	88	63	37	28	134	22	25	27	593

AVERAGE TEMPERATURE (DEGREES CENTIGRADE)

OBS.	JAN.	FEB.	MAR.	APR.	MAY	JUN.	JUL.	AUG.	SEP.	OCT.	NOV.	DEC.	ANNUAL
229	13.9	15.3	17.8	19.9	21.2	20.3	19.4	19.4	18.9	17.5	15.9	14.7	17.8
230	17.0	18.7	20.6	24.3	26.6	31.0	31.4	31.4	30.4	27.8	22.4	18.0	25.0
231	11.7	13.6	15.4	17.5	19.9	20.9	19.2	19.2	19.1	17.8	14.9	12.3	16.8
232	22.0	23.4	25.3	28.0	28.6	27.5	26.5	26.2	25.9	24.8	23.8	22.7	25.4
233	11.3	13.4	15.1	17.8	21.6	23.3	22.5	21.6	21.1	17.9	14.2	12.3	17.7
234	18.8	20.2	22.8	26.7	28.3	28.5	27.4	27.7	27.0	24.7	20.6	19.2	24.3
235	15.6	17.7	19.8	23.1	24.6	24.2	23.3	23.3	22.6	20.5	17.8	15.9	20.7
236	13.3	15.2	17.6	20.9	24.2	26.2	24.7	24.4	23.6	21.1	16.4	13.6	20.1
237	19.6	21.5	21.6	22.2	24.4	27.7	25.7	27.6	25.8	27.0	24.6	22.3	24.2
238	.	.	.	.	.	.	.	.	.	.	.	.	.
239	14.6	15.9	17.5	21.1	23.9	29.1	28.0	28.2	27.9	24.7	19.2	15.6	22.1
240	9.8	14.4	16.7	23.0	28.0	32.5	32.6	32.5	28.1	21.4	16.0	10.6	22.1

CLIMATOLOGICAL DATA FOR MEXICO

GEOGRAPHY

OBS.	LOCATION	YEARS	LATITUDE	LONGITUDE	ELEVATION -M-	LIFEZONE	CCTX	ID	ONMAP
241	SALINA CRUZ	15	16.20 N	95.20 W	56.1	TD	EH53	24	YES
242	SALINAS	15	22.62 N	101.72 W	2098.5	.	.	450	NO
243	SALTILLO	15	25.43 N	101.00 W	1608.7	.	.	527	YES
244	SALTO EL	15	23.77 N	105.37 W	2537.5	WD	EC64	475	YES
245	SALVATIERRA	15	20.22 N	100.88 W	1760.5	SD	EW63	297	NO
246	SAN ANDRES TUXTLA	15	18.43 N	95.20 W	360.0	TM	EW33	94	NO
247	SAN BARTOLO	15	24.67 N	103.22 W	1686.5	.	.	504	NO
248	SAN BARTOLOME	15	16.33 N	92.53 W	709.0	.	.	29	YES
249	SAN BLAS	15	26.07 N	108.75 W	71.0	.	.	560	NO
250	SAN BLAS	15	21.53 N	105.32 W	7.0	TD	EW53	413	YES
251	SAN BUENAVENTURA	15	29.83 N	107.50 W	1535.6	WT	SC85	636	YES
252	SAN CARLOS	15	24.57 N	98.97 W	1399.6	.	.	500	YES

AVERAGE PRECIPITATION (MM)

OBS.	JAN.	FEB.	MAR.	APR.	MAY	JUN.	JUL.	AUG.	SEP.	OCT.	NOV.	DEC.	ANNUAL
241	3	1	0	5	147	334	137	161	224	99	5	1	1040
242	.	.	.	.	.	.	.	.	.	.	.	.	.
243	13	14	6	18	23	49	71	36	51	25	19	15	341
244	37	34	8	2	36	153	193	189	129	61	23	60	925
245	10	4	7	8	37	117	182	166	165	44	10	16	766
246	30	17	48	13	54	222	282	313	450	441	179	63	2114
247	21	8	1	9	17	80	106	70	67	32	19	14	445
248	9	3	11	29	214	387	174	222	364	106	15	9	1543
249	5	8	3	1	3	5	115	115	90	30	2	40	415
250	14	5	0	0	1	211	404	329	337	111	24	28	1464
251	8	17	3	11	9	26	88	108	46	29	9	21	376
252	46	8	25	49	112	88	123	99	173	43	20	9	797

AVERAGE TEMPERATURE (DEGREES CENTIGRADE)

OBS.	JAN.	FEB.	MAR.	APR.	MAY	JUN.	JUL.	AUG.	SEP.	OCT.	NOV.	DEC.	ANNUAL
241	24.8	25.2	26.0	27.4	28.3	27.1	27.7	27.6	26.7	26.7	26.0	25.4	26.6
242	9.8	12.2	11.8	15.0	18.6	18.7	18.0	19.5	19.1	13.9	13.3	11.7	15.1
243	.	.	.	.	.	.	.	.	.	.	.	.	.
244	6.3	6.5	8.0	10.6	12.9	22.2	15.4	15.5	15.2	12.8	9.3	7.2	11.4
245	16.2	17.5	19.6	21.6	23.4	22.9	21.8	21.3	21.0	20.0	17.9	16.5	20.0
246	20.5	22.3	24.0	26.8	28.1	27.0	25.9	25.6	25.5	24.2	22.1	20.9	24.4
247	.	.	.	.	.	.	.	.	.	.	.	.	.
248	.	.	.	.	.	.	.	.	.	.	.	.	.
249	.	.	.	.	.	.	.	.	.	.	.	.	.
250	20.9	20.2	20.1	21.7	24.0	26.5	27.3	27.5	27.3	26.7	23.8	21.8	24.0
251	8.8	10.6	12.7	17.0	20.2	25.4	25.3	23.7	22.7	19.1	12.6	8.9	17.2
252	.	.	.	.	.	.	.	.	.	.	.	.	.

CLIMATOLOGICAL DATA FOR MEXICO

GEOGRAPHY

OBS.	LOCATION	YEARS	LATITUDE	LONGITUDE	ELEVATION -M-	LIFEZONE	CCTX	ID	ONMAP
253	SAN CIRO	15	21.63 N	99.83 W	882.7	SD	EW63	420	NO
254	SAN CRISTOBAL	15	16.73 N	92.63 W	2127.5	WM	EC52	39	YES
255	SAN CRISTOBAL	15	18.35 N	95.80 W	125.0	.	.	90	NO
256	SAN D CURUCUPASEO	15	19.30 N	101.15 W	1019.9	TV	EW62	174	NO
257	SAN DIEGO UNION	15	21.47 N	100.87 W	2079.3	WD	EC73	407	NO
258	SAN FERNANDO	15	24.85 N	98.15 W	43.0	.	.	510	NO
259	SAN IGNACIO	15	23.93 N	106.43 W	150.0	TV	EW64	483	NO
260	SAN J DE GUADALUPE	15	24.62 N	102.63 W	1569.7	ST	SW83	501	YES
261	SAN J IXCAQUISTLA	15	18.47 N	97.82 W	1849.5	WD	EC63	95	YES
262	SAN JUAN DEL RIO	15	24.75 N	104.43 W	1699.6	ST	SC84	505	NO
263	SAN JUAN DEL RIO	15	20.38 N	100.00 W	.	WD	EC73	319	NO
264	SAN LUIS DE PAZ	15	21.27 N	100.53 W	1932.4	WT	EC83	396	NO

AVERAGE PRECIPITATION (MM)

OBS.	JAN.	FEB.	MAR.	AFR.	MAY	JUN.	JUL.	AUG.	SEP.	OCT.	NOV.	DEC.	ANNUAL
253	28	13	6	9	54	92	170	53	172	119	34	5	753
254	7	2	11	35	129	249	142	158	247	152	23	15	1171
255	.	.	.	.	.	.	.	.	.	.	.	.	.
256	11	4	4	4	24	178	184	176	210	68	22	12	895
257	21	6	13	6	44	99	111	75	120	35	5	24	558
258	40	22	14	31	77	113	29	69	103	99	31	7	635
259	14	12	6	0	9	57	226	175	173	62	9	44	789
260	7	9	4	3	11	53	87	58	71	34	20	14	370
261	5	6	13	12	91	173	105	121	164	76	13	1	782
262	8	8	1	3	26	62	144	86	75	29	7	10	459
263	12	8	9	15	43	115	108	82	126	44	19	7	587
264	12	6	7	2	44	65	67	35	82	22	14	16	371

AVERAGE TEMPERATURE (DEGREES CENTIGRADE)

OBS.	JAN.	FEB.	MAR.	APR.	MAY	JUN.	JUL.	AUG.	SEP.	OCT.	NOV.	DEC.	ANNUAL
253	15.2	18.7	21.1	23.7	25.8	25.5	23.6	24.3	24.3	21.1	17.6	15.5	21.3
254	12.6	12.9	14.1	15.4	15.9	15.7	15.8	15.6	15.4	14.9	12.8	12.9	14.5
255	20.3	23.2	25.2	26.6	28.1	26.3	26.6	26.3	26.4	25.1	22.5	21.0	24.8
256	23.3	24.4	25.7	26.8	27.3	26.3	25.2	25.0	24.8	24.7	24.5	23.7	25.1
257	12.6	14.2	16.4	18.3	21.4	20.0	18.9	19.2	18.5	16.9	14.1	13.1	17.0
258	.	.	.	.	.	.	.	.	.	.	.	.	.
259	19.5	20.4	21.8	24.2	26.9	29.0	28.3	28.3	27.5	27.2	23.8	21.3	24.8
260	18.9	20.4	21.5	22.3	23.8	23.8	23.8	23.8	22.8	22.6	21.5	19.4	22.0
261	14.0	16.1	16.7	18.1	19.9	18.9	18.0	17.8	17.5	16.6	15.1	14.5	17.0
262	12.8	14.3	16.1	19.1	20.9	23.1	21.4	20.1	20.0	18.4	15.7	13.5	17.9
263	13.1	15.0	17.0	19.0	20.5	20.2	18.9	19.2	18.8	16.9	14.8	13.9	17.3
264	12.6	14.1	16.7	17.8	20.1	19.7	19.4	19.4	18.5	16.6	13.8	13.3	16.8

CLIMATOLOGICAL DATA FOR MEXICO

GEOGRAPHY

OBS.	LOCATION	YEARS	LATITUDE	LONGITUDE	ELEVATION -M-	LIFEZONE	CCTX	ID	ONMAP
265	SAN LUIS POTOSI	15	22.15 N	100.97 W	1876.7	ST	EC83	444	NO
266	SAN MARCOS	15	16.70 N	99.35 W	909.8	SM	EC54	36	NO
267	SAN PEDRO	15	22.47 N	102.35 W	2188.5	.	.	448	YES
268	SAN RAFAEL	15	19.22 N	98.82 W	2529.2	WM	EC52	165	NO
269	SANCARLOSYAUTEPEC	15	16.50 N	96.10 W	999.7	SD	EW63	30	NO
270	SANLUISCOLORADO	15	32.47 N	114.83 W	39.9	SO	SW95	665	NO
271	SANMIGUELALLENDE	15	20.95 N	100.75 W	1851.7	SD	EC73	368	NO
272	SANTA ANA	15	30.57 N	111.12 W	685.8	ST	SW85	646	NO
273	SANTIAGO DE LA PEN	15	21.97 N	97.40 W	10.1	TD	EW53	436	NO
274	SANTIAGO VALLE	15	20.40 N	101.42 W	1779.4	SD	EW73	323	NO
275	SANTIAGOPAPASQUIAR	15	25.03 N	105.43 W	1739.5	WT	SC84	515	NO
276	SANTIAGUILLO	15	19.13 N	95.82 W	4.0	TD	EH43	153	YES

AVERAGE PRECIPITATION (MM)

OBS.	JAN.	FEB.	MAR.	APR.	MAY	JUN.	JUL.	AUG.	SEP.	OCT.	NOV.	DEC.	ANNUAL
265	11	5	10	5	30	72	57	43	85	18	9	14	361
266	0	0	0	0	11	209	180	180	398	94	4	0	1077
267	19	10	3	2	36	69	176	85	125	22	27	7	580
268	11	9	13	41	86	179	270	225	221	72	31	21	1179
269	0	0	4	10	57	242	87	126	166	87	11	0	790
270	7	9	4	3	1	0	9	12	6	8	3	14	76
271	9	7	7	9	28	115	104	94	110	25	14	17	539
272	12	22	14	6	4	20	108	87	47	21	16	27	385
273	47	31	24	23	101	201	185	163	314	222	65	34	1410
274	4	4	8	7	22	25	171	151	141	49	37	29	725
275	12	9	3	11	15	68	136	104	78	29	10	23	488
276	24	19	14	17	42	210	247	283	346	250	96	48	1595

AVERAGE TEMPERATURE (DEGREES CENTIGRADE)

OBS.	JAN.	FEB.	MAR.	APR.	MAY	JUN.	JUL.	AUG.	SEP.	OCT.	NOV.	DEC.	ANNUAL
265	12.9	15.0	17.0	20.4	21.5	20.9	19.5	19.6	18.4	17.2	14.8	13.9	17.6
266	13.3	15.6	18.0	20.4	22.3	21.0	19.8	19.1	19.1	17.6	15.2	14.0	18.0
267	.	.	.	.	.	.	.	.	.	.	.	.	.
268	11.1	12.0	13.5	14.8	15.1	14.5	13.4	13.6	13.5	13.2	12.2	11.8	13.2
269	19.6	21.5	22.8	24.0	25.4	24.7	24.0	24.2	23.5	23.3	22.1	21.2	23.0
270	11.9	14.7	18.3	21.5	25.4	29.7	34.4	34.0	30.2	23.5	16.8	12.4	22.7
271	13.3	15.6	18.0	20.4	22.3	21.0	19.8	19.1	19.1	17.6	15.2	14.0	18.0
272	11.7	13.6	15.9	18.7	23.1	28.6	30.0	29.4	28.2	24.1	18.2	12.3	21.1
273	18.8	20.4	22.1	25.4	27.6	28.0	27.6	27.7	26.6	25.3	21.7	19.3	24.2
274	16.4	18.4	20.8	22.7	23.9	23.2	21.2	21.6	21.2	20.4	18.5	16.6	20.4
275	10.6	12.9	15.2	18.2	21.0	23.3	21.9	21.4	20.7	18.2	14.3	11.4	17.4
276	24.2	23.9	24.4	26.4	27.8	28.9	28.2	28.2	28.7	27.9	26.2	24.6	26.6

CLIMATOLOGICAL DATA FOR MEXICO

GEOGRAPHY

OBS.	LOCATION	YEARS	LATITUDE	LONGITUDE	ELEVATION -M-	LIFEZONE	CCTX	ID	ONMAP
277	SANTO DOMINGO	15	16.53 N	94.60 W	89.9	TD	EH53	31	NO
278	SANVICENTE JAUMAVE	15	23.42 N	99.32 W	734.9	SM	EW53	466	NO
279	SIERRA MOJADA	15	27.28 N	103.70 W	1255.8	ST	SW84	584	YES
280	SILAO	15	20.93 N	101.42 W	1776.7	SD	EW63	367	NO
281	SIMEJOVEL	15	17.15 N	92.63 W	249.9	TD	EW42	51	NO
282	SINALOA	15	25.83 N	108.22 W	54.9	.	.	550	NO
283	SOLEDAD DOBLADO	15	19.07 N	96.40 W	182.9	TV	EH63	147	NO
284	SOMBRERETE	15	23.32 N	103.62 W	2349.4	WD	EC74	464	YES
285	SOTO LA MARINA	15	23.77 N	98.20 W	25.0	SM	EW53	476	NO
286	SUAHUAYO	15	20.03 N	102.72 W	1574.6	.	.	271	NO
287	SULTEPEC	15	18.83 N	99.85 W	2335.4	.	.	123	NO
288	TACAMBARO	15	19.18 N	101.35 W	1754.4	.	.	159	NO

AVERAGE PRECIPITATION (MM)

OBS.	JAN.	FEB.	MAR.	APR.	MAY	JUN.	JUL.	AUG.	SEP.	OCT.	NOV.	DEC.	ANNUAL
277	0	1	2	5	87	357	197	182	364	170	9	2	1376
278	27	13	12	31	85	227	195	141	201	88	37	19	1076
279	17	9	8	7	23	46	80	67	84	29	9	14	394
280	21	10	10	6	30	151	182	166	180	47	17	21	840
281	72	30	46	42	159	220	162	220	232	188	106	94	1572
282	13	23	2	0	4	21	172	163	83	44	13	60	597
283	8	2	0	3	19	119	182	186	164	125	26	11	844
284	16	10	1	5	23	74	160	86	111	47	21	29	582
285	62	13	2	23	118	175	139	138	221	91	112	14	1108
286	17	4	9	2	34	136	170	159	102	36	7	8	685
287	17	21	17	22	136	304	311	316	339	176	59	20	1739
288	.	.	.	.	.	.	.	.	.	.	.	.	.

AVERAGE TEMPERATURE (DEGREES CENTIGRADE)

OBS.	JAN.	FEB.	MAR.	APR.	MAY	JUN.	JUL.	AUG.	SEP.	OCT.	NOV.	DEC.	ANNUAL
277	24.3	25.4	27.1	27.8	28.7	27.7	27.9	28.0	27.8	26.6	25.6	24.5	26.8
278	15.8	17.6	19.6	22.3	24.7	25.4	24.5	24.7	24.1	21.7	17.8	16.1	21.2
279	13.2	15.7	18.6	22.6	26.0	28.3	27.2	26.7	24.5	21.8	17.3	12.9	24.2
280	14.1	15.9	17.7	19.9	21.5	21.6	20.7	20.4	20.1	18.7	16.5	15.2	18.5
281	20.3	22.4	24.4	25.9	26.4	25.6	25.3	25.2	25.3	24.0	21.9	21.3	24.0
282	.	.	.	.	.	.	.	.	.	.	.	.	.
283	22.2	23.5	25.7	28.6	30.0	29.5	28.4	27.3	27.4	26.8	24.3	21.7	26.3
284	10.0	12.3	14.0	17.3	19.6	21.0	18.9	18.8	17.8	16.1	13.0	11.4	15.8
285	16.9	18.5	21.0	24.4	26.9	27.9	27.2	28.1	26.9	24.4	20.5	17.1	23.3
286	.	.	.	.	.	.	.	.	.	.	.	.	.
287	.	.	.	.	.	.	.	.	.	.	.	.	.
288	18.6	20.1	21.5	23.4	24.0	23.5	22.8	22.1	21.9	21.9	21.1	19.5	21.7

CLIMATOLOGICAL DATA FOR MEXICO

GEOGRAPHY

OBS.	LOCATION	YEARS	LATITUDE	LONGITUDE	ELEVATION -M-	LIFEZONE	CCTX	ID	ONMAP
289	TAMAZULA GORDIANO	15	19.65 N	103.17 W	1284.7	SD	EW63	216	NO
290	TAMPICO	15	22.20 N	97.85 W	18.0	TD	EW54	445	NO
291	TANTOYUCA	15	21.35 N	98.22 W	217.0	SM	EW53	401	NO
292	TAPACHULA	15	14.90 N	92.27 W	167.9	TM	EW31	7	YES
293	TAPALPA	15	20.00 N	103.75 W	1564.5	WD	EC63	265	NO
294	TAXCO	15	18.55 N	99.60 W	1754.4	SM	EW53	102	NO
295	TEAPA	15	17.55 N	92.95 W	79.9	TM	EW22	66	NO
296	TECOLUTLA	10	20.50 N	97.02 W	3.0	SM	EW53	332	NO
297	TEHUACAN	15	18.47 N	97.38 W	1675.5	ST	EW83	96	NO
298	TEMOSACHIC	15	28.97 N	107.83 W	1857.5	WT	SC85	621	YES
299	TENANCINGO	15	19.03 N	99.55 W	2079.3	WM	EC43	142	NO
300	TENOSIQUE	15	17.48 N	91.42 W	60.0	TM	EH32	60	NO

AVERAGE PRECIPITATION (MM)

OBS.	JAN.	FEB.	MAR.	APR.	MAY	JUN.	JUL.	AUG.	SEP.	OCT.	NOV.	DEC.	ANNUAL
289	12	3	0	1	25	200	172	191	171	125	42	34	982
290	52	23	12	10	50	199	148	151	332	177	56	42	1176
291	25	56	18	38	86	202	159	154	211	114	59	52	1172
292	7	6	32	73	312	473	298	331	453	405	88	11	2490
293	25	9	6	6	56	129	111	119	128	107	29	53	777
294	1	5	9	23	77	269	301	342	329	90	5	3	1455
295	366	249	207	114	198	377	344	375	572	508	309	345	3963
296	62	51	68	14	48	200	142	123	281	284	85	74	1431
297	2	3	4	17	67	95	71	57	195	32	4	6	477
298	9	4	7	7	12	27	109	129	65	30	15	40	454
299	9	15	20	29	105	276	325	295	332	113	27	4	1550
300	132	88	55	43	178	276	172	232	339	367	175	156	2214

AVERAGE TEMPERATURE (DEGREES CENTIGRADE)

OBS.	JAN.	FEB.	MAR.	APR.	MAY	JUN.	JUL.	AUG.	SEP.	OCT.	NOV.	DEC.	ANNUAL
289	19.0	20.0	21.5	23.4	24.9	25.3	24.0	24.0	24.0	23.5	21.5	19.9	22.6
290	18.8	20.3	22.0	24.8	26.9	27.9	28.1	28.3	27.3	25.6	22.3	19.3	24.3
291	17.9	18.8	22.7	25.4	26.0	26.4	25.5	26.3	25.8	23.9	20.1	16.8	23.0
292	25.2	25.6	26.4	27.2	26.6	25.3	25.8	25.6	25.2	25.1	25.2	25.2	25.7
293	13.0	14.0	16.3	18.1	19.8	19.7	19.3	18.7	18.1	17.6	15.8	14.5	17.1
294	19.2	20.7	22.1	23.9	24.7	22.3	21.1	20.8	20.7	20.7	19.8	19.6	21.3
295	21.8	22.8	24.1	26.3	27.6	27.1	26.8	26.9	26.3	25.2	23.2	22.5	25.0
296	19.4	21.2	21.0	22.8	25.1	25.3	25.2	25.6	24.7	24.0	22.1	19.2	23.0
297	15.8	16.8	18.5	20.0	21.2	20.8	19.5	19.9	20.1	18.2	16.7	16.1	18.6
298	4.3	6.6	8.7	11.7	16.6	21.0	21.8	21.2	18.7	14.0	8.2	4.8	13.0
299	12.6	14.1	15.9	17.3	18.5	17.9	17.4	17.1	16.7	16.8	15.3	13.4	16.1
300	23.7	24.8	26.5	29.1	30.1	29.2	28.5	28.8	28.3	27.2	25.1	24.2	27.1

CLIMATOLOGICAL DATA FOR MEXICO

GEOGRAPHY

OBS.	LOCATION	YEARS	LATITUDE	LONGITUDE	ELEVATION -M-	LIFEZONE	CCTX	ID	ONMAP
301	TEOCALTICHE	15	21.42 N	102.62 W	1799.5	.	.	406	NO
302	TEPEHUANES	15	25.33 N	105.72 W	1786.4	WD	SC74	525	NO
303	TEPIC	15	21.52 N	104.88 W	917.8	SM	EW53	411	NO
304	TEXCOCO	15	19.52 N	98.87 W	2215.3	WD	EC63	207	NO
305	TEZIUTLAN	15	19.80 N	97.35 W	2442.4	WM	EC43	235	NO
306	TICUF EL	15	18.25 N	101.67 W	25.0	TD	EH53	86	NO
307	TIERRA BLANCA	15	18.33 N	96.33 W	61.0	TD	EH43	89	NO
308	TIERRA COLORADA	15	17.17 N	99.60 W	297.8	TD	EH52	52	NO
309	TIJUANA	15	32.48 N	117.03 W	29.0	WT	SC83	667	NO
310	TLACOLULA	15	16.95 N	98.48 W	.	ST	EW83	46	NO
311	TLACOTALPAN	15	18.23 N	97.95 W	31.1	TD	EH53	85	NO
312	TLACOTOPEC	15	18.68 N	97.65 W	1974.5	.	.	118	NO

AVERAGE PRECIPITATION (MM)

OBS.	JAN.	FEB.	MAR.	APR.	MAY	JUN.	JUL.	AUG.	SEP.	OCT.	NOV.	DEC.	ANNUAL
301	19	5	2	3	34	136	151	184	192	29	20	8	782
302	19	12	5	2	13	70	139	100	104	31	14	32	542
303	31	21	1	0	2	170	344	288	202	75	9	53	1196
304	6	7	12	22	60	135	153	150	139	55	20	9	768
305	59	42	55	62	55	235	210	195	330	347	118	44	1752
306	30	0	0	0	0	255	84	199	568	75	149	51	1411
307	25	11	3	14	46	301	295	299	264	183	65	21	1527
308	5	2	1	1	42	255	263	189	323	144	196	15	1240
309	54	57	43	29	20	3	0	2	5	21	18	66	315
310	3	1	5	11	73	99	76	76	106	36	3	4	492
311	46	10	10	6	34	117	124	93	265	204	125	43	1079
312	6	2	0	7	58	263	302	266	337	81	24	2	1349

AVERAGE TEMPERATURE (DEGREES CENTIGRADE)

OBS.	JAN.	FEB.	MAR.	APR.	MAY	JUN.	JUL.	AUG.	SEP.	OCT.	NOV.	DEC.	ANNUAL
301	.	.	.	.	.	.	.	.	.	.	.	.	.
302	9.3	12.2	13.4	16.4	19.5	23.0	22.0	21.2	20.3	17.4	13.1	9.9	16.5
303	17.2	17.9	18.6	20.7	21.5	23.7	23.5	23.5	23.5	23.0	20.2	18.0	20.9
304	12.5	14.2	16.4	17.9	19.1	18.4	17.6	17.5	17.3	15.8	13.9	12.7	16.1
305	11.4	12.5	14.9	17.4	18.1	16.3	15.8	15.9	15.6	14.2	12.4	11.6	14.7
306	27.0	27.1	28.4	28.8	30.1	29.5	30.0	29.6	27.8	28.4	27.4	26.9	28.4
307	22.1	23.5	25.7	27.7	29.1	28.7	27.5	28.0	27.6	26.7	24.8	23.2	26.2
308	26.4	27.1	28.2	28.9	29.9	28.4	27.6	27.7	27.3	27.3	26.7	26.6	27.7
309	12.6	13.3	14.1	15.2	16.7	18.4	21.1	21.4	19.6	17.5	15.6	13.0	16.5
310	17.9	19.4	20.6	21.9	22.7	21.9	21.2	21.1	21.1	20.5	18.5	18.0	20.4
311	24.9	26.0	26.7	28.1	29.2	28.5	28.3	28.3	27.9	27.4	25.8	25.3	27.2
312	.	.	.	.	.	.	.	.	.	.	.	.	.

CLIMATOLOGICAL DATA FOR MEXICO

GEOGRAPHY

OBS.	LOCATION	YEARS	LATITUDE	LONGITUDE	ELEVATION -M-	LIFEZONE	CCTX	ID	ONMAP
313	TLAXCALA	15	19.32 N	98.23 W	2251.6	WD	EC73	177	NO
314	TLAXIACO	15	17.25 N	97.67 W	1936.4	WM	EC53	56	NO
315	TLAYACAPAN	15	18.95 N	98.98 W	1633.7	SD	EW63	133	YES
316	TOLIMAN	15	20.92 N	99.93 W	1509.7	ST	EW83	366	YES
317	TOLUCA	15	19.28 N	99.65 W	2674.3	WD	EC63	172	NO
318	TOMATLAN	15	19.93 N	105.25 W	38.1	.	.	252	NO
319	TONALA	15	16.08 N	93.75 W	54.9	TD	EH42	23	NO
320	TOPOLO BAMPO	15	25.60 N	109.05 W	3.0	TT	SW84	541	NO
321	TULA	15	20.05 N	99.35 W	2035.5	.	.	276	NO
322	TULANCINGO	15	20.08 N	98.37 W	2180.5	WD	EC73	280	NO
323	TUXPAN	15	20.95 N	97.40 W	14.0	TD	EW53	369	NO
324	TUXPAN	15	21.95 N	105.30 W	.	.	.	433	NO

AVERAGE PRECIPITATION (MM)

OBS.	JAN.	FEB.	MAR.	APR.	MAY	JUN.	JUL.	AUG.	SEP.	OCT.	NOV.	DEC.	ANNUAL
313	6	2	9	19	72	126	135	150	136	48	7	6	717
314	1	2	16	18	78	207	180	206	193	91	8	4	1003
315	0	2	0	2	58	183	213	183	199	69	11	4	924
316	14	3	8	6	27	38	76	53	84	28	10	5	352
317	10	9	11	26	63	135	168	144	151	49	20	7	791
318	27	19	0	0	5	122	205	224	283	78	11	29	1004
319	2	3	13	9	122	299	301	270	374	199	29	6	1628
320	7	7	7	0	2	6	40	101	56	74	7	55	360
321	18	14	10	16	62	94	128	124	131	56	13	9	674
322	8	12	11	24	47	95	74	66	128	89	16	10	578
323	40	26	26	38	87	201	174	168	306	223	53	39	1382
324	.	.	.	.	.	.	.	.	.	.	.	.	.

AVERAGE TEMPERATURE (DEGREES CENTIGRADE)

OBS.	JAN.	FEB.	MAR.	APR.	MAY	JUN.	JUL.	AUG.	SEP.	OCT.	NOV.	DEC.	ANNUAL
313	12.9	14.2	16.4	17.1	17.9	17.1	16.3	16.2	16.2	15.7	14.5	13.7	15.7
314	13.8	15.0	16.0	17.4	18.9	18.7	18.2	18.1	18.2	17.0	14.8	14.2	16.7
315	16.9	18.0	20.1	21.1	22.0	20.9	19.8	19.4	19.2	18.9	17.9	17.2	19.3
316	14.7	16.0	17.1	20.6	22.1	22.7	22.1	21.6	20.5	18.2	15.7	14.7	18.8
317	9.3	11.1	13.0	14.3	14.9	14.4	13.4	13.5	13.4	12.8	11.3	10.3	12.7
318	.	.	.	.	.	.	.	.	.	.	.	.	.
319	27.2	27.4	28.5	29.4	29.1	27.1	27.3	27.5	27.2	27.2	27.8	27.2	27.7
320	18.6	19.7	20.1	22.2	25.0	29.1	29.8	29.7	29.7	27.8	24.0	19.7	24.6
321	.	.	.	.	.	.	.	.	.	.	.	.	.
322	11.4	12.8	14.8	16.7	17.4	16.7	16.1	15.9	15.7	14.2	12.6	12.1	14.7
323	19.5	21.1	22.4	25.6	27.6	28.3	27.9	27.9	27.3	25.4	22.3	19.8	24.6
324	19.5	21.3	22.0	23.9	26.0	28.6	27.8	29.0	29.1	28.2	24.8	21.5	25.1

CLIMATOLOGICAL DATA FOR MEXICO

GEOGRAPHY

OBS.	LOCATION	YEARS	LATITUDE	LONGITUDE	ELEVATION -M-	LIFEZONE	CCTX	ID	ONMAP
325	TUXTLA GUTIERREZ	15	16.75 N	93.10 W	535.8	TV	EW63	40	NO
326	UNION DE TULA	15	19.98 N	104.25 W	1384.7	SD	EW62	263	NO
327	UNION LA	15	17.97 N	101.80 W	38.1	TD	EH52	75	YES
328	URES	15	29.43 N	110.40 W	431.9	SD	SW75	631	YES
329	URUAPAN	15	19.42 N	101.97 W	1610.6	SM	EW43	190	NO
330	VALLADOLID	15	20.68 N	88.22 W	21.9	TD	EW52	355	NO
331	VALLARTA PUERTO	15	20.62 N	105.25 W	1.8	TD	EW53	348	YES
332	VALLE DE BRAVO	15	19.23 N	100.12 W	2025.4	.	.	167	NO
333	VENUSTIANOCARRANZA	15	17.82 N	95.82 W	.	TM	EW33	72	NO
334	VERACRUZ	15	19.20 N	96.13 W	15.8	TD	EW43	163	NO
335	VICTORIA	15	23.73 N	99.13 W	321.0	SD	EW64	474	NO
336	VICTORIA VILLA	15	21.35 N	100.22 W	.	ST	EW83	400	NO

AVERAGE PRECIPITATION (MM)

OBS.	JAN.	FEB.	MAR.	APR.	MAY	JUN.	JUL.	AUG.	SEP.	OCT.	NOV.	DEC.	ANNUAL
325	0	1	1	6	75	235	175	155	201	81	4	6	942
326	67	9	5	0	30	113	179	145	167	59	26	18	818
327	7	23	0	0	12	190	236	210	319	86	18	22	1123
328	10	9	22	10	8	20	156	125	68	37	19	50	535
329	16	21	7	4	32	273	347	329	406	179	37	31	1683
330	55	28	27	74	120	148	140	160	179	145	49	56	1181
331	17	55	2	0	9	207	347	359	332	138	15	7	1488
332	25	5	13	18	86	383	493	538	382	180	78	53	2254
333	39	28	18	36	71	286	370	419	406	319	101	60	2153
334	23	16	7	20	53	245	346	300	345	153	88	25	1622
335	36	24	20	39	126	121	104	68	198	108	43	15	903
336	12	10	15	11	34	87	83	56	92	25	12	6	444

AVERAGE TEMPERATURE (DEGREES CENTIGRADE)

OBS.	JAN.	FEB.	MAR.	APR.	MAY	JUN.	JUL.	AUG.	SEP.	OCT.	NOV.	DEC.	ANNUAL
325	21.6	23.0	24.7	26.6	27.2	25.9	25.4	25.5	25.1	24.3	22.6	20.7	24.4
326	20.1	18.9	20.8	21.9	23.3	23.4	21.7	21.4	21.6	21.7	20.6	19.9	21.4
327	26.5	26.4	26.5	27.7	28.7	28.9	28.7	28.5	27.9	28.0	27.5	26.6	27.7
328	15.0	16.6	18.0	20.6	24.5	28.3	29.7	28.8	28.4	24.3	19.0	16.0	22.4
329	16.0	17.0	19.6	21.5	22.4	21.8	21.1	20.8	20.3	19.7	17.5	16.6	19.5
330	22.1	23.0	24.7	26.6	27.2	26.5	26.7	26.5	26.3	25.2	23.1	22.7	25.0
331	22.7	22.5	23.0	24.5	26.2	27.3	27.8	27.9	28.2	28.1	26.3	24.0	25.7
332	.	.	.	.	.	.	.	.	.	.	.	.	.
333	21.2	22.3	23.1	26.0	27.7	27.2	26.6	26.6	26.4	25.1	22.7	21.4	24.7
334	21.2	21.9	22.8	24.9	26.7	27.3	27.0	27.2	26.9	26.1	23.8	22.0	24.8
335	15.7	18.1	21.3	24.5	26.4	27.0	27.4	27.9	25.9	23.3	19.6	15.5	23.7
336	15.3	17.2	18.9	20.9	22.5	21.9	20.9	20.8	20.6	19.0	16.7	15.9	19.2

CLIMATOLOGICAL DATA FOR MEXICO

GEOGRAPHY

OBS.	LOCATION	YEARS	LATITUDE	LONGITUDE	ELEVATION -M-	LIFEZONE	CCTX	ID	ONMAP
337	VILLA ALDAMA	15	26.50 N	100.43 W	468.8	ST	SW84	568	YES
338	VILLA ALDAMA	15	26.50 N	100.43 W	468.8	ST	SW84	569	NO
339	VILLA SANTIAGO	15	25.43 N	100.13 W	445.0	.	.	528	NO
340	VILLAHERMOSA	15	17.98 N	92.92 W	10.1	TD	EW42	78	NO
341	XILITLA	15	21.40 N	99.00 W	.	SW	EW23	404	NO
342	ZACAPU	15	19.75 N	101.75 W	1999.5	WD	EC63	227	NO
343	ZACATECAS	15	22.78 N	102.57 W	2611.2	WT	EC83	453	NO
344	ZACATLAN	15	19.93 N	97.93 W	.	.	.	253	NO
345	ZAMORA	15	19.98 N	102.30 W	1632.5	SD	EW63	264	NO
346	ZAPOTITLAN	15	18.55 N	94.82 W	4.0	TM	EW22	103	YES
347	ZARAGOZA	15	29.75 N	107.43 W	2499.4	.	.	635	NO
348	ZARAGOZA	15	23.83 N	99.75 W	1370.7	.	.	481	NO

AVERAGE PRECIPITATION (MM)

OBS.	JAN.	FEB.	MAR.	APR.	MAY	JUN.	JUL.	AUG.	SEP.	OCT.	NOV.	DEC.	ANNUAL
337	25	9	14	21	31	38	58	36	77	34	27	11	381
338	25	9	14	21	31	38	58	36	77	34	27	11	381
339	.	.	.	.	.	.	.	.	.	.	.	.	.
340	140	100	46	46	89	204	194	195	272	292	142	181	1902
341	66	56	71	65	113	345	358	321	744	263	100	37	2538
342	14	5	8	8	30	102	199	192	168	67	23	19	836
343	10	5	2	2	23	52	90	59	76	22	15	11	368
344	33	12	16	35	40	157	199	100	387	114	133	19	1245
345	16	7	13	7	33	158	181	162	153	56	14	19	819
346	169	104	99	50	80	363	376	366	549	535	328	374	3393
347	14	20	0	0	15	4	94	158	42	40	11	16	413
348	22	16	15	18	56	65	120	111	166	38	13	9	649

AVERAGE TEMPERATURE (DEGREES CENTIGRADE)

OBS.	JAN.	FEB.	MAR.	APR.	MAY	JUN.	JUL.	AUG.	SEP.	OCT.	NOV.	DEC.	ANNUAL
337	12.3	16.0	20.4	22.2	24.9	27.2	28.7	28.9	27.1	23.0	18.8	12.9	21.9
338	12.3	16.0	20.4	22.2	24.9	27.2	28.7	28.9	27.1	23.0	18.8	12.9	21.9
339	13.8	16.6	18.8	22.7	25.4	27.7	28.1	28.5	26.4	22.6	17.0	12.8	21.7
340	22.2	23.8	24.9	26.8	28.2	28.2	28.0	28.4	27.6	26.7	24.7	22.9	26.0
341	15.6	17.6	19.5	22.5	23.5	24.2	23.9	24.0	23.1	21.4	18.3	16.4	20.8
342	11.4	12.9	15.4	17.4	19.1	19.1	17.7	17.2	16.9	15.2	13.0	11.9	15.6
343	9.3	10.5	12.7	14.9	16.7	16.0	14.1	14.6	13.8	13.1	11.3	10.1	13.1
344	.	.	.	.	.	.	.	.	.	.	.	.	.
345	16.2	17.6	19.7	21.9	23.2	22.6	20.9	20.8	20.8	20.3	18.7	17.2	20.0
346	22.0	22.7	23.3	24.4	26.0	26.5	26.2	26.2	26.3	25.7	24.1	23.0	24.7
347	.	.	.	.	.	.	.	.	.	.	.	.	.
348	.	.	.	.	.	.	.	.	.	.	.	.	.

CLIMATOLOGICAL DATA FOR MEXICO

GEOGRAPHY

OBS.	LOCATION	YEARS	LATITUDE	LONGITUDE	ELEVATION -M-	LIFEZONE	CCTX	ID	ONMAP
349	ZIMAPAN	15	20.75 N	99.38 W	1719.7	ST	EW83	360	NO
350	ZINAPECUARO	15	19.83 N	100.80 W	1839.5	SD	EW63	241	NO
351	ZIRANDARO	15	18.48 N	100.98 W	192.9	TV	EH63	98	YES
352	ZITACUARO	15	19.37 N	100.38 W	2063.5	WD	EC63	185	NO

AVERAGE PRECIPITATION (MM)

OBS.	JAN.	FEB.	MAR.	APR.	MAY	JUN.	JUL.	AUG.	SEP.	OCT.	NOV.	DEC.	ANNUAL
349	10	2	6	9	36	78	41	47	87	43	9	7	375
350	15	6	4	5	25	137	144	193	149	43	15	22	756
351	4	1	0	4	13	157	240	191	183	57	16	11	877
352	9	12	8	6	37	173	216	185	209	65	28	15	963

AVERAGE TEMPERATURE (DEGREES CENTIGRADE)

OBS.	JAN.	FEB.	MAR.	APR.	MAY	JUN.	JUL.	AUG.	SEP.	OCT.	NOV.	DEC.	ANNUAL
349	16.0	18.2	20.3	21.4	21.6	20.4	20.2	20.0	19.3	18.5	16.4	15.5	19.0
350	15.4	16.4	18.1	20.2	21.6	21.4	19.9	19.1	19.1	18.1	17.0	16.0	18.5
351	26.5	28.2	30.2	31.4	33.2	31.4	29.3	28.6	28.2	29.0	28.0	26.7	29.2
352	14.3	15.5	17.4	19.3	20.4	19.0	17.4	17.2	17.4	17.2	15.8	14.7	17.1

MOZAMBIQUE

Agriculture

Of 78,303,000 ha in total area, Mozambique was estimated in 1978 to have 3,080,000 ha of arable or permanent cropland, 44,000,000 ha in permanent pasture, 19,400,000 ha in forest and woodland, and 10,073,000 ha in other terrain categories. In 1978, approximately 68,000 ha were irrigated. Of the 1979 population, 65.4% or 2,539,000 individuals of the economically active population were engaged in agriculture. The total population in 1979 was 10,199,000. According to FAO statistics for 1979, there were 20,000 asses, 1,380,000 cattle, 110,000 pigs, 105,000 sheep, 330,000 goats, 17,500,000 chickens, and 565,000 ducks.[2] Agricultural production is summarized in Table 1. Conventional and energetic yield data on most of these will be found in Appendixes I and II.

Climate

Data for Wernstedt[3] are presented in the Climatological Data Table (see Introduction for explanation). Based on the means for data points cited in Wernstedt, the mean annual precipitation is 10.2 dm (mean of 90 sites) and the mean annual temperature 23.96°C (mean of 90 sites). The averages would suggest that much of the country could be classified as Subtropical Moist Forest Life Zone.[4]

Productivity

The vegetation map depicts the major vegetation types in Mozambique. With reasonable management, these units might achieve annual productivities of:

Vegetation type	MT/ha/yr
Broadleaf evergreen forest	18—24
Deciduous forest	12—18
Grassland	3—12
Savanna	3—12
Mangrove	12—24
Swamp and marsh	6—24

Economics

Food consumption per capita was 1906 cal/day, energy consumption 138 kg coal equivalent per capita in 1977. In 1970, the caloric supply was 94.0% of requirements. The total population growth rate was 1.9% in 1970. Population density was 10 persons per square kilometer in 1970, with 55.0% of the total population urban. Life expectancy at birth of those born in 1970 was 41 years. In 1977, agriculture accounted for 56.3% of GDP, manufacturing for 6.1%, other industry for 6.2%, and services for 31.4%. In 1970, food and beverage constituted 57.2% of exports, nonfood agriculture for 23.1%, fuels, minerals, and metal for 10.3%, machines for 4.7%, and other manufacturing for 4.8% of total merchandise exports. Food and beverage constituted 13.5% of imports, nonfood agriculture 0.9%, fuel and lubricants 8.4%, nonfuel minerals and metals 1.4%, machinery and equipment 35.2%, and other manufactured goods constituted 40.6% of total merchandise imports.[10]

Energetics

Mozambique has 400 million ton coal (80 technically and economically recoverable), and a hydroelectric potential of 11,920 MW. Importing less than 50% (26 to 50%) of its commercial energy demands as oil, Mozambique is expected to be experiencing fuelwood problems by the year 2000.[10]

Table 1
AGRICULTURAL PRODUCTION IN MOZAMBIQUE

Crop	Area planted (ha)	Yield (kg/ha)	Total production (MT)
Wheat	5,000	600	3,000
Rice paddy	65,000	538	35,000
Corn	600,000	583	350,000
Millet	20,000	250	5,000
Sorghum	250,000	720	180,000
Potatoes	6,000	6,333	38,000
Sweet potatoes	9,000	4,706	40,000
Cassava	450,000	5,556	2,500,000
Peanuts in shell	200,000	400	80,000
Sunflower seed	10,000	1,050	11,000
Sesame seed	7,000	429	3,000
Seed cotton	121,000	372	45,000
Cottonseed			30,000
Coconuts			400,000
Copra			75,000
Tomatoes	1,000	10,000	13,000
Pumpkins, squash, gourds		2,429	1,000
Dry onions	1,000	5,000	3,000
Sugarcane	50,000	42,000	2,100,000
Sugar (centrifugal, raw)			200,000
Oranges			18,000
Tang, mand., clement., satsuma			1,000
Lemons and limes			1,000
Grapefruit amd pomelo			16,000
Mangoes			30,000
Pineapples			13,000
Bananas			65,000
Papayas			36,000
Cashew nuts			150,000
Green coffee	1,000	714	1,000
Tea	11,000	1,455	16,000
Tobacco leaves	3,000	962	3,000
Jute	6,000	667	4,000
Sisal	50,000	360	18,000
Cotton lint			15,000

From FAO 1979 Production Yearbook, Vol. 33, Food and Agriculture Organization, Rome, 1980.

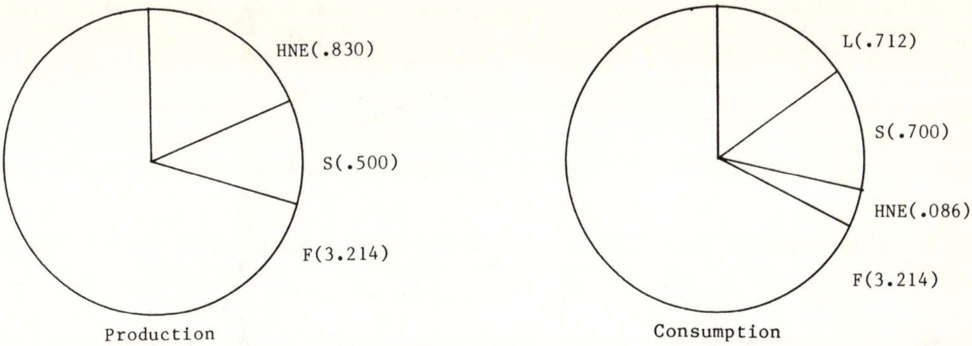

1978 Energy production and consumption in Mozambique; fuel types listed in million metric tons coal equivalents (MMTCE). 1 MMTCE = 27.78×10^{12} Btu. L = liquid fuel, S = solid fuel, HNE = hydro, nuclear, and electric fuel, F = firewood, NG = natural gas. (Data from United Nations, World Energy Supplies, 1973—1978, Statistical Papers Series J, 22, Department of International Economic and Social Affairs, Statistical Office, New York, 1979.)

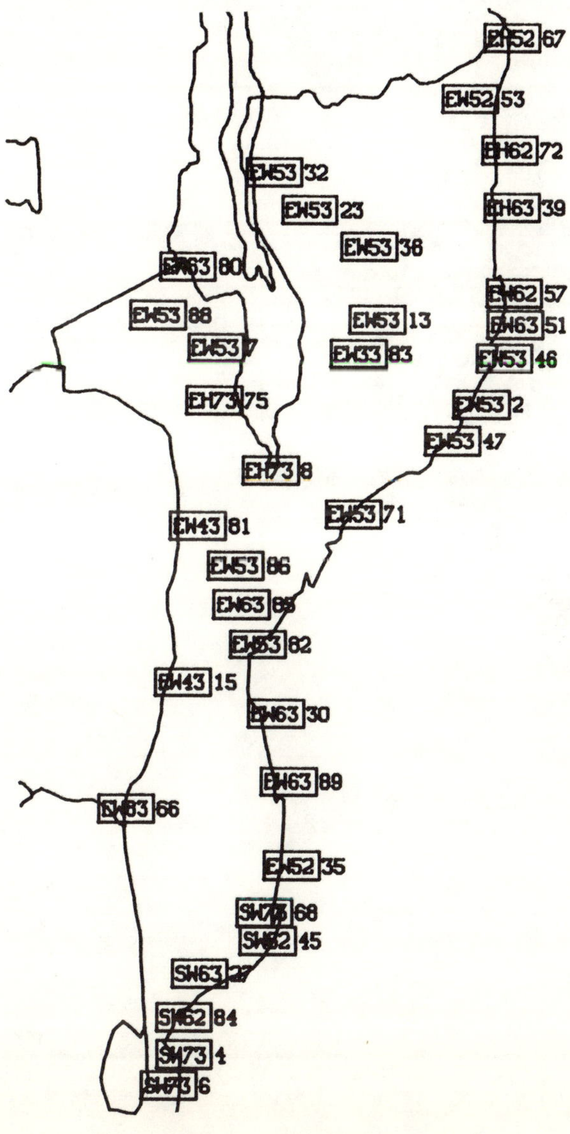

Selected weather stations for Mozambique.

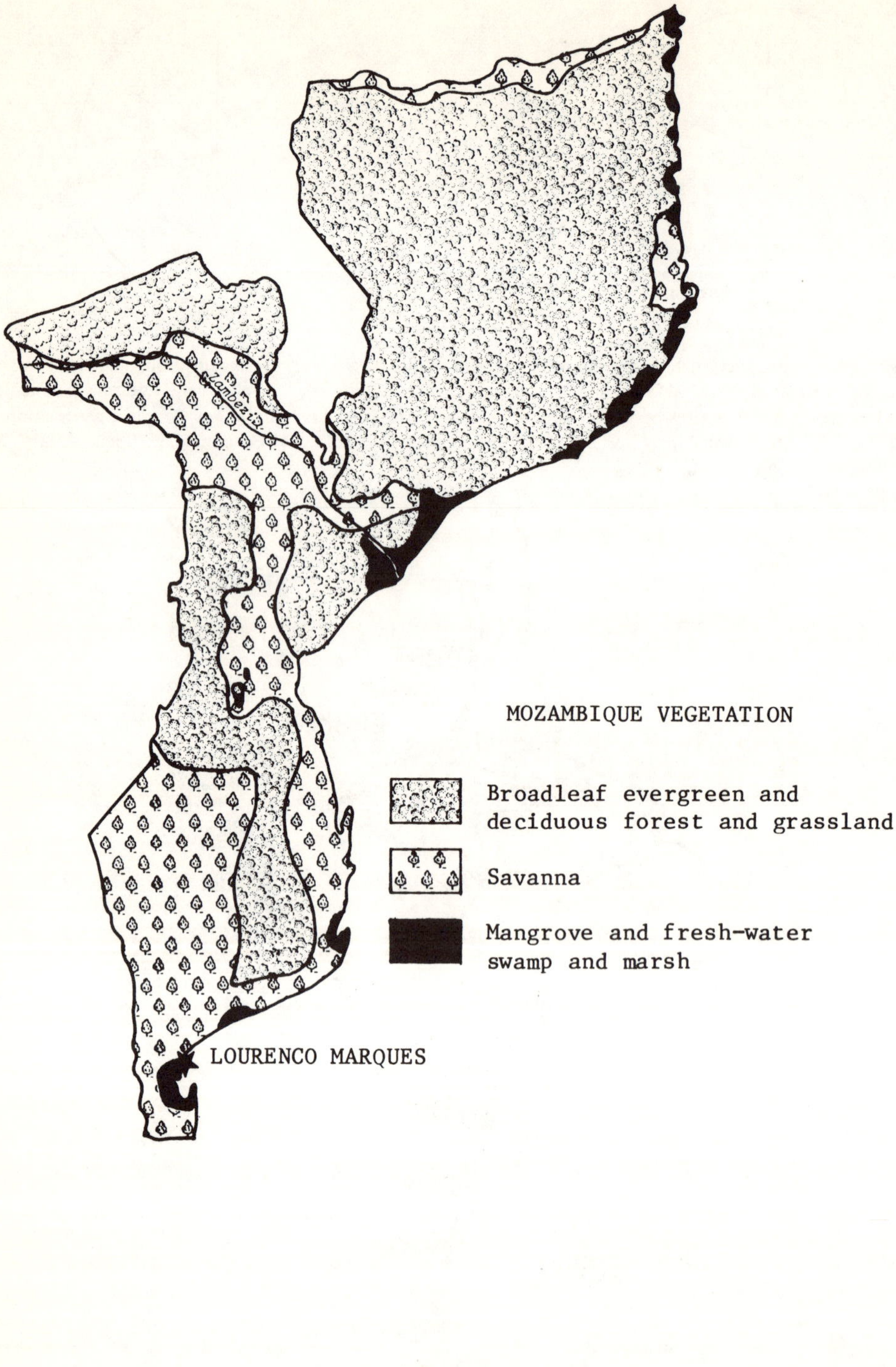
Zambezi
MOZAMBIQUE VEGETATION
Broadleaf evergreen and deciduous forest and grassland
Savanna
Mangrove and fresh-water swamp and marsh
LOURENCO MARQUES

CLIMATOLOGICAL DATA FOR MOZAMBIQUE

GEOGRAPHY

OBS.	LOCATION	YEARS	LATITUDE	LONGITUDE	ELEVATION -M-	LIFEZONE	CCTX	ID	ONMAP
1	ALTO MOLOCUE	30	15.63 S	37.68 E	562.7	SM	EW53	1	NO
2	ANTONIO ENES	30	16.22 S	39.90 E	61.0	TD	EW53	2	YES
3	BEIRA	30	19.83 S	34.85 E	7.0	TD	EW53	3	NO
4	BELLA VISTA	30	26.33 S	32.68 E	14.9	SD	SW73	4	YES
5	CANICADO	30	24.33 S	33.08 E	32.9	TV	SW73	5	NO
6	CATUANE	26	26.83 S	32.27 E	36.9	SD	SW73	6	YES
7	CAZULA	14	15.38 S	33.63 E	596.8	TD	EW53	7	YES
8	CHEMBA	24	17.23 S	34.90 E	100.0	TV	EH73	8	YES
9	CHIBUTO	25	24.68 S	33.53 E	89.9	TV	SW73	9	NO
10	CHINDE	14	18.57 S	36.47 E	4.0	TD	EW53	10	NO
11	CHIUTA	14	15.55 S	33.30 E	1332.6	TD	EW53	11	NO
12	CHOBELA	15	25.00 S	32.73 E	39.9	TV	SW73	12	NO

AVERAGE PRECIPITATION (MM)

OBS.	JAN.	FEB.	MAR.	APR.	MAY	JUN.	JUL.	AUG.	SEP.	OCT.	NOV.	DEC.	ANNUAL
1	299	292	249	77	30	24	25	27	15	16	96	224	1373
2	223	201	173	94	48	52	33	22	10	10	62	146	1074
3	274	209	255	103	50	38	35	28	20	39	130	233	1412
4	126	114	78	34	21	22	16	14	26	49	86	105	692
5	114	103	87	40	24	18	16	12	17	29	72	82	613
6	103	85	77	44	31	17	14	12	30	45	74	96	629
7	254	248	217	74	14	21	11	10	8	21	130	188	1196
8	148	145	103	30	13	12	9	8	5	11	90	132	706
9	102	120	105	41	43	30	19	18	25	34	83	102	721
10	172	194	170	109	61	80	55	26	26	7	48	186	1135
11	272	237	172	48	9	14	9	3	4	16	114	183	1081
12	107	123	123	42	22	22	13	11	23	38	101	86	711

AVERAGE TEMPERATURE (DEGREES CENTIGRADE)

OBS.	JAN.	FEB.	MAR.	APR.	MAY	JUN.	JUL.	AUG.	SEP.	OCT.	NOV.	DEC.	ANNUAL
1	25.2	25.2	24.5	23.5	21.0	18.8	18.4	19.8	22.2	25.1	25.8	25.4	22.9
2	27.8	27.9	27.4	26.4	24.8	22.8	22.0	22.6	24.4	27.0	27.2	27.8	25.7
3	27.5	27.6	26.9	25.4	22.9	20.9	20.4	21.2	23.2	24.9	26.3	26.7	24.5
4	26.0	25.8	25.0	23.8	21.3	19.0	18.7	19.8	21.2	23.0	24.1	25.5	22.8
5	27.8	27.4	26.5	24.9	22.3	20.0	19.7	22.5	23.9	25.6	27.7	28.8	24.5
6	26.2	26.0	25.3	23.7	21.3	18.8	18.7	20.1	21.9	23.4	24.4	25.6	23.0
7	26.2	25.5	25.1	24.8	23.3	21.2	21.2	23.3	25.8	29.3	27.6	26.3	25.0
8	29.2	28.6	28.0	27.1	24.3	21.8	21.7	23.6	26.6	29.6	30.0	29.4	26.7
9	27.6	27.3	26.4	25.2	22.7	20.4	20.0	21.7	23.7	25.7	26.6	27.2	24.5
10	28.1	28.2	27.6	26.6	24.3	22.2	22.0	22.6	24.2	26.2	27.0	27.4	25.5
11	23.7	23.9	22.4	23.6	22.2	20.3	20.7	22.7	24.7	27.5	25.7	25.0	23.6
12	26.7	26.4	25.4	24.4	21.6	19.3	19.3	20.7	22.2	24.3	25.6	26.5	23.5

CLIMATOLOGICAL DATA FOR MOZAMBIQUE

GEOGRAPHY

OBS.	LOCATION	YEARS	LATITUDE	LONGITUDE	ELEVATION -M-	LIFEZONE	CCTX	ID	ONMAP
13	ENTRE-RIOS	30	14.95 S	37.42 E	624.5	SM	EW53	13	YES
14	ERREGO	18	16.03 S	37.18 E	532.8	SM	EW43	14	NO
15	ESPUNGABERA	30	20.47 S	32.77 E	823.6	SM	EW43	15	YES
16	FINGOE	30	15.17 S	31.88 E	856.5	SM	EW53	16	NO
17	FUNHALOURO	18	23.08 S	34.38 E	115.8	TV	EW73	17	NO
18	FURANCUNGO	21	14.90 S	33.60 E	1275.6	SM	EW53	18	NO
19	INHAMBANE	30	23.87 S	35.38 E	14.0	TV	EW63	19	NO
20	INHAMINGA	24	18.38 S	35.03 E	315.8	TD	EW53	20	NO
21	INHAMUSSUA	30	23.90 S	35.23 E	4.0	SD	EW63	21	NO
22	INHARRIME	30	24.48 S	35.02 E	43.0	TV	SW62	22	NO
23	LATUNDE	13	13.32 S	35.83 E	1027.8	SM	EW53	23	YES
24	LOURENCO MARQUES	30	25.97 S	32.60 E	59.1	SD	SW63	24	NO

AVERAGE PRECIPITATION (MM)

OBS.	JAN.	FEB.	MAR.	APR.	MAY	JUN.	JUL.	AUG.	SEP.	OCT.	NOV.	DEC.	ANNUAL
13	284	258	176	30	3	2	2	1	5	6	53	181	1001
14	350	271	313	137	40	53	32	27	20	22	146	253	1664
15	308	282	243	78	28	47	33	36	39	64	133	242	1531
16	244	259	168	35	4	8	5	4	1	8	94	233	1064
17	99	107	66	24	18	10	6	6	16	29	66	85	530
18	287	263	183	61	8	7	4	5	1	20	146	247	1232
19	171	132	134	70	72	49	41	33	25	26	100	108	960
20	210	157	171	46	24	36	20	17	14	24	103	193	1015
21	143	129	102	57	48	45	28	25	33	27	76	124	836
22	119	108	109	61	67	62	41	30	30	35	89	91	845
23	314	237	271	104	26	8	2	4	8	18	68	217	1277
24	130	106	135	55	22	23	12	9	28	41	92	101	753

AVERAGE TEMPERATURE (DEGREES CENTIGRADE)

OBS.	JAN.	FEB.	MAR.	APR.	MAY	JUN.	JUL.	AUG.	SEP.	OCT.	NOV.	DEC.	ANNUAL
13	23.4	23.5	23.2	22.6	20.8	18.6	18.2	19.2	21.8	24.0	24.4	23.6	21.9
14	25.0	24.8	24.0	23.2	21.1	19.1	18.9	20.4	23.4	26.1	26.1	25.2	23.1
15	22.0	21.9	21.2	20.4	18.8	16.5	16.4	17.5	19.4	21.4	21.6	21.6	19.9
16	23.6	23.4	22.8	22.2	20.1	17.7	17.0	19.3	22.6	24.6	25.4	24.0	21.9
17	27.9	27.5	26.8	25.7	23.0	20.6	20.5	22.1	24.4	26.8	27.1	27.6	25.0
18	21.0	20.9	20.5	20.0	18.4	16.2	16.0	17.7	20.6	23.0	22.3	21.7	19.8
19	26.7	26.7	26.8	24.6	22.4	19.5	20.0	19.6	22.2	23.8	24.7	26.0	23.7
20	27.0	27.0	26.1	25.0	23.0	20.8	20.2	21.7	24.3	26.6	27.1	27.0	24.6
21	25.9	25.8	25.3	23.0	20.9	19.0	18.3	18.8	20.5	22.7	24.4	25.4	22.5
22	26.6	26.7	26.1	25.0	22.8	20.5	20.2	21.2	23.0	24.9	25.6	26.6	24.1
23	22.2	22.4	22.0	22.3	20.2	18.2	17.8	19.0	21.2	24.4	24.2	22.9	21.3
24	24.8	25.2	25.5	23.3	21.0	19.2	18.7	20.4	19.7	21.6	23.7	23.7	22.2

CLIMATOLOGICAL DATA FOR MOZAMBIQUE

GEOGRAPHY

OBS.	LOCATION	YEARS	LATITUDE	LONGITUDE	ELEVATION -M-	LIFEZONE	CCTX	ID	ONMAP
25	LUGELA	14	16.43 S	36.75 E	292.9	TD	EW43	25	NO
26	MABOTE	17	22.05 S	34.12 E	143.0	TV	EW72	26	NO
27	MACIA	30	25.03 S	33.10 E	55.8	SD	SW63	27	YES
28	MACOMIA	17	12.25 S	40.13 E	342.9	TD	EW52	28	NO
29	MAGANJA DA COSTA	30	17.30 S	37.53 E	69.8	TD	EW53	29	NO
30	MAMBONE	23	20.95 S	35.02 E	4.0	TV	EW63	30	YES
31	MANHICA	30	25.40 S	32.80 E	35.1	SD	SW62	31	NO
32	MANIAMBA	19	12.77 S	35.00 E	1092.7	SM	EW53	32	YES
33	MANJACAZE	23	24.72 S	33.88 E	64.9	TV	SW73	33	NO
34	MARROMEU	14	18.28 S	35.95 E	20.1	TV	EW63	34	NO
35	MASSINGA	30	23.32 S	35.40 E	356.9	SM	EW52	35	YES
36	MAUA	12	13.87 S	37.22 E	593.8	TD	EW53	36	YES

AVERAGE PRECIPITATION (MM)

OBS.	JAN.	FEB.	MAR.	APR.	MAY	JUN.	JUL.	AUG.	SEP.	OCT.	NOV.	DEC.	ANNUAL
25	345	323	299	120	54	54	30	26	16	39	171	256	1734
26	117	130	89	21	17	5	2	8	10	4	61	113	576
27	105	163	123	58	42	41	25	23	31	40	75	103	828
28	236	225	273	124	23	7	7	4	5	14	44	232	1193
29	250	238	218	100	63	60	48	38	14	15	77	186	1306
30	148	120	179	58	24	24	21	24	5	15	88	160	865
31	120	138	121	63	32	40	30	24	39	53	108	105	874
32	281	284	307	146	30	18	2	3	1	19	80	228	1398
33	119	141	82	43	33	39	18	15	33	34	72	71	700
34	177	154	161	68	46	31	22	17	14	15	84	162	950
35	204	187	175	89	60	59	48	28	26	33	107	154	1172
36	307	270	280	50	10	4	0	4	3	10	72	264	1274

AVERAGE TEMPERATURE (DEGREES CENTIGRADE)

OBS.	JAN.	FEB.	MAR.	APR.	MAY	JUN.	JUL.	AUG.	SEP.	OCT.	NOV.	DEC.	ANNUAL
25	26.8	26.4	25.7	24.9	22.8	20.7	20.8	22.0	24.6	27.3	27.5	26.8	24.7
26	25.9	25.5	25.4	24.8	23.1	21.3	21.6	22.1	23.4	23.4	25.2	25.4	24.0
27	25.3	23.5	24.2	23.1	20.4	18.4	18.2	19.6	21.5	23.5	24.0	25.2	22.2
28	25.9	26.2	25.8	25.4	24.9	23.3	23.9	23.6	24.4	25.4	26.0	26.1	25.1
29	27.4	27.4	26.6	25.1	23.7	21.3	21.5	22.5	24.3	26.8	27.4	27.8	25.2
30	27.5	26.7	26.7	25.3	22.8	20.5	20.0	21.1	23.1	25.2	26.4	27.0	24.4
31	26.1	25.9	25.2	24.1	21.7	19.3	19.2	20.4	22.0	23.9	24.7	25.7	23.2
32	21.7	21.6	21.4	20.8	19.8	17.8	17.6	18.9	20.9	23.6	23.8	22.2	20.8
33	27.1	26.8	25.9	24.9	22.6	20.3	19.6	21.2	22.7	24.8	25.9	26.5	24.0
34	27.3	27.6	26.5	25.5	24.0	22.2	21.5	23.0	24.4	26.6	27.2	27.4	25.3
35	25.7	25.8	25.0	24.0	22.3	20.2	19.6	20.4	21.6	23.6	24.3	25.1	23.1
36	25.0	24.9	24.5	24.5	22.3	20.6	20.8	22.4	24.8	28.0	28.2	25.9	24.3

CLIMATOLOGICAL DATA FOR MOZAMBIQUE

GEOGRAPHY

OBS.	LOCATION	YEARS	LATITUDE	LONGITUDE	ELEVATION -M-	LIFEZONE	CCTX	ID	ONMAP
37	MAUE	11	14.75 S	34.37 E	1188.7	SD	EW63	37	NO
38	MECONTA	17	15.00 S	39.85 E	235.0	TV	EW62	38	NO
39	MECUFI	10	13.28 S	40.55 E	10.1	TV	EH63	39	YES
40	MEMBA	21	14.18 S	40.53 E	14.0	TV	EW62	40	NO
41	MILANGE	21	16.10 S	35.78 E	744.6	SM	EW43	41	NO
42	MOAMBA	30	25.60 S	32.23 E	110.0	TV	SW73	42	NO
43	MOCIMBOA DA PRAIA	17	11.32 S	40.37 E	27.1	TV	EW62	43	NO
44	MOCUBA	18	16.83 S	36.98 E	133.8	TD	EW53	44	NO
45	MOCUMBI	30	24.50 S	34.80 E	44.8	TV	SW62	45	YES
46	MOGINCUAL	22	15.53 S	40.42 E	35.1	TD	EW53	46	YES
47	MOMA	20	16.77 S	39.22 E	4.0	TD	EW53	47	YES
48	MONTEPUEZ	30	13.13 S	39.03 E	533.7	TV	EW63	48	NO

AVERAGE PRECIPITATION (MM)

OBS.	JAN.	FEB.	MAR.	APR.	MAY	JUN.	JUL.	AUG.	SEP.	OCT.	NOV.	DEC.	ANNUAL
37	232	249	166	34	12	3	2	2	2	9	68	188	966
38	247	206	196	38	10	12	8	9	5	14	41	204	987
39	186	154	194	82	18	9	3	4	4	3	21	120	798
40	195	184	154	52	9	12	3	5	4	10	24	127	777
41	292	296	268	152	68	74	36	30	27	63	158	367	1831
42	124	101	72	50	16	12	8	7	27	39	90	88	633
43	188	173	164	177	37	19	7	7	10	11	36	83	912
44	249	245	188	86	25	33	24	25	14	28	102	178	1196
45	109	125	87	52	48	45	37	22	33	41	80	102	779
46	291	234	188	121	56	36	19	25	9	6	110	120	1216
47	229	220	212	68	88	50	35	24	12	5	44	109	1096
48	246	216	200	64	9	3	2	2	1	8	41	152	943

AVERAGE TEMPERATURE (DEGREES CENTIGRADE)

OBS.	JAN.	FEB.	MAR.	APR.	MAY	JUN.	JUL.	AUG.	SEP.	OCT.	NOV.	DEC.	ANNUAL
37	21.7	21.6	21.2	20.6	18.3	15.5	15.9	17.5	20.5	23.6	23.4	22.3	20.2
38	25.9	26.2	26.0	25.6	24.4	23.4	23.2	23.7	24.7	25.6	26.6	26.5	25.2
39	28.4	28.4	28.3	28.1	27.0	25.1	24.4	24.8	25.5	26.6	27.3	29.0	26.9
40	27.4	27.2	27.3	27.0	25.4	24.0	23.5	23.6	24.6	25.8	27.0	27.5	25.8
41	24.5	24.4	23.6	23.1	21.2	20.0	19.5	20.7	22.8	25.4	25.1	24.7	22.9
42	27.1	27.0	26.4	24.8	22.0	19.6	19.5	21.3	23.2	24.8	25.7	27.0	24.0
43	27.0	27.1	27.1	26.5	25.5	24.1	23.4	23.7	24.5	25.4	26.8	27.5	25.7
44	27.4	27.1	26.3	25.3	22.9	20.6	20.3	21.5	24.5	27.4	28.1	27.2	24.9
45	26.0	26.7	25.8	24.8	22.8	20.7	20.1	21.1	23.0	24.8	25.1	26.3	24.0
46	27.7	27.6	27.2	26.5	24.8	23.1	22.6	22.9	24.5	26.2	27.4	27.9	25.7
47	28.0	28.1	27.6	26.9	24.9	23.1	22.7	23.1	24.6	26.3	27.6	28.3	25.9
48	25.2	25.3	25.2	24.8	23.3	21.6	21.1	22.0	23.6	25.2	26.7	26.4	24.2

CLIMATOLOGICAL DATA FOR MOZAMBIQUE

GEOGRAPHY

OBS.	LOCATION	YEARS	LATITUDE	LONGITUDE	ELEVATION -M-	LIFEZONE	CCTX	ID	ONMAP
49	MOPEIA	12	18.00 S	35.50 E	50.9	TV	EH63	49	NO
50	MORRUMBENE	30	23.67 S	35.37 E	20.1	SD	EW63	50	NO
51	MOSSURIL	30	15.02 S	40.67 E	9.8	TV	EW63	51	YES
52	MUECATE	30	14.90 S	39.63 E	27.7	TD	EW53	52	NO
53	MUEDA	18	11.67 S	39.55 E	846.7	SM	EW52	53	YES
54	MUNGARI	11	17.17 S	33.55 E	534.9	TV	EW73	54	NO
55	MUTARARA	15	17.38 S	35.05 E	88.1	TV	EW73	55	NO
56	MUTUALI	13	14.88 S	37.05 E	569.7	TV	EW63	56	NO
57	NACALA	18	14.55 S	40.63 E	24.1	TV	EW62	57	YES
58	NACOONGO	10	24.30 S	35.18 E	29.9	TD	SW52	58	NO
59	NAMAACHA	27	25.98 S	32.02 E	522.7	SD	SW63	59	NO
60	NAMAPA	21	13.82 S	39.82 E	199.9	TV	EW63	60	NO

AVERAGE PRECIPITATION (MM)

OBS.	JAN.	FEB.	MAR.	AFR.	MAY	JUN.	JUL.	AUG.	SEP.	OCT.	NOV.	DEC.	ANNUAL
49	166	182	158	57	37	24	16	13	10	18	97	174	951
50	142	145	151	60	57	47	27	24	19	27	88	112	899
51	218	219	150	90	28	33	14	15	10	8	28	136	950
52	314	269	264	94	17	11	12	10	8	24	77	245	1345
53	268	217	234	125	24	4	1	4	4	4	41	176	1101
54	151	150	56	14	7	9	1	1	2	6	68	145	609
55	150	131	140	37	11	9	28	7	9	4	65	126	718
56	245	204	199	38	7	2	5	5	6	9	66	206	991
57	194	194	170	57	11	9	3	7	6	6	33	121	809
58	163	175	117	61	99	84	49	32	47	27	92	115	1061
59	141	167	157	56	31	22	20	15	43	63	129	116	959
60	229	211	158	80	9	3	1	5	5	17	35	152	904

AVERAGE TEMPERATURE (DEGREES CENTIGRADE)

OBS.	JAN.	FEB.	MAR.	APR.	MAY	JUN.	JUL.	AUG.	SEP.	OCT.	NOV.	DEC.	ANNUAL
49	28.9	28.7	27.9	26.8	24.2	22.4	21.8	23.4	25.4	28.4	29.0	28.9	26.3
50	26.5	26.4	25.6	24.2	21.8	20.1	19.4	20.4	22.1	23.9	24.6	25.6	23.4
51	27.4	27.2	27.0	25.1	24.4	22.0	22.2	22.6	24.0	25.9	27.4	27.8	25.3
52	26.4	25.8	25.4	25.4	23.8	22.3	21.7	22.5	24.4	26.4	27.2	27.0	24.8
53	23.3	23.3	22.9	22.0	21.2	20.0	19.6	19.8	21.2	22.5	23.6	23.7	21.9
54	25.4	25.2	25.0	24.3	22.0	20.2	19.9	21.8	24.3	26.7	26.2	25.5	23.9
55	28.3	27.9	27.2	26.2	24.2	21.8	21.5	23.0	25.6	28.1	28.8	28.2	25.9
56	25.9	25.9	25.4	24.7	22.7	20.7	20.7	22.4	25.0	27.9	28.0	26.4	24.6
57	27.0	26.7	26.9	26.3	25.5	24.3	24.0	24.2	24.8	26.1	26.9	27.0	25.8
58	26.4	26.4	25.8	24.4	22.1	20.1	19.6	20.6	22.1	24.3	24.9	26.1	23.6
59	23.1	23.0	22.3	21.2	19.1	16.5	17.2	18.5	19.5	21.1	21.6	22.8	20.6
60	27.2	26.7	26.8	26.4	24.9	23.4	23.3	24.0	25.6	27.2	28.4	28.1	26.0

CLIMATOLOGICAL DATA FOR MOZAMBIQUE

GEOGRAPHY

OBS.	LOCATION	YEARS	LATITUDE	LONGITUDE	ELEVATION -M-	LIFEZONE	CCTX	ID	ONMAP
61	NAMARROI	11	15.95 S	36.87 E	602.9	SM	EW43	61	NO
62	NAMETIL	25	15.72 S	39.35 E	171.0	TD	EH53	62	NO
63	NAMPULA	19	15.13 S	39.25 E	431.9	TD	EW53	63	NO
64	NOVA FREIXO	30	14.80 S	36.53 E	586.7	TV	EW63	64	NO
65	NOVA SOFALA	30	20.15 S	34.73 E	9.8	TD	EW53	65	NO
66	PAFURI	15	22.43 S	31.33 E	214.9	TT	EW83	66	YES
67	PALMA	10	10.77 S	40.50 E	60.0	TD	EH52	67	YES
68	PANDA	30	24.05 S	34.72 E	150.0	TV	SW73	68	YES
69	PEBANE	30	17.27 S	38.15 E	24.7	TD	EW52	69	NO
70	PORTO AMELIA	17	12.97 S	40.50 E	49.1	TV	EW62	70	NO
71	QUELIMANE	30	17.88 S	36.88 E	6.1	TD	EW53	71	YES
72	QUISSANGA	16	12.43 S	40.48 E	42.1	TV	EH62	72	YES

AVERAGE PRECIPITATION (MM)

OBS.	JAN.	FEB.	MAR.	APR.	MAY	JUN.	JUL.	AUG.	SEP.	OCT.	NOV.	DEC.	ANNUAL
61	302	317	329	120	56	54	40	25	15	71	187	246	1764
62	266	225	194	86	23	21	17	16	9	17	63	171	1108
63	247	232	212	80	31	26	13	14	11	23	53	165	1108
64	246	205	146	25	7	2	1	2	5	15	50	186	889
65	204	214	164	47	23	15	17	13	15	16	114	186	1025
66	75	54	46	23	3	8	1	3	8	10	41	67	337
67	153	173	170	208	100	32	15	6	13	14	35	84	1005
68	119	124	98	43	39	24	19	12	22	36	76	112	724
69	224	216	203	107	72	86	59	30	13	8	56	152	1226
70	161	150	175	120	30	8	14	5	10	3	30	92	799
71	240	261	278	105	85	65	52	36	18	14	76	169	1400
72	241	166	174	152	39	18	25	3	3	23	22	102	967

AVERAGE TEMPERATURE (DEGREES CENTIGRADE)

OBS.	JAN.	FEB.	MAR.	APR.	MAY	JUN.	JUL.	AUG.	SEP.	OCT.	NOV.	DEC.	ANNUAL
61	25.9	25.1	24.6	24.2	22.2	19.8	19.2	20.1	22.6	25.6	25.6	25.4	23.3
62	27.9	27.7	27.4	26.4	24.5	22.8	22.3	23.6	25.9	28.2	29.6	29.0	26.3
63	25.9	25.4	25.5	24.7	23.3	21.5	21.3	22.4	24.5	27.0	27.4	27.0	24.6
64	25.7	25.8	25.4	24.6	22.0	20.0	19.8	21.6	24.6	27.3	28.0	26.5	24.2
65	27.2	27.5	26.7	25.4	23.2	21.2	20.6	21.4	23.0	24.9	26.0	26.8	24.5
66	28.6	28.2	27.4	25.8	22.4	19.7	19.8	21.9	24.5	27.2	27.8	28.6	25.2
67	27.8	27.7	27.7	26.9	26.1	24.8	24.3	24.6	25.3	26.2	27.2	27.9	26.4
68	26.9	26.6	25.8	24.5	22.6	20.6	20.3	21.4	23.3	25.1	25.9	26.7	24.1
69	27.3	27.1	27.0	26.2	24.2	22.6	21.9	22.3	24.2	25.8	26.8	27.4	25.2
70	27.1	27.0	27.1	26.6	25.4	23.9	23.6	24.1	24.9	26.2	27.1	27.7	25.9
71	27.5	27.4	26.8	25.7	23.2	21.1	20.7	21.5	23.7	25.8	27.0	27.3	24.8
72	27.2	27.2	27.0	26.9	25.9	24.8	24.1	24.8	25.6	26.5	27.2	27.5	26.2

CLIMATOLOGICAL DATA FOR MOZAMBIQUE

GEOGRAPHY

OBS.	LOCATION	YEARS	LATITUDE	LONGITUDE	ELEVATION -M-	LIFEZONE	CCTX	ID	ONMAP
73	QUISSICO	30	24.72 S	34.75 E	150.0	SD	SW62	73	NO
74	RIBAUE	30	14.98 S	38.27 E	534.6	TD	EW53	74	NO
75	TETE	19	16.18 S	33.58 E	139.0	TV	EH73	75	YES
76	UMBELUZI	30	26.05 S	32.38 E	11.9	SD	SW73	76	NO
77	VILA COUTINHO	23	14.58 S	34.30 E	1299.7	SD	EW63	77	NO
78	VILA DE MANICA	30	18.93 S	32.87 E	722.7	SM	EW53	78	NO
79	VILA FONTES	15	17.82 S	35.38 E	25.0	TV	EW63	79	NO
80	VILA GAMITO	11	14.17 S	32.98 E	957.7	SD	EW63	80	YES
81	VILA GOUBEIA	30	18.07 S	33.17 E	610.8	SM	EW43	81	YES
82	VILA GUILHERME	14	19.88 S	34.58 E	10.1	TD	EW53	82	YES
83	VILA JUNQUEIRO	14	15.47 S	36.98 E	733.7	SW	EW33	83	YES
84	VILA LUISA	30	25.73 S	32.68 E	25.0	TV	SW62	84	YES

AVERAGE PRECIPITATION (MM)

OBS.	JAN.	FEB.	MAR.	APR.	MAY	JUN.	JUL.	AUG.	SEP.	OCT.	NOV.	DEC.	ANNUAL
73	163	109	107	84	75	59	40	34	35	38	99	116	958
74	277	278	210	51	18	14	7	8	8	12	70	191	1145
75	148	126	76	11	4	2	10	3	7	7	33	91	519
76	123	115	104	48	18	17	14	6	30	49	82	108	715
77	256	222	167	47	10	3	2	2	2	12	68	188	978
78	241	176	152	46	16	13	9	12	16	32	107	209	1030
79	211	152	178	54	27	31	20	13	9	10	99	164	967
80	218	200	156	39	11	0	0	2	1	20	118	182	947
81	401	342	227	105	26	21	11	16	14	26	153	348	1690
82	161	224	192	58	19	23	12	10	20	32	107	194	1051
83	339	356	402	181	58	67	40	49	19	53	184	293	2042
84	133	115	129	66	30	28	16	18	33	43	93	95	800

AVERAGE TEMPERATURE (DEGREES CENTIGRADE)

OBS.	JAN.	FEB.	MAR.	APR.	MAY	JUN.	JUL.	AUG.	SEP.	OCT.	NOV.	DEC.	ANNUAL
73	25.9	26.1	25.4	24.1	21.9	20.3	19.9	20.4	21.9	23.8	24.4	25.5	23.3
74	25.9	25.4	25.1	24.2	22.3	19.8	19.8	21.2	23.6	25.4	26.6	25.9	23.8
75	28.6	28.2	27.4	27.3	25.0	22.4	22.2	23.8	26.9	29.6	30.1	29.9	26.8
76	26.5	26.4	25.4	23.7	21.0	18.4	18.2	20.0	21.8	23.8	24.7	26.0	23.0
77	21.4	21.2	20.8	20.0	17.9	16.0	15.8	17.1	20.2	22.7	22.9	22.1	19.8
78	24.6	24.6	23.5	22.0	18.9	16.2	15.9	17.2	20.5	23.5	24.2	24.2	21.2
79	27.3	27.6	26.5	25.9	24.0	21.4	21.2	22.7	24.6	27.6	28.0	28.1	25.4
80	23.6	23.5	23.3	22.9	21.3	18.3	18.4	20.2	23.9	26.2	25.6	24.0	22.6
81	24.1	24.0	23.4	22.8	20.3	17.9	17.8	19.0	21.9	23.4	24.7	24.6	22.0
82	25.7	27.7	26.8	25.4	22.8	20.5	20.1	21.7	23.6	25.9	26.9	29.9	24.7
83	24.1	24.1	23.4	22.3	20.4	18.0	18.0	19.3	22.7	25.0	24.9	24.8	22.2
84	26.5	26.5	25.8	24.5	22.3	20.1	19.7	20.7	22.0	23.6	24.9	26.2	23.6

CLIMATOLOGICAL DATA FOR MOZAMBIQUE

GEOGRAPHY

OBS.	LOCATION	YEARS	LATITUDE	LONGITUDE	ELEVATION -M-	LIFEZONE	CCTX	ID	ONMAP
85	VILA MACHADO	30	19.27 S	34.20 E	57.0	TV	EW63	85	YES
86	VILA PAIRADE ANDRA	21	18.68 S	34.07 E	299.9	SM	EW53	86	YES
87	VILA PERY	30	19.12 S	33.47 E	730.6	SM	EW53	87	NO
88	VILA VASCO DE GAMA	14	14.90 S	32.27 E	1169.5	SM	EW53	88	YES
89	VILANCULAS	22	22.00 S	35.33 E	20.1	TV	EW63	89	YES
90	ZUMBO	17	15.62 S	30.45 E	342.9	TV	EH64	90	NO

AVERAGE PRECIPITATION (MM)

OBS.	JAN.	FEB.	MAR.	APR.	MAY	JUN.	JUL.	AUG.	SEP.	OCT.	NOV.	DEC.	ANNUAL
85	170	151	121	52	20	17	11	13	11	25	102	228	922
86	251	230	204	76	28	29	31	28	24	34	150	279	1365
87	225	208	165	57	20	21	16	20	16	34	110	191	1081
88	320	329	228	70	18	25	17	13	4	11	124	304	1465
89	174	144	124	49	36	21	26	20	18	21	65	132	830
90	206	178	106	10	1	0	0	2	0	8	81	174	766

AVERAGE TEMPERATURE (DEGREES CENTIGRADE)

OBS.	JAN.	FEB.	MAR.	APR.	MAY	JUN.	JUL.	AUG.	SEP.	OCT.	NOV.	DEC.	ANNUAL
85	28.3	27.9	27.0	26.2	23.9	21.5	21.1	22.3	24.7	27.0	28.1	27.7	25.5
86	26.2	25.8	24.8	23.8	21.1	18.7	18.3	20.1	23.0	26.3	26.1	26.1	23.4
87	24.4	24.2	23.2	22.2	20.0	17.6	17.2	18.8	21.2	23.6	23.8	23.9	21.7
88	21.3	21.2	20.8	20.2	18.4	15.8	16.1	17.6	20.9	24.3	23.5	21.8	20.2
89	27.2	27.2	26.4	25.0	22.6	20.6	19.8	21.0	22.7	24.7	25.7	26.5	24.1
90	27.1	27.0	27.0	26.7	24.1	21.3	21.0	23.4	26.7	30.7	30.1	17.0	26.1

NEPAL

Agriculture

Of 14,080,000 ha in total area, Nepal was estimated in 1978 to have 2,319,000 ha of arable or permanent cropland, 1,700,000 ha in permanent pasture, 4,450,000 ha in forest and woodland, and 5,211,000 ha in other terrain categories. In 1978, approximately 210,000 ha were irrigated. Of the 1979 population, 92.7% or 6,162,000 individuals of the economically active population were engaged in agriculture. The total population in 1979 was 13,938,000. According to FAO statistics for 1979, there were 6,850,000 cattle, 4,150,000 buffalo, 350,000 pigs, 2,360,000 sheep, 2,480,000 goats, and 21,500,000 chickens.[2] Agricultural production is summarized in Table 1. Conventional and energetic yield data on most of these will be found in Appendixes I and II.

Climate

Data from Wernstedt[3] are presented in the Climatological Data Table (see Introduction for explanation). Based on the means for data points cited in Wernstedt, the mean annual precipitation is 16.8 dm (mean of 94 sites) and the mean annual temperature 19.5°C (mean of 94 sites). The averages would suggest that much of the country could be classified as Subtropical Moist Forest Life Zone.[4]

Productivity

Vegetation types are roughly indicated in the vegetation map. Depending on climate and edaphic factors, I estimate the annual productivity potentials, assuming reasonable management, as follows:

Vegetation type	MT/ha/yr
Grazing land	3—6
Arable land	3—12
Scrub	3—12
Forest	6—24
Barren highland	0—3

Economics

Food consumption per capita was 2002 cal/day, energy consumption 11 kg coal equivalent per capita in 1977. In 1970, the caloric supply was 94.0% of requirements. The total population growth rate was 2.0% in 1970. Population density was 81 persons per square kilometer in 1970, with 3.9% of the total population urban. Life expectancy at birth of those born in 1970 was 42 years. In 1977, with GNP at $110 per capita, agriculture accounted for 62.3% of GDP, manufacturing for 10.4%, other industry for 2.0%, and services for 25.3%.[10]

Energetics

Importing more than 75% of its commercial energy demands as oil, Nepal is expected to be experiencing fuelwood problems by the year 2000.[10]

Table 1
AGRICULTURAL PRODUCTION IN NEPAL

Crop	Area planted (ha)	Yield (kg/ha)	Total production (MT)
Wheat	356,000	1,166	415,000
Rice	1,265,000	1,852	250,000
Barley	26,000	885	23,000
Corn	450,000	1,778	800,000
Millet	120,000	1,167	140,000
Potatoes	55,000	5,455	300,000
Chick peas	38,000	485	33,000
Sugarcane	20,000	16,850	337,000
Sugar (centrifugal, raw)			29,000
Tobacco leaves	8,000	750	6,000
Jute and substitutes	45,000	1,000	45,000

From FAO 1979 Production Yearbook, Vol. 33, Food and Agriculture Organization, Rome, 1980.

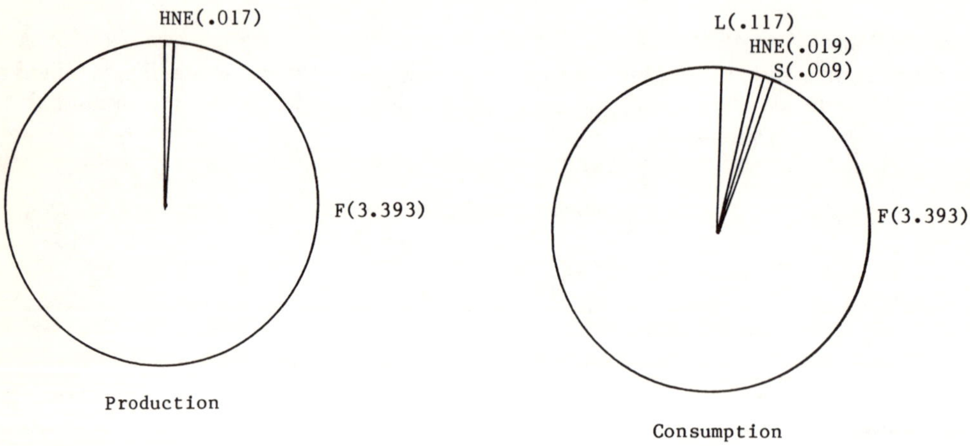

1978 Energy production and consumption in Nepal; fuel types listed in million metric tons coal equivalents (MMTCE). 1 MMTCE = 27.78 × 10^{12} Btu. L = liquid fuel, S = solid fuel, HNE = hydro, nuclear, and electric fuel. F = firewood, NG = natural gas. (Data from United Nations, World Energy Supplies, 1973—1978, Statistical Papers Series J, 22, Department of International Economic and Social Affairs, Statistical Office, New York, 1979.)

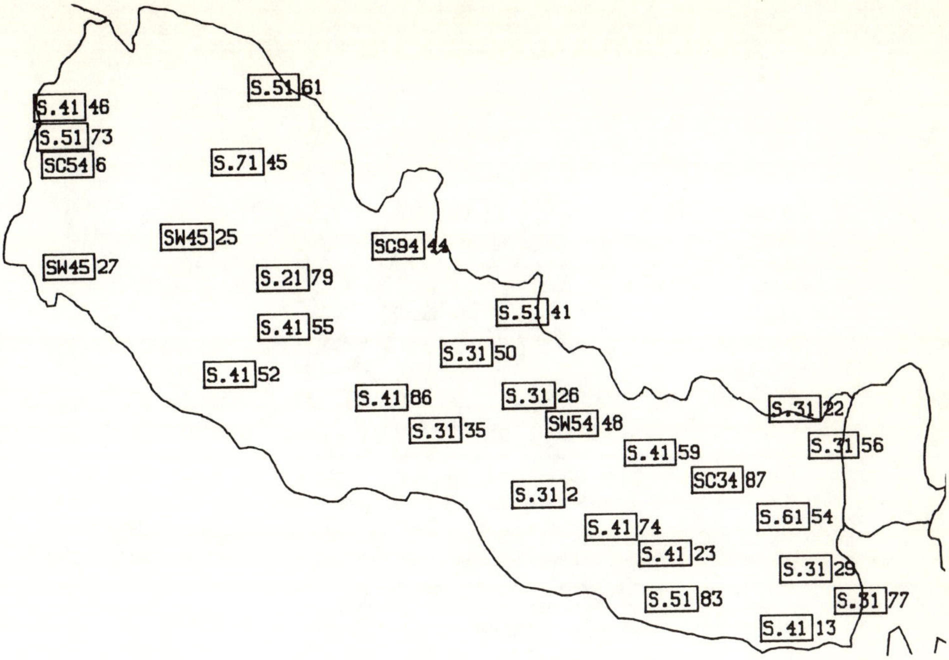

Selected weather stations for Nepal.

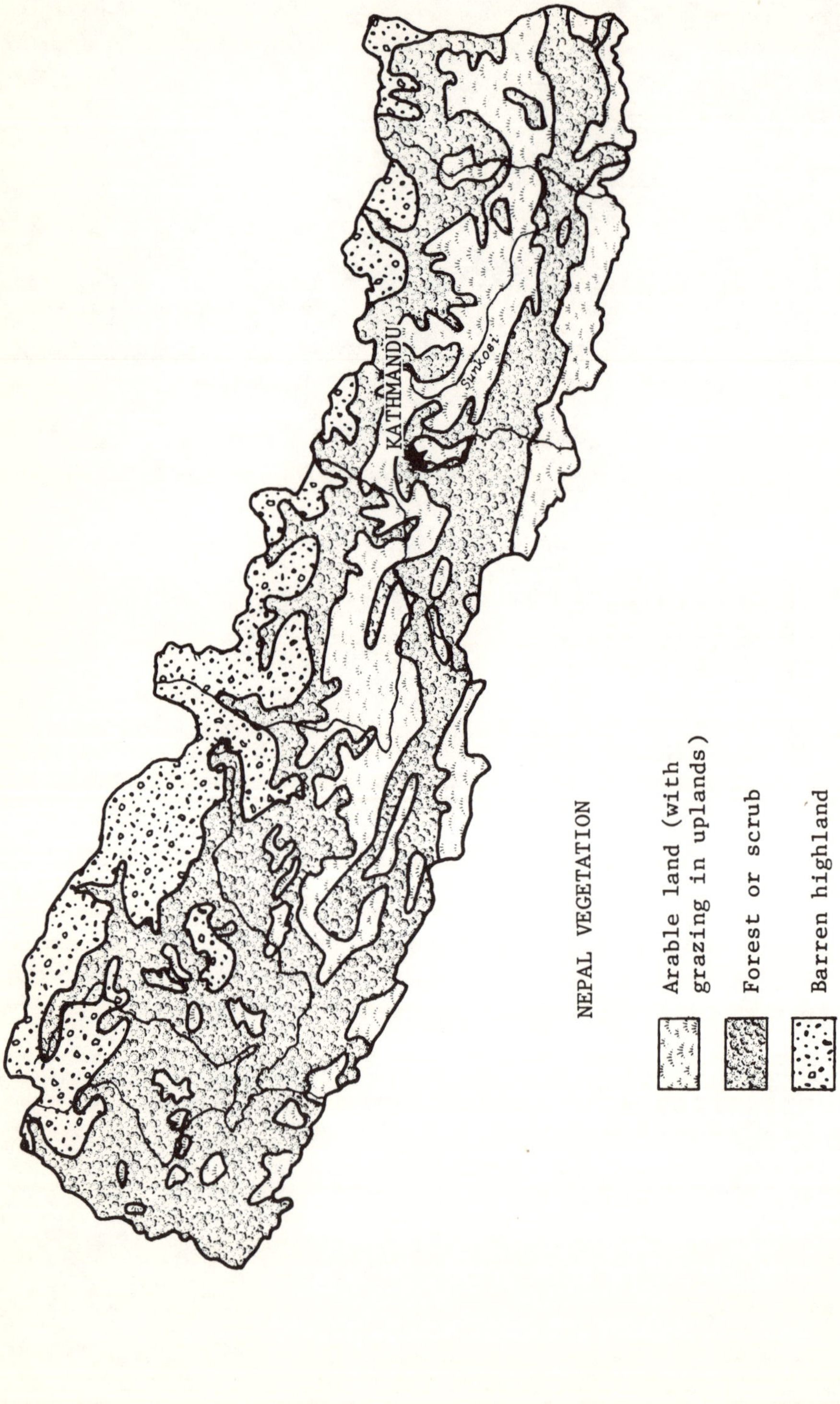
KATHMANDU
Sunkosi
NEPAL VEGETATION
Arable land (with grazing in uplands)
Forest or scrub
Barren highland

CLIMATOLOGICAL DATA FOR NEPAL

GEOGRAPHY

OBS.	LOCATION	YEARS	LATITUDE	LONGITUDE	ELEVATION -M-	LIFEZONE	CCTX	ID	ONMAP
1	AISYALUKHARKA	21	27.35 N	86.75 E	2878.2	.	.	1	NO
2	AMLEKHAGANJ	15	27.28 N	85.00 E	302.7	.	.	2	YES
3	ANGBUNG	19	27.27 N	87.72 E	1211.6	.	.	3	NO
4	ARUGHAT BAZAR	12	28.05 N	84.82 E	605.6	.	.	4	NO
5	BAIRIA	14	27.00 N	85.38 E	151.8	.	.	5	NO
6	BANDELDURA	6	29.30 N	80.58 E	1836.4	WM	SC54	6	YES
7	BANDIPUR	14	27.93 N	84.40 W	1060.7	.	.	7	NO
8	BARAKSHETRA	8	26.87 N	87.17 W	145.7	TM	SW24	8	NO
9	BENI BAZAR	14	28.35 N	82.57 E	914.7	.	.	9	NO
10	BHOJPUR	8	27.18 N	87.05 E	1666.3	WM	SC54	10	NO
11	BIJAYA PUR RASKOT	12	29.23 N	81.63 E	1514.6	.	.	11	NO
12	BIJUWAR PATHAN	13	28.10 N	82.87 E	889.7	.	.	12	NO

AVERAGE PRECIPITATION (MM)

OBS.	JAN.	FEB.	MAR.	APR.	MAY	JUN.	JUL.	AUG.	SEP.	OCT.	NOV.	DEC.	ANNUAL
1	18	9	23	81	181	472	612	591	311	109	13	7	2427
2	12	5	36	39	94	383	591	575	373	74	11	2	2195
3	19	12	47	124	234	288	271	230	165	59	17	2	1468
4	25	16	51	50	91	364	599	623	320	82	9	9	2239
5	24	11	20	29	64	261	405	365	253	58	10	5	1505
6	68	59	70	39	54	170	410	327	171	81	9	39	1497
7	35	12	40	48	144	319	480	478	167	64	6	2	1795
8	22	10	34	73	122	373	815	612	498	85	13	3	2660
9	31	20	34	30	62	212	335	382	188	59	7	10	1370
10	19	13	31	57	118	200	292	221	162	32	12	2	1159
11	54	43	62	28	53	93	197	192	98	47	6	18	891
12	27	31	23	30	44	235	316	257	119	47	3	9	1141

AVERAGE TEMPERATURE (DEGREES CENTIGRADE)

OBS.	JAN.	FEB.	MAR.	APR.	MAY	JUN.	JUL.	AUG.	SEP.	OCT.	NOV.	DEC.	ANNUAL
1	.	.	.	.	.	.	.	.	.	.	.	.	.
2	.	.	.	.	.	.	.	.	.	.	.	.	.
3	.	.	.	.	.	.	.	.	.	.	.	.	.
4	.	.	.	.	.	.	.	.	.	.	.	.	.
5	.	.	.	.	.	.	.	.	.	.	.	.	.
6	8.0	9.6	13.4	17.6	20.3	21.7	20.1	20.1	19.1	16.5	12.4	9.4	15.7
7	.	.	.	.	.	.	.	.	.	.	.	.	.
8	16.9	19.5	23.8	27.8	28.7	28.7	27.9	27.5	27.0	25.1	21.3	18.3	24.4
9	.	.	.	.	.	.	.	.	.	.	.	.	.
10	10.0	12.2	15.7	19.0	20.1	21.0	20.9	21.0	20.3	16.8	14.5	11.9	17.0
11	.	.	.	.	.	.	.	.	.	.	.	.	.
12	.	.	.	.	.	.	.	.	.	.	.	.	.

CLIMATOLOGICAL DATA FOR NEPAL

GEOGRAPHY

OBS.	LOCATION	YEARS	LATITUDE	LONGITUDE	ELEVATION -M-	LIFEZONE	CCTX	ID	ONMAP
13	BIRATNAGAR	15	26.47 N	87.28 E	75.9	.	.	13	YES
14	BUTWAL	9	27.70 N	83.47 E	262.7	TM	SW35	14	NO
15	CHAINPUR EAST	8	27.28 N	87.33 E	1328.6	SM	SW54	15	NO
16	CHAINPUR WEST	13	29.55 N	81.22 E	1303.6	.	.	16	NO
17	CHAPKOT	12	27.92 N	83.82 E	605.6	.	.	17	NO
18	CHARIKOT	11	27.67 N	86.05 E	1980.3	.	.	18	NO
19	CHATRA	21	26.82 N	87.15 E	151.8	.	.	19	NO
20	CHAURI-KHARKA	20	27.70 N	86.73 E	2437.2	.	.	20	NO
21	CHAUTARA	6	27.78 N	85.72 E	1514.6	SW	SW34	21	NO
22	CHEPUA	11	27.77 N	87.42 E	0.0	.	.	22	YES
23	CHISAPANI BAZAR	13	26.92 N	86.17 E	734.6	.	.	23	YES
24	CHISPANI GARHI	13	27.55 N	85.13 E	1727.3	.	.	24	NO

AVERAGE PRECIPITATION (MM)

OBS.	JAN.	FEB.	MAR.	APR.	MAY	JUN.	JUL.	AUG.	SEP.	OCT.	NOV.	DEC.	ANNUAL
13	21	5	11	34	117	293	469	414	300	95	4	0	1763
14	15	17	17	26	55	453	660	855	313	46	12	6	2475
15	15	12	25	66	148	219	306	312	184	42	28	1	1358
16	66	55	72	33	31	161	360	393	217	76	6	21	1491
17	35	28	49	39	104	294	461	391	178	85	2	13	1679
18	21	29	38	50	127	290	545	588	370	97	13	3	2171
19	19	15	30	53	117	361	611	518	345	117	10	3	2199
20	18	11	43	49	112	362	620	587	346	66	14	6	2234
21	17	28	31	98	53	344	526	568	275	62	13	3	2018
22	40	67	150	108	221	396	489	433	352	146	39	9	2450
23	24	7	25	19	60	277	449	398	187	87	10	5	1548
24	26	21	40	71	135	367	516	525	207	76	8	4	1996

AVERAGE TEMPERATURE (DEGREES CENTIGRADE)

OBS.	JAN.	FEB.	MAR.	APR.	MAY	JUN.	JUL.	AUG.	SEP.	OCT.	NOV.	DEC.	ANNUAL
13	.	.	.	.	.	.	.	.	.	.	.	.	.
14	18.0	20.1	24.8	29.3	31.5	30.3	28.8	28.6	28.3	26.1	22.4	18.8	25.6
15	12.9	15.2	18.3	22.1	23.2	23.9	23.6	23.1	22.5	20.2	17.1	14.0	19.7
16	.	.	.	.	.	.	.	.	.	.	.	.	.
17	.	.	.	.	.	.	.	.	.	.	.	.	.
18	.	.	.	.	.	.	.	.	.	.	.	.	.
19	.	.	.	.	.	.	.	.	.	.	.	.	.
20	.	.	.	.	.	.	.	.	.	.	.	.	.
21	11.8	14.1	17.2	20.8	22.6	23.1	22.6	22.2	21.9	19.7	15.9	13.4	18.8
22	.	.	.	.	.	.	.	.	.	.	.	.	.
23	.	.	.	.	.	.	.	.	.	.	.	.	.
24	.	.	.	.	.	.	.	.	.	.	.	.	.

CLIMATOLOGICAL DATA FOR NEPAL

GEOGRAPHY

OBS.	LOCATION	YEARS	LATITUDE	LONGITUDE	ELEVATION -M-	LIFEZONE	CCTX	ID	ONMAP
25	DAILEKH	6	28.85 N	81.72 E	1303.6	SM	SW45	25	YES
26	DHADING	14	27.87 N	84.93 E	1439.6	.	.	26	YES
27	DHANAGARHI	14	28.68 N	80.60 E	279.8	TD	SW45	27	YES
28	DHANKUTA	21	26.98 N	87.35 E	1059.5	.	.	28	NO
29	DHARAN BAZAR	21	26.82 N	87.48 E	302.7	.	.	29	YES
30	DHULIKHEL	22	27.62 N	85.55 E	1665.4	.	.	30	NO
31	DOLAL GHAT	20	27.63 N	85.72 E	1371.6	.	.	31	NO
32	DUMUHAN	21	26.98 N	87.35 E	1059.5	.	.	32	NO
33	DUNAI	12	28.93 N	82.92 E	2802.0	.	.	33	NO
34	DWARPA	10	27.22 N	86.85 E	1817.5	.	.	34	NO
35	GIRWARI	12	27.67 N	84.05 E	302.7	.	.	35	YES
36	GODAVARI	13	27.70 N	85.33 E	1514.6	.	.	36	NO

AVERAGE PRECIPITATION (MM)

OBS.	JAN.	FEB.	MAR.	APR.	MAY	JUN.	JUL.	AUG.	SEP.	OCT.	NOV.	DEC.	ANNUAL
25	24	24	37	42	54	212	434	510	169	45	8	10	1569
26	34	27	60	65	130	359	606	606	271	89	5	5	2257
27	33	15	25	3	30	220	547	408	278	88	4	9	1660
28	6	7	34	43	56	184	243	172	78	52	3	5	883
29	16	9	25	42	144	306	573	660	395	126	13	3	2312
30	22	19	33	49	98	246	390	394	197	59	7	2	1516
31	11	9	24	36	54	176	236	267	105	28	8	2	947
32	6	7	34	43	56	184	243	172	78	52	3	5	883
33	53	48	70	56	60	68	228	202	101	49	13	33	981
34	11	6	47	60	144	309	365	330	213	103	9	5	1602
35	23	7	15	31	80	487	659	719	293	95	4	3	2416
36	32	18	39	48	112	301	459	500	225	49	2	4	1789

AVERAGE TEMPERATURE (DEGREES CENTIGRADE)

OBS.	JAN.	FEB.	MAR.	APR.	MAY	JUN.	JUL.	AUG.	SEP.	OCT.	NOV.	DEC.	ANNUAL
25	10.5	12.9	16.4	20.6	23.4	23.9	22.9	22.7	22.2	19.0	15.2	11.7	18.5
26	.	.	.	.	.	.	.	.	.	.	.	.	.
27	14.7	16.9	22.8	28.6	32.2	32.5	29.7	28.9	28.6	25.6	20.0	15.6	24.7
28	.	.	.	.	.	.	.	.	.	.	.	.	.
29	.	.	.	.	.	.	.	.	.	.	.	.	.
30	.	.	.	.	.	.	.	.	.	.	.	.	.
31	.	.	.	.	.	.	.	.	.	.	.	.	.
32	.	.	.	.	.	.	.	.	.	.	.	.	.
33	.	.	.	.	.	.	.	.	.	.	.	.	.
34	.	.	.	.	.	.	.	.	.	.	.	.	.
35	.	.	.	.	.	.	.	.	.	.	.	.	.
36	.	.	.	.	.	.	.	.	.	.	.	.	.

CLIMATOLOGICAL DATA FOR NEPAL

GEOGRAPHY

OBS.	LOCATION	YEARS	LATITUDE	LONGITUDE	ELEVATION -M-	LIFEZONE	CCTX	ID	ONMAP
37	GORKHA	7	28.00 N	84.45 E	1060.7	SM	SW44	37	NO
38	GULARIA	13	28.17 N	81.35 E	151.8	.	.	38	NO
39	HARIHARPUR GARHI	14	27.33 N	85.50 E	302.7	.	.	39	NO
40	ILAM	14	26.92 N	87.92 E	1060.7	.	.	40	NO
41	JAGAT SETIBAS	13	28.37 N	84.88 E	3002.0	.	.	41	YES
42	JAJARKOT	13	28.70 N	82.20 E	1211.6	.	.	42	NO
43	JHAWANI	12	27.58 N	84.53 E	1514.6	.	.	43	NO
44	JOMOSOM	8	28.78 N	83.72 E	2799.0	WT	SC94	44	YES
45	JUMLA	13	29.30 N	82.20 E	2423.2	.	.	45	YES
46	KAKERPAKHA	14	29.65 N	80.50 E	841.6	.	.	46	YES
47	KATAI	12	29.00 N	81.02 E	1362.5	.	.	47	NO
48	KATHMANDU	10	27.70 N	85.33 E	1287.5	SM	SW54	48	YES

AVERAGE PRECIPITATION (MM)

OBS.	JAN.	FEB.	MAR.	APR.	MAY	JUN.	JUL.	AUG.	SEP.	OCT.	NOV.	DEC.	ANNUAL
37	28	6	24	62	111	282	441	474	159	50	24	10	1671
38	30	18	27	11	30	127	338	385	205	65	2	9	1247
39	17	15	32	36	119	553	852	753	383	112	4	0	2876
40	12	7	20	54	129	267	420	297	206	86	5	3	1506
41	32	43	93	58	51	175	311	281	152	58	8	8	1270
42	35	29	34	55	24	269	464	406	177	40	1	30	1564
43	17	16	17	20	83	291	402	506	262	80	11	3	1708
44	32	29	28	17	5	8	42	47	25	12	2	3	250
45	33	36	42	28	29	49	193	163	78	32	2	8	693
46	40	32	58	44	38	245	438	385	215	56	9	14	1574
47	62	42	35	32	56	327	605	530	280	67	2	19	2057
48	15	41	23	58	122	246	373	345	155	38	8	3	1427

AVERAGE TEMPERATURE (DEGREES CENTIGRADE)

OBS.	JAN.	FEB.	MAR.	APR.	MAY	JUN.	JUL.	AUG.	SEP.	OCT.	NOV.	DEC.	ANNUAL
37	12.9	15.3	19.6	23.4	24.8	25.0	24.0	23.7	23.3	20.7	17.1	14.0	20.3
38	.	.	.	.	.	.	.	.	.	.	.	.	.
39	.	.	.	.	.	.	.	.	.	.	.	.	.
40	.	.	.	.	.	.	.	.	.	.	.	.	.
41	.	.	.	.	.	.	.	.	.	.	.	.	.
42	.	.	.	.	.	.	.	.	.	.	.	.	.
43	.	.	.	.	.	.	.	.	.	.	.	.	.
44	4.5	5.9	8.9	11.9	15.1	18.7	19.4	18.9	17.0	11.8	8.3	5.6	12.2
45	.	.	.	.	.	.	.	.	.	.	.	.	.
46	.	.	.	.	.	.	.	.	.	.	.	.	.
47	.	.	.	.	.	.	.	.	.	.	.	.	.
48	9.7	12.2	15.7	19.3	22.1	24.1	23.9	23.7	23.0	19.4	14.9	11.3	18.3

CLIMATOLOGICAL DATA FOR NEPAL

GEOGRAPHY

OBS.	LOCATION	YEARS	LATITUDE	LONGITUDE	ELEVATION -M-	LIFEZONE	CCTX	ID	ONMAP
49	KHOTANG BAZAR	11	27.03 N	86.83 E	1211.6	.	.	49	NO
50	KUNCHA	14	28.13 N	84.35 E	908.6	.	.	50	YES
51	KURLEGAT	21	27.13 N	86.42 E	605.6	.	.	51	NO
52	KUSUM	13	28.02 N	82.12 E	151.8	.	.	52	YES
53	LAHAN	14	26.73 N	86.50 E	150.9	.	.	53	NO
54	LEGUA GHAT	15	27.13 N	87.28 E	302.7	.	.	54	YES
55	LIBAN GAON	13	28.30 N	82.63 E	1362.5	.	.	55	YES
56	LUNGTHUNG	21	27.55 N	87.78 E	1817.5	.	.	56	YES
57	MACHUAGHAT	21	26.97 N	87.17 E	181.7	.	.	57	NO
58	MANEBHANJANG	22	27.20 N	86.45 E	1514.6	.	.	58	NO
59	MELUNG	11	27.52 N	86.05 E	1514.6	.	.	59	YES
60	MEMENJAGAT	20	27.20 N	87.93 E	1817.5	.	.	60	NO

AVERAGE PRECIPITATION (MM)

OBS.	JAN.	FEB.	MAR.	APR.	MAY	JUN.	JUL.	AUG.	SEP.	OCT.	NOV.	DEC.	ANNUAL
49	12	10	36	29	96	183	273	226	131	59	4	1	1060
50	27	22	60	73	173	451	569	500	266	87	15	7	2250
51	35	11	32	51	54	156	256	212	121	35	6	1	970
52	41	18	28	15	21	253	501	461	212	87	1	6	1644
53	23	18	24	28	66	277	319	375	190	112	8	2	1442
54	11	4	18	47	79	134	173	193	118	26	7	1	811
55	30	35	38	55	61	238	384	400	198	61	3	10	1513
56	18	24	74	78	126	345	507	504	372	94	19	6	2167
57	16	14	29	46	99	240	377	306	177	64	9	2	1379
58	22	19	43	59	103	210	287	258	131	41	6	3	1182
59	15	20	36	59	119	305	364	507	219	65	9	4	1722
60	16	13	48	127	235	354	481	417	273	109	17	6	2095

AVERAGE TEMPERATURE (DEGREES CENTIGRADE)

OBS.	JAN.	FEB.	MAR.	APR.	MAY	JUN.	JUL.	AUG.	SEP.	OCT.	NOV.	DEC.	ANNUAL
49	.	.	.	.	.	.	.	.	.	.	.	.	.
50	.	.	.	.	.	.	.	.	.	.	.	.	.
51	.	.	.	.	.	.	.	.	.	.	.	.	.
52	.	.	.	.	.	.	.	.	.	.	.	.	.
53	.	.	.	.	.	.	.	.	.	.	.	.	.
54	.	.	.	.	.	.	.	.	.	.	.	.	.
55	.	.	.	.	.	.	.	.	.	.	.	.	.
56	.	.	.	.	.	.	.	.	.	.	.	.	.
57	.	.	.	.	.	.	.	.	.	.	.	.	.
58	.	.	.	.	.	.	.	.	.	.	.	.	.
59	.	.	.	.	.	.	.	.	.	.	.	.	.
60	.	.	.	.	.	.	.	.	.	.	.	.	.

CLIMATOLOGICAL DATA FOR NEPAL

GEOGRAPHY

OBS.	LOCATION	YEARS	LATITUDE	LONGITUDE	ELEVATION -M-	LIFEZONE	CCTX	ID	ONMAP
61	MUGU	.	29.75 N	82.55 E	3785.9	.	.	61	YES
62	MULGHAT	20	26.92 N	87.30 E	302.7	.	.	62	NO
63	MUNGA	21	27.03 N	87.23 E	1211.6	.	.	63	NO
64	MUSIKOT	13	28.17 N	83.27 E	1817.5	.	.	64	NO
65	NAMCHE BAZAR	.	27.83 N	86.72 E	3448.8	.	.	65	NO
66	NAWAKOT	12	27.92 N	85.17 E	908.6	.	.	94	NO
67	NAWALPUR	11	27.80 N	85.62 E	1817.5	.	.	66	NO
68	NEPAL THOK	22	27.45 N	85.82 E	756.5	.	.	67	NO
69	NEPALGANJ	45	28.10 N	81.62 E	182.9	TD	SW55	68	NO
70	NIJGARHI	12	27.20 N	85.17 E	150.9	.	.	69	NO
71	NUM	11	27.55 N	87.28 E	1362.5	.	.	70	NO
72	PAKARNAS	21	27.43 N	86.57 E	2271.4	.	.	71	NO

AVERAGE PRECIPITATION (MM)

OBS.	JAN.	FEB.	MAR.	APR.	MAY	JUN.	JUL.	AUG.	SEP.	OCT.	NOV.	DEC.	ANNUAL
61	117	55	112	77	65	78	177	185	110	68	9	43	1096
62	11	5	26	30	83	167	255	194	89	45	8	3	916
63	16	5	33	42	92	201	317	284	185	66	9	3	1253
64	22	13	36	39	78	343	368	311	179	50	7	9	1455
65	43	20	31	26	39	129	226	206	140	57	11	24	952
66	22	21	31	44	76	288	429	483	263	61	7	3	1728
67	18	24	35	59	80	362	586	698	362	73	7	3	2307
68	16	10	34	51	69	128	192	208	151	92	6	14	971
69	21	23	10	8	31	157	311	325	218	40	7	9	1161
70	20	10	24	31	60	346	490	548	319	73	8	3	1932
71	66	37	107	139	397	634	634	619	551	216	78	32	3510
72	15	13	34	42	80	229	449	489	283	75	6	0	1715

AVERAGE TEMPERATURE (DEGREES CENTIGRADE)

OBS.	JAN.	FEB.	MAR.	APR.	MAY	JUN.	JUL.	AUG.	SEP.	OCT.	NOV.	DEC.	ANNUAL
61	.	.	.	.	.	.	.	.	.	.	.	.	.
62	.	.	.	.	.	.	.	.	.	.	.	.	.
63	.	.	.	.	.	.	.	.	.	.	.	.	.
64	.	.	.	.	.	.	.	.	.	.	.	.	.
65	.	.	.	.	.	.	.	.	.	.	.	.	.
66	.	.	.	.	.	.	.	.	.	.	.	.	.
67	.	.	.	.	.	.	.	.	.	.	.	.	.
68	.	.	.	.	.	.	.	.	.	.	.	.	.
69	15.6	17.8	23.3	28.9	31.9	31.9	29.7	28.9	28.9	26.1	20.8	16.4	25.0
70	.	.	.	.	.	.	.	.	.	.	.	.	.
71	.	.	.	.	.	.	.	.	.	.	.	.	.
72	.	.	.	.	.	.	.	.	.	.	.	.	.

CLIMATOLOGICAL DATA FOR NEPAL

GEOGRAPHY

OBS.	LOCATION	YEARS	LATITUDE	LONGITUDE	ELEVATION -M-	LIFEZONE	CCTX	ID	ONMAP
73	PANGTHANGDOMA	17	27.68 N	87.82 E	2817.3	.	.	72	NO
74	PATAN WEST	14	29.47 N	80.53 E	1211.6	.	.	73	YES
75	PATHARKOT	14	27.08 N	85.67 E	226.8	.	.	74	YES
76	PEPALKOTE	13	29.62 N	85.87 E	1362.5	.	.	75	NO
77	POKHARA	11	28.18 N	84.00 E	832.7	SW	SW24	76	NO
78	RANIBIRTA	13	26.63 N	87.98 E	151.8	.	.	77	YES
79	RIRIBAZAR	14	27.95 N	83.43 E	605.6	.	.	78	NO
80	RUKUMKOT	11	28.60 N	82.63 E	1590.4	.	.	79	YES
81	SALLYAN	9	28.38 N	82.17 E	1665.4	SM	SW54	80	NO
82	SILGARHI	14	29.27 N	80.98 E	1059.5	.	.	81	NO
83	SINDHULI GARHI	15	27.28 N	85.97 E	1363.4	.	.	82	NO
84	SIRHA	22	26.65 N	86.22 E	75.9	.	.	83	YES

AVERAGE PRECIPITATION (MM)

OBS.	JAN.	FEB.	MAR.	APR.	MAY	JUN.	JUL.	AUG.	SEP.	OCT.	NOV.	DEC.	ANNUAL
73	26	41	89	74	87	223	424	327	195	81	28	17	1612
74	40	30	48	34	49	263	377	310	180	51	11	20	1413
75	22	6	23	31	86	292	515	440	289	91	14	1	1810
76	63	54	55	63	68	339	579	625	352	94	16	28	2336
77	33	30	49	87	210	632	834	862	536	162	12	14	3461
78	10	5	20	33	127	411	697	514	321	164	7	1	2310
79	34	14	30	34	61	219	342	314	157	63	1	11	1280
80	47	70	64	56	95	406	727	759	391	105	10	17	2747
81	33	31	46	20	35	202	263	279	81	58	1	41	1090
82	62	37	53	27	46	179	284	248	148	78	50	20	1232
83	38	8	49	72	165	430	572	615	282	104	18	7	2360
84	22	8	20	22	88	227	351	354	181	69	14	1	1357

AVERAGE TEMPERATURE (DEGREES CENTIGRADE)

OBS.	JAN.	FEB.	MAR.	APR.	MAY	JUN.	JUL.	AUG.	SEP.	OCT.	NOV.	DEC.	ANNUAL
73	.	.	.	.	.	.	.	.	.	.	.	.	.
74	.	.	.	.	.	.	.	.	.	.	.	.	.
75	.	.	.	.	.	.	.	.	.	.	.	.	.
76	.	.	.	.	.	.	.	.	.	.	.	.	.
77	13.3	15.7	19.6	23.7	24.8	25.6	25.6	25.5	23.1	21.4	17.5	14.1	20.8
78	.	.	.	.	.	.	.	.	.	.	.	.	.
79	.	.	.	.	.	.	.	.	.	.	.	.	.
80	.	.	.	.	.	.	.	.	.	.	.	.	.
81	11.2	13.5	17.3	22.0	25.1	24.0	23.0	22.5	21.6	18.8	15.4	12.5	18.9
82	.	.	.	.	.	.	.	.	.	.	.	.	.
83	.	.	.	.	.	.	.	.	.	.	.	.	.
84	.	.	.	.	.	.	.	.	.	.	.	.	.

CLIMATOLOGICAL DATA FOR NEPAL

GEOGRAPHY

OBS.	LOCATION	YEARS	LATITUDE	LONGITUDE	ELEVATION -M-	LIFEZONE	CCTX	ID	ONMAP
85	SUNDARIJAL	26	27.78 N	85.43 E	1575.5	.	.	84	NO
86	SURKHET	13	28.60 N	81.62 E	605.6	.	.	85	NO
87	TANSEN	12	27.87 N	83.55 E	756.5	.	.	86	YES
88	TAPLEJUNG	8	27.35 N	86.67 E	1767.5	WW	SC34	87	YES
89	TAPLETHOK	8	27.48 N	87.78 W	1371.6	SW	SW33	88	NO
90	THIRPU	11	29.32 N	81.77 E	1211.6	.	.	89	NO
91	TIMURE	10	28.28 N	85.38 E	1675.5	WM	SC54	90	NO
92	TRIBENI	21	26.93 N	87.15 E	151.8	.	.	91	NO
93	TULSI	14	27.03 N	85.92 E	302.7	.	.	92	NO
94	UDAIPUR GARHI	17	26.93 N	86.52 E	1389.6	.	.	93	NO

AVERAGE PRECIPITATION (MM)

OBS.	JAN.	FEB.	MAR.	APR.	MAY	JUN.	JUL.	AUG.	SEP.	OCT.	NOV.	DEC.	ANNUAL
85	18	19	24	75	116	344	602	680	300	53	6	0	2237
86	44	35	28	36	50	315	721	739	364	77	1	19	2429
87	35	21	39	34	60	227	438	437	192	78	0	8	1569
88	21	17	70	107	218	387	448	429	255	47	24	4	2027
89	27	25	72	127	164	408	540	535	395	69	20	7	2389
90	56	24	49	21	30	38	82	103	53	23	3	16	493
91	25	28	59	33	31	120	265	266	137	38	5	7	1014
92	20	11	24	46	109	272	501	427	281	70	8	0	1769
93	16	4	24	76	70	302	393	414	222	75	12	2	1610
94	27	10	27	29	142	346	470	498	302	115	18	7	1991

AVERAGE TEMPERATURE (DEGREES CENTIGRADE)

OBS.	JAN.	FEB.	MAR.	APR.	MAY	JUN.	JUL.	AUG.	SEP.	OCT.	NOV.	DEC.	ANNUAL
85	.	.	.	.	.	.	.	.	.	.	.	.	.
86	.	.	.	.	.	.	.	.	.	.	.	.	.
87	.	.	.	.	.	.	.	.	.	.	.	.	.
88	8.8	10.6	13.9	17.4	18.6	20.5	20.9	20.8	20.0	16.6	13.0	10.5	16.0
89	12.4	14.2	17.3	20.2	21.6	22.9	22.9	22.6	21.9	19.9	16.4	13.7	18.8
90	.	.	.	.	.	.	.	.	.	.	.	.	.
91	9.9	11.0	14.6	18.9	20.9	22.4	21.9	21.7	20.7	17.3	13.0	10.3	16.9
92	.	.	.	.	.	.	.	.	.	.	.	.	.
93	.	.	.	.	.	.	.	.	.	.	.	.	.
94	.	.	.	.	.	.	.	.	.	.	.	.	.

NICARAGUA*

Agriculture

Of 13,000,000 ha in total area, Nicaragua was estimated in 1978 to have 1,511,000 ha of arable or permanent cropland, 3,384,000 ha in permanent pasture, 6,282,000 ha in forest and woodland, and 698,000 ha in other terrain categories. In 1978, approximately 76,000 ha were irrigated. Of the 1979 population, 43.7% or 344,000 individuals of the economically active population were engaged in agriculture. The total population in 1979 was 2,649,000. According to FAO statistics for 1979, there were 280,000 horses, 44,000 mules, 7000 asses, 2,846,000 cattle, 725,000 pigs, 2000 sheep, 7000 goats, and 4,620,000 chickens.[2] Agricultural production is summarized in Table 1. Conventional and energetic yield data on most of these will be found in Appendixes I and II.

Climate

Data for Wernstedt[3] are presented in the Climatological Data Table (see Introduction for explanation). Based on the means for data points cited in Wernstedt, the mean annual precipitation is 18.4 dm (mean of 35 sites) and the mean annual temperature 24.4°C (mean of 11 sites). The averages would suggest that much of the country could be classified as Tropical Dry Forest Life Zone.[4]

Productivity

Major vegetation types of Nicaragua are indicated in the vegetation map. Productivity potentials, assuming reasonable management, are

Vegetation type	MT/ha/yr
Tree crops	12—36
Field crops	6—12
Pasture	3—12
Brush	6—12
Forest	18—36
Dense woodland	12—24
Open woodland	18—24
Open woodland	12—24
Savanna	6—12
Marsh and swamp	6—36

Economics

Food consumption per capita was 2446 cal/day, energy consumption 540 kg coal equivalent per capita in 1977. In 1970, the caloric supply was 106.0% of requirements. The total population growth rate was 3.0% in 1970. Population density was 15 persons per square kilometer in 1970, with 47.2% of the total population urban. Life expectancy at birth of those born in 1970 was 50.4 years. In 1977, with GNP at $870 per capita, agriculture accounted for 22.8% of GDP, manufacturing for 19.7%, other industry for 6.3%, and services for 51.2%. In 1977, food and beverage constituted 56.0% of exports, nonfood agriculture for 26.1%, fuels, minerals, and metal for 1.2%, machines for 0.6%, and other manufacturing for 16.0% of total merchandise exports. Food and beverage constituted 7.7% of imports, nonfood agriculture 0.7%, fuel and lubricants 13.8%, nonfuel minerals and metals 1.1%, machinery and equipment 31.0%, and other manufactured goods constituted 45.6% of total merchandise imports.[10]

* Maps and tables follow text.

Table 1
AGRICULTURAL PRODUCTION IN NICARAGUA

Crop	Area planted (ha)	Yield (kg/ha)	Total production (MT)
Rice paddy	20,000	1,050	21,000
Corn	180,000	931	168,000
Sorghum	55,000	1,058	58,000
Potatoes		4,250	2,000
Cassava	7,000	4,046	26,000
Dry beans	69,000	781	54,000
Peanuts in shell	5,000	1,696	8,000
Sesame seed	8,000	675	6,000
Seed cotton	173,000	1,788	310,000
Cottonseed	185,000		
Coconuts	1,000		
Cabbages	6,000	1,656	11,000
Tomatoes	1,000	41,311	25,000
Onions	1,000	4,000	6,000
Sugar cane	41,000	60,976	2,500,000
Sugar (centrifugal raw)			223,000
Sugar (noncentrifugal)			10,000
Oranges			57,000
Pineapples			36,000
Bananas			160,000
Plantains			83,000
Green coffee	85,000	621	53,000
Cocoa beans	4,000	100	
Tobacco leaves	2,000	1,671	3,000
Cotton lint			109,000

From FAO 1979 Food Production Yearbook, Vol. 33, Food and Agriculture Organization, Rome, 1980.

Energetics

Nicaragua has a hydroelectric potential of 3600 MW. Importing more than 75% of its commercial energy demands as oil, Nicaragua is not expected to be experiencing fuelwood problems by the year 2000.[10]

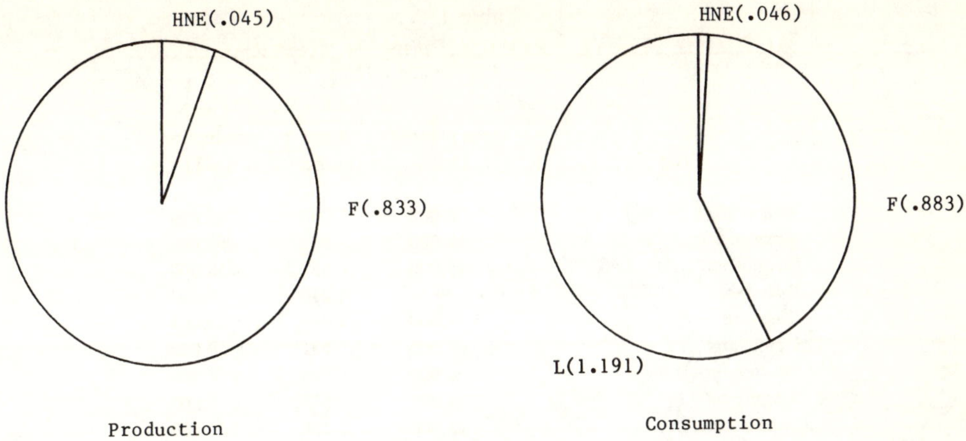

1978 Energy Production and Consumption in Nicaragua; fuel types listed in million metric tons coal equivalents (MMTCE). 1 MMTCE = 27.78 × 10^{12} Btu. L = liquid fuel, S = solid fuel, HNE = hydro, nuclear, and electric fuel. F = firewood, NG = natural gas. (Data from United Nations, World Energy Supplies, 1973–1978, Statistical Papers Series J, 22, Department of International Economic and Social Affairs, Statistical Office, New York, 1979.)

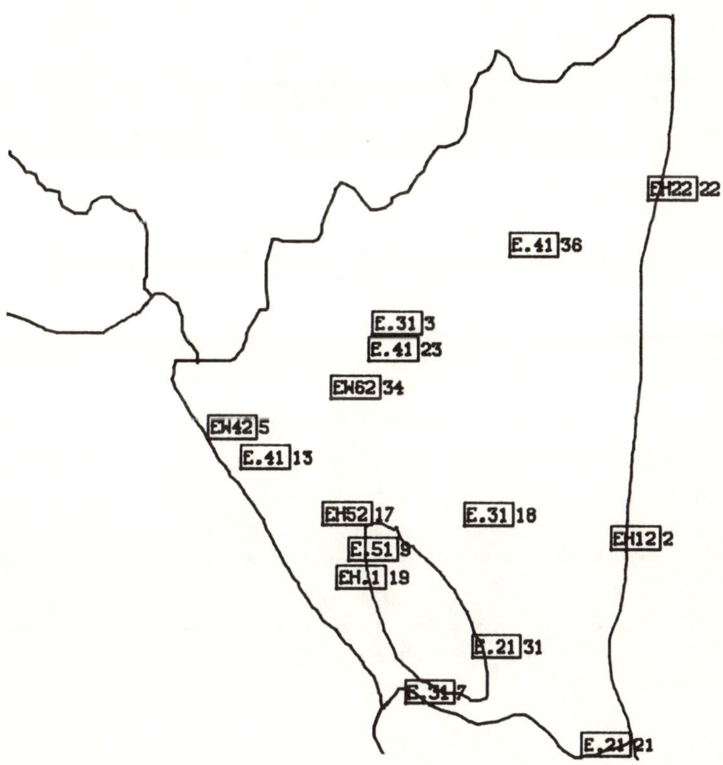

Selected weather stations for Nicaragua.

NICARAGUA VEGETATION

Cultivated area--tree crops; primarily coffee

Cultivated area--field crops; most continuous areas in the west; largely intermixed with pasture, brush or forest in the east

Dense woodland; mostly broadleaf evergreen, with some pine in northern part

Open woodland; mostly deciduous, with some evergreen; also small areas of brush, pasture and cropland

Savanna; intermixed with cultivated plots, forest and brush in the west, and with scattered clusters of pine in the northeast

Marsh and swamp

CLIMATOLOGICAL DATA FOR NICARAGUA

GEOGRAPHY

OBS.	LOCATION	YEARS	LATITUDE	LONGITUDE	ELEVATION -M-	LIFEZONE	CCTX	ID	ONMAP
1	APANAS	.	13.17 N	86.00 W	.	.	.	1	NO
2	BLUEFIELDS	14	12.00 N	83.72 W	8.5	TW	EH12	2	YES
3	CARATERA	10	13.22 N	85.75 W	.	.	.	3	YES
4	CARDENAS	25	11.20 N	85.52 W	.	.	.	4	NO
5	CHINANDEGA	14	12.62 N	87.15 W	.	TD	EW42	5	YES
6	CIUDAD RIVAS	26	11.43 N	85.85 W	.	.	.	6	NO
7	COLON	23	11.12 N	85.47 W	.	.	.	7	YES
8	CONDEGA	.	13.35 N	86.42 W	.	.	.	8	NO
9	CUIDAD GRANADA	79	11.93 N	85.95 W	.	.	.	9	YES
10	ESTELI	12	13.08 N	86.38 W	.	.	.	10	NO
11	JINOTEGA	10	13.10 N	86.00 W	.	.	.	11	NO
12	JUIGALPA	24	12.12 N	85.37 W	.	TD	EH52	12	NO

AVERAGE PRECIPITATION (MM)

OBS.	JAN.	FEB.	MAR.	APR.	MAY	JUN.	JUL.	AUG.	SEP.	OCT.	NOV.	DEC.	ANNUAL
1	.	.	.	.	.	.	.	.	.	.	.	.	.
2	253	104	71	98	349	506	801	619	317	335	363	359	4175
3	111	49	34	37	225	354	420	247	265	287	167	119	2315
4	81	34	28	60	204	388	341	296	295	326	195	107	2354
5	1	1	5	3	204	314	187	225	434	515	95	4	1988
6	11	5	4	5	193	294	176	176	296	326	78	23	1587
7	80	47	17	19	186	338	327	257	269	299	191	154	2183
8	.	.	.	.	.	.	.	.	.	.	.	.	.
9	5	2	6	21	153	286	163	171	271	289	76	23	1467
10	2	3	7	25	108	198	69	79	184	222	36	23	956
11	39	23	11	11	140	185	148	123	164	218	57	44	1163
12	10	4	2	8	180	234	151	157	301	266	71	26	1410

AVERAGE TEMPERATURE (DEGREES CENTIGRADE)

OBS.	JAN.	FEB.	MAR.	APR.	MAY	JUN.	JUL.	AUG.	SEP.	OCT.	NOV.	DEC.	ANNUAL
1	17.1	17.5	18.2	18.5	19.7	19.5	19.2	19.5	19.5	19.2	18.2	17.7	18.7
2	25.4	26.3	28.3	28.8	26.4	26.3	25.2	26.1	26.2	26.4	25.3	24.7	26.4
3	.	.	.	.	.	.	.	.	.	.	.	.	.
4	.	.	.	.	.	.	.	.	.	.	.	.	.
5	23.2	23.6	25.5	26.9	26.1	25.8	26.3	26.1	25.6	25.4	24.3	23.3	25.2
6	.	.	.	.	.	.	.	.	.	.	.	.	.
7	.	.	.	.	.	.	.	.	.	.	.	.	.
8	20.2	20.3	20.6	20.6	23.4	22.0	21.9	21.5	21.5	21.1	20.3	20.2	21.1
9	.	.	.	.	.	.	.	.	.	.	.	.	.
10	.	.	.	.	.	.	.	.	.	.	.	.	.
11	.	.	.	.	.	.	.	.	.	.	.	.	.
12	25.3	26.4	27.4	27.7	28.0	26.7	25.7	26.0	25.6	26.0	25.6	25.1	26.3

CLIMATOLOGICAL DATA FOR NICARAGUA

GEOGRAPHY

OBS.	LOCATION	YEARS	LATITUDE	LONGITUDE	ELEVATION -M-	LIFEZONE	CCTX	ID	ONMAP
13	LEON	13	12.45 N	86.87 W	109.7	.	.	13	YES
14	LOVAGO	10	11.97 N	85.20 W	.	.	.	14	NO
15	MANAGUA	28	12.15 N	86.28 E	.	TD	EH52	15	NO
16	MATAGALPA	10	12.88 N	85.95 W	.	.	.	16	NO
17	MERCEDES (LAS)	14	12.13 N	86.17 W	55.8	TD	EH52	17	YES
18	MUHAN	10	12.13 N	84.97 W	.	.	.	18	YES
19	NANDAIME	.	11.77 N	86.05 W	.	.	.	19	YES
20	PANALOYA	25	12.07 N	85.93 W	.	.	.	20	NO
21	PILARES (LOS)	24	10.83 N	83.97 W	.	.	.	21	YES
22	PUERTO CABEZAS	11	14.00 N	83.40 W	13.7	TM	EH22	22	YES
23	REINA (LA)	10	13.07 N	85.78 W	.	.	.	23	YES
24	ROBLES (LOS)	10	13.13 N	86.03 W	.	SM	EW52	24	NO

AVERAGE PRECIPITATION (MM)

OBS.	JAN.	FEB.	MAR.	APR.	MAY	JUN.	JUL.	AUG.	SEP.	OCT.	NOV.	DEC.	ANNUAL
13	0	0	0	19	145	320	132	153	226	400	111	86	1592
14	28	13	3	7	118	273	168	146	212	236	92	30	1326
15	3	1	3	11	147	211	136	110	216	292	44	10	1185
16	31	13	13	18	136	250	182	155	244	229	68	21	1360
17	5	2	3	5	93	260	134	141	191	228	56	10	1128
18	61	53	21	27	168	316	290	291	239	268	222	107	2063
19	.	.	.	.	.	.	.	.	.	.	.	.	.
20	9	1	3	6	112	186	126	143	219	233	48	23	1109
21	237	143	63	89	247	436	429	373	335	264	271	327	3214
22	156	90	63	39	173	413	306	336	271	303	237	170	2557
23	43	25	31	20	199	242	196	174	204	253	94	37	1518
24	46	29	11	20	128	185	198	157	180	168	67	48	1237

AVERAGE TEMPERATURE (DEGREES CENTIGRADE)

OBS.	JAN.	FEB.	MAR.	APR.	MAY	JUN.	JUL.	AUG.	SEP.	OCT.	NOV.	DEC.	ANNUAL
13	.	.	.	.	.	.	.	.	.	.	.	.	.
14	.	.	.	.	.	.	.	.	.	.	.	.	.
15	26.3	27.2	28.6	29.3	29.4	27.2	26.9	27.2	26.9	26.5	26.3	26.1	27.3
16	.	.	.	.	.	.	.	.	.	.	.	.	.
17	26.0	26.8	28.0	28.9	29.0	26.9	26.5	26.7	26.6	26.2	25.9	25.7	27.0
18	.	.	.	.	.	.	.	.	.	.	.	.	.
19	27.0	27.3	27.4	28.5	29.1	27.5	27.5	27.4	27.4	27.1	26.9	26.7	27.4
20	.	.	.	.	.	.	.	.	.	.	.	.	.
21	.	.	.	.	.	.	.	.	.	.	.	.	.
22	24.9	26.3	26.2	27.8	27.3	27.6	27.5	27.1	27.0	26.3	25.6	25.7	26.9
23	.	.	.	.	.	.	.	.	.	.	.	.	.
24	16.9	17.5	18.3	19.0	20.0	19.7	19.4	19.8	20.0	19.8	18.5	17.8	18.9

CLIMATOLOGICAL DATA FOR NICARAGUA

GEOGRAPHY

OBS.	LOCATION	YEARS	LATITUDE	LONGITUDE	ELEVATION -M-	LIFEZONE	CCTX	ID	ONMAP
25	SAN ANTONIO	56	12.58 N	87.08 W	.	.	.	25	NO
26	SAN CARLOS	28	11.12 N	84.78 W	.	.	.	26	NO
27	SAN FRANCISCO	10	12.97 N	85.88 W	.	.	.	27	NO
28	SAN JORGE	26	11.45 N	85.80 W	.	.	.	28	NO
29	SAN JUAN DEL NORTE	24	10.93 N	83.70 W	.	.	.	29	NO
30	SAN MARCOS	30	11.83 N	86.20 W	.	.	.	30	NO
31	SAN MIGUELITO	25	11.38 N	84.90 W	.	.	.	31	YES
32	SAN RAFAEL	10	13.18 N	86.10 W	.	.	.	32	NO
33	SAN UBALDO	25	11.83 N	85.33 W	.	.	.	33	NO
34	SEBACO	10	12.85 N	86.10 W	.	SD	EW62	34	YES
35	SERENA (LA)	10	13.17 N	86.38 W	.	.	.	35	NO
36	SIUNA	23	13.67 N	84.58 W	152.4	.	.	36	YES

AVERAGE PRECIPITATION (MM)

OBS.	JAN.	FEB.	MAR.	APR.	MAY	JUN.	JUL.	AUG.	SEP.	OCT.	NOV.	DEC.	ANNUAL
25	1	1	3	24	247	339	194	265	406	513	79	7	2076
26	78	33	16	19	161	282	277	250	221	231	163	126	1857
27	78	67	36	27	201	281	296	211	284	283	96	60	1920
28	10	5	1	2	134	263	146	136	238	342	68	20	1366
29	436	286	122	171	388	606	808	580	354	546	899	787	5983
30	17	6	8	9	221	320	162	185	322	434	79	39	1800
31	73	31	16	19	169	419	427	376	379	357	156	108	2529
32	65	27	18	16	114	187	175	149	188	225	84	67	1315
33	18	4	2	13	154	233	190	191	282	268	91	34	1481
34	3	5	5	10	152	194	107	72	177	216	29	2	972
35	10	6	11	21	130	149	78	82	169	166	29	2	853
36	73	61	26	35	210	364	273	234	282	241	120	87	1998

AVERAGE TEMPERATURE (DEGREES CENTIGRADE)

OBS.	JAN.	FEB.	MAR.	APR.	MAY	JUN.	JUL.	AUG.	SEP.	OCT.	NOV.	DEC.	ANNUAL
25	.	.	.	.	.	.	.	.	.	.	.	.	.
26	.	.	.	.	.	.	.	.	.	.	.	.	.
27	.	.	.	.	.	.	.	.	.	.	.	.	.
28	.	.	.	.	.	.	.	.	.	.	.	.	.
29	.	.	.	.	.	.	.	.	.	.	.	.	.
30	.	.	.	.	.	.	.	.	.	.	.	.	.
31	.	.	.	.	.	.	.	.	.	.	.	.	.
32	.	.	.	.	.	.	.	.	.	.	.	.	.
33	.	.	.	.	.	.	.	.	.	.	.	.	.
34	22.2	22.9	23.9	24.7	24.9	23.3	23.1	23.7	23.9	23.3	22.6	22.3	23.4
35	.	.	.	.	.	.	.	.	.	.	.	.	.
36	.	.	.	.	.	.	.	.	.	.	.	.	.

CLIMATOLOGICAL DATA FOR NICARAGUA

GEOGRAPHY

OBS.	LOCATION	YEARS	LATITUDE	LONGITUDE	ELEVATION -M-	LIFEZONE	CCTX	ID	ONMAP
37	TIPITAPA	19	12.20 N	86.10 W	.	.	.	37	NO

AVERAGE PRECIPITATION (MM)

OBS.	JAN.	FEB.	MAR.	APR.	MAY	JUN.	JUL.	AUG.	SEP.	OCT.	NOV.	DEC.	ANNUAL
37	4	1	4	4	118	186	127	108	200	230	59	13	1054

AVERAGE TEMPERATURE (DEGREES CENTIGRADE)

OBS.	JAN.	FEB.	MAR.	APR.	MAY	JUN.	JUL.	AUG.	SEP.	OCT.	NOV.	DEC.	ANNUAL
37	.	.	.	.	.	.	.	.	.	.	.	.	.

NIGER*

Agriculture

Of 126,700,000 ha in total area, Niger was estimated in 1978 to have 3,112,000 ha of arable or permanent cropland, 9,300,000 ha in permanent pasture, 11,000,000 ha in forest and woodland, and 103,258,000 ha in other terrain categories. In 1978, approximately 34,000 ha were irrigated. Of the 1979 population, 88.5% or 1,419,000 individuals of the economically active population were engaged in agriculture. The total population in 1979 was 5,150,000. According to FAO statistics for 1979, there were 230,000 horses, 440,000 asses, 2,995,000 cattle, 330,000 camels, 29,000 pigs, 2,500,000 sheep, 6,400,000 goats, and 7,600,000 chickens.[2] Agricultural production is summarized in Table 1. Conventional and energetic yield data on most of these will be found in Appendixes I and II.

Climate

Data from Wernstedt[3] are presented in the Climatological Data Table (see Introduction for explanation). Based on the means for data points cited in Wernstedt, the mean annual precipitation is 4.7 dm (mean of 37 sites) and the mean annual temperature 28.0°C (mean of 10 sites). The averages would suggest that much of the country could be classified as Tropical Thorn Woodland Life Zone.[4]

Productivity

The vegetation map shows the isohyets for Niger. For a tropical climate, I often equate the annual productivity with the decimeters of precipitation or roughly 0 to 4 in. (0 to 1 dm), 0 to 1 MT/ha/yr; 1 to 2 dm, 1 to 2 MT/ha/yr; 2 to 4 dm, 2 to 4 MT/ha/yr; 4 to 6 dm, 4 to 6 MT/ha/yr; 6 to 8 dm, 6 to 8 MT/ha/yr. Because of the extremely hot temperatures, I suggest that the lower ends of these ranges prevail. The vegetation map also shows the distribution of some major crops of Niger.

Economics

Food consumption per capita was 2139 cal/day, energy consumption 37 kg coal equivalent per capita in 1977. In 1970, the caloric supply was 83.0% of requirements. The total population growth rate was 3.3% in 1970. Population density was 3 persons per square kilometer in 1970, with 8.4% of the total population urban. Life expectancy at birth of those born in 1970 was 38.5 years. In 1977, with GNP at $190 per capita, agriculture accounted for 43.3% of GDP, manufacturing for 10.7%, other industry for 14.1%, and services for 32.0%. In 1970, food and beverage constituted 91.3% of exports, nonfood agriculture for 4.8%, fuels, minerals, and metal for 0.2%, machines for 1.4%, and other manufacturing for 2.4% of total merchandise exports. Food and beverage constituted 14.2% of imports, nonfood agriculture 2.3%, fuel and lubricants 4.0%, nonfuel minerals and metals 1.9%, machinery and equipment 26.5%, and other manufactured goods constituted 51.2% of total merchandise imports.[10]

Energetics

Importing more than 75% of its commercial energy demands as oil, Niger is expected to be experiencing fuelwood problems by the year 2000. Oil has been discovered in this oil-importing nation.[10]

Table 1
AGRICULTURAL PRODUCTION IN NIGER

Crop	Area planted (ha)	Yield (kg/ha)	Total production (MT)
Wheat	2,000	1,056	2,000
Rice paddy	25,000	960	24,000
Corn	7,000	1,143	8,000
Millet	2,800,000	445	1,246,000
Sorghum	810,000	427	346,000
Sweet potatoes	4,000	7,500	30,000
Cassava	25,000	8,400	210,000
Seed cotton	9,000	444	4,000
Cottonseed			4,000
Tomatoes	1,000	14,444	13,000
Dry onions	2,000	27,083	65,000
Dates			6,000
Sugar cane	4,000	38,750	155,000
Tobacco leaves	1,000	477	1,000
Cotton lint			2,000

From FAO 1979 Production Yearbook, Vol. 33, Food and Agriculture Organization, Rome, 1980.

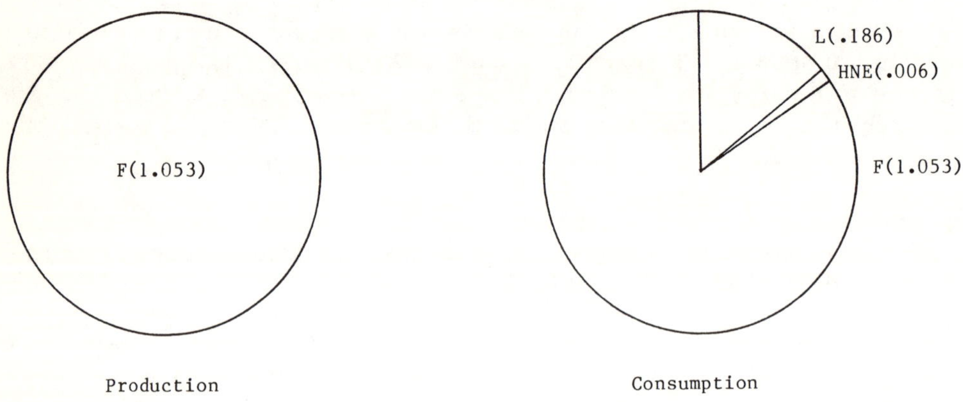

1978 Energy production and consumption in Niger; fuel types listed in million metric tons coal equivalents (MMTCE). 1 MMTCE = 27.78×10^{12} Btu. L = liquid fuel, S = solid fuel, HNE = hydro, nuclear, and electric fuel, F = firewood, NG = natural gas. (Data from United Nations, World Energy Supplies, 1973—1978, Statistical Papers Series J, 22, Department of International Economic and Social Affairs, Statistical Office, New York, 1979.)

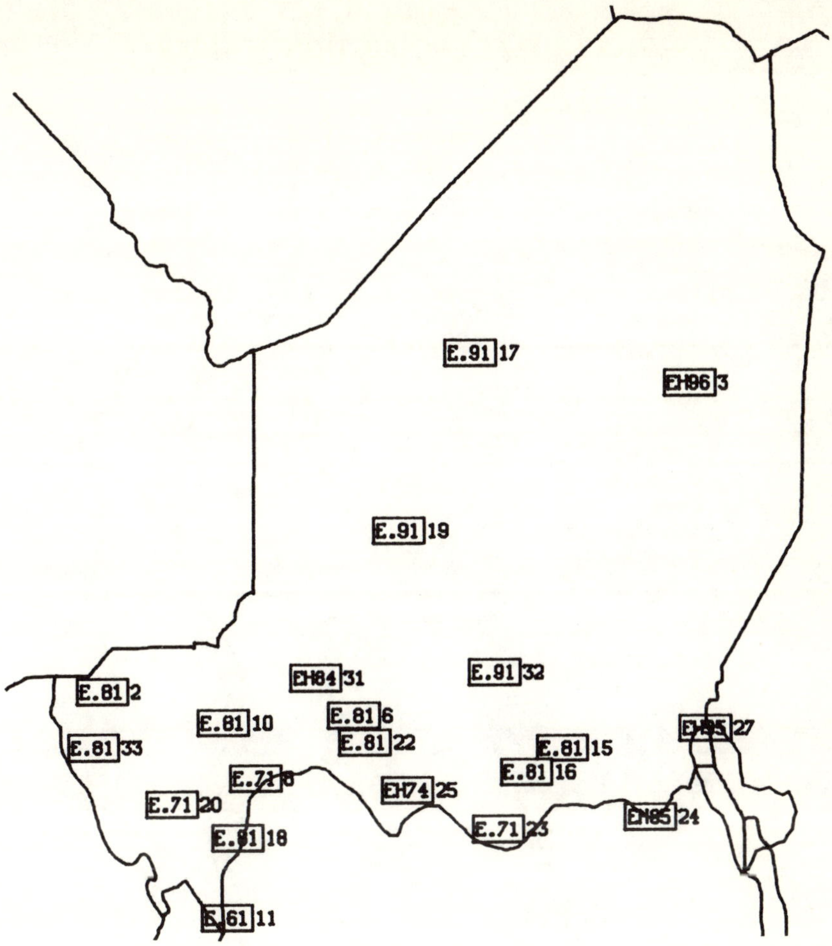

Selected weather stations for Niger.

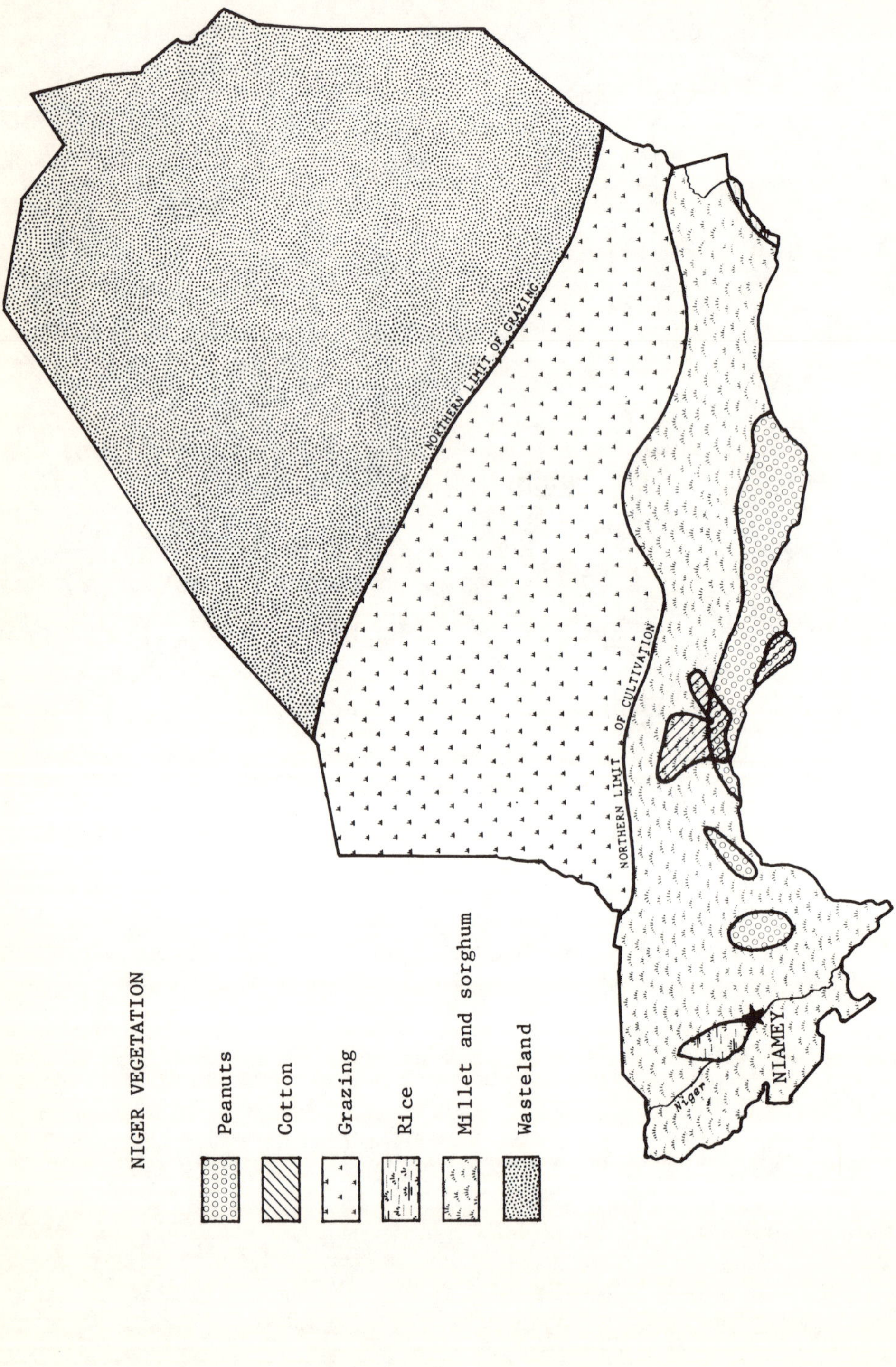
NIGER VEGETATION
Peanuts
Cotton
Grazing
Rice
Millet and sorghum
Wasteland
NORTHERN LIMIT OF GRAZING
NORTHERN LIMIT OF CULTIVATION
NIAMEY
Niger

CLIMATOLOGICAL DATA FOR NIGER

GEOGRAPHY

OBS.	LOCATION	YEARS	LATITUDE	LONGITUDE	ELEVATION -M-	LIFEZONE	CCTX	ID	ONMAP
1	AGADES	35	16.98 N	7.98 E	497.7	TX	EH95	1	NO
2	AYOROU	10	14.73 N	0.92 E	.	.	.	2	YES
3	BILMA	34	18.68 N	12.92 E	359.1	TO	EH96	3	YES
4	BIRNI N'GAOURE	10	13.08 N	2.90 E	.	.	.	4	NO
5	BIRNI-N'KONI	22	13.80 N	5.25 E	271.9	TV	EH74	5	NO
6	BOUZA	10	14.42 N	6.03 E	.	.	.	6	YES
7	DIFFA	10	13.32 N	12.62 E	.	.	.	7	NO
8	DOGONDOUTCHI	26	13.63 N	4.03 E	229.8	.	.	8	YES
9	DOSSO	19	13.03 N	3.18 E	199.9	.	.	9	NO
10	FILINGUE	18	14.33 N	3.38 E	299.9	.	.	10	YES
11	GAYA	18	11.87 N	3.48 E	160.0	.	.	11	YES
12	GAZAOUA	10	13.53 N	7.92 E	.	.	.	12	NO

AVERAGE PRECIPITATION (MM)

OBS.	JAN.	FEB.	MAR.	APR.	MAY	JUN.	JUL.	AUG.	SEP.	OCT.	NOV.	DEC.	ANNUAL
1	0	0	0	1	6	10	43	89	15	1	0	0	165
2	0	0	1	5	9	48	102	163	38	10	4	0	380
3	0	0	0	0	1	1	3	12	4	2	0	0	23
4	0	0	3	12	40	98	192	216	105	8	4	0	678
5	0	0	0	3	26	73	141	235	115	9	0	0	602
6	0	0	0	3	21	43	144	192	61	16	1	0	481
7	0	0	0	1	6	25	86	172	46	10	1	0	347
8	0	1	3	8	31	73	151	235	101	14	1	0	617
9	0	1	1	17	42	104	147	254	126	10	0	0	702
10	0	1	0	4	20	48	109	194	79	5	1	1	459
11	0	0	6	16	69	113	186	297	146	17	3	0	851
12	0	0	0	0	31	85	181	266	114	4	0	0	681

AVERAGE TEMPERATURE (DEGREES CENTIGRADE)

OBS.	JAN.	FEB.	MAR.	APR.	MAY	JUN.	JUL.	AUG.	SEP.	OCT.	NOV.	DEC.	ANNUAL
1	20.1	22.8	27.1	30.6	33.6	33.1	31.8	30.3	30.9	29.5	24.8	20.5	27.9
2	.	.	.	.	.	.	.	.	.	.	.	.	.
3	17.2	20.2	24.4	28.8	32.0	33.2	33.0	32.8	31.2	27.7	23.0	18.0	26.8
4	.	.	.	.	.	.	.	.	.	.	.	.	.
5	24.2	26.7	30.4	32.4	32.9	31.2	28.2	26.9	27.8	29.5	27.4	24.5	28.5
6	.	.	.	.	.	.	.	.	.	.	.	.	.
7	.	.	.	.	.	.	.	.	.	.	.	.	.
8	.	.	.	.	.	.	.	.	.	.	.	.	.
9	.	.	.	.	.	.	.	.	.	.	.	.	.
10	.	.	.	.	.	.	.	.	.	.	.	.	.
11	.	.	.	.	.	.	.	.	.	.	.	.	.
12	.	.	.	.	.	.	.	.	.	.	.	.	.

CLIMATOLOGICAL DATA FOR NIGER

GEOGRAPHY

OBS.	LOCATION	YEARS	LATITUDE	LONGITUDE	ELEVATION -M-	LIFEZONE	CCTX	ID	ONMAP
13	GOTHEYE	10	13.87 N	1.57 E	.	.	.	13	NO
14	GOUDOUMARIA	10	13.70 N	11.17 E	.	.	.	14	NO
15	GOURE	13	14.00 N	10.23 E	449.9	.	.	15	YES
16	GUIDIMOUNI	10	13.70 N	9.50 E	.	.	.	16	YES
17	IFEROUANE	20	19.07 N	8.40 E	.	.	.	17	YES
18	ILLELA	10	12.88 N	3.68 E	.	.	.	18	YES
19	INGALL	10	16.78 N	6.93 E	.	.	.	19	YES
20	KODO	10	13.30 N	2.35 E	210.0	.	.	20	YES
21	KOLO	10	12.93 N	1.95 E	.	.	.	21	NO
22	MADAOUA	13	14.08 N	6.25 E	329.8	.	.	22	YES
23	MAGARIA	11	12.98 N	8.93 E	360.0	.	.	23	YES
24	MAINE-SOROA	20	13.13 N	11.98 E	338.9	TT	EH85	24	YES

AVERAGE PRECIPITATION (MM)

OBS.	JAN.	FEB.	MAR.	APR.	MAY	JUN.	JUL.	AUG.	SEP.	OCT.	NOV.	DEC.	ANNUAL
13	0	1	3	0	21	7	138	191	71	12	1	0	445
14	0	0	0	0	7	30	109	180	82	6	1	1	416
15	0	0	0	1	9	24	135	171	52	3	0	0	394
16	0	0	0	0	11	35	153	211	68	7	0	0	485
17	0	0	0	3	6	6	11	30	7	1	0	0	64
18	0	0	0	1	1	47	164	168	62	15	1	0	459
19	0	0	0	3	6	13	74	138	26	0	0	0	260
20	0	0	0	9	32	79	141	175	85	13	1	0	535
21	0	0	0	1	44	77	132	191	91	14	0	0	550
22	0	0	1	6	16	46	127	194	84	3	0	0	477
23	0	0	0	6	36	44	183	249	94	5	0	0	617
24	0	0	0	2	19	31	107	179	65	9	0	0	412

AVERAGE TEMPERATURE (DEGREES CENTIGRADE)

OBS.	JAN.	FEB.	MAR.	APR.	MAY	JUN.	JUL.	AUG.	SEP.	OCT.	NOV.	DEC.	ANNUAL
13	.	.	.	.	.	.	.	.	.	.	.	.	.
14	.	.	.	.	.	.	.	.	.	.	.	.	.
15	.	.	.	.	.	.	.	.	.	.	.	.	.
16	.	.	.	.	.	.	.	.	.	.	.	.	.
17	.	.	.	.	.	.	.	.	.	.	.	.	.
18	.	.	.	.	.	.	.	.	.	.	.	.	.
19	.	.	.	.	.	.	.	.	.	.	.	.	.
20	.	.	.	.	.	.	.	.	.	.	.	.	.
21	.	.	.	.	.	.	.	.	.	.	.	.	.
22	.	.	.	.	.	.	.	.	.	.	.	.	.
23	.	.	.	.	.	.	.	.	.	.	.	.	.
24	22.1	24.5	28.6	31.2	32.4	31.2	28.4	26.5	28.2	28.9	26.1	22.2	27.6

CLIMATOLOGICAL DATA FOR NIGER

GEOGRAPHY

OBS.	LOCATION	YEARS	LATITUDE	LONGITUDE	ELEVATION -M-	LIFEZONE	CCTX	ID	ONMAP
25	MARADI	24	13.47 N	7.10 E	268.8	TV	EH74	25	YES
26	MYRRIAH	11	13.72 N	9.15 E	370.0	.	.	26	NO
27	N'GUIGMI	34	14.25 N	13.12 E	289.0	TX	EH95	27	YES
28	NIAMEY	34	13.52 N	2.10 E	215.8	TV	EH74	28	NO
29	OUALLAM	10	14.32 N	2.08 E	.	.	.	29	NO
30	SAY	28	13.12 N	2.32 E	229.8	.	.	30	NO
31	TAHOUA	35	14.90 N	5.25 E	385.9	TT	EH84	31	YES
32	TANOUT	13	14.97 N	8.87 E	399.9	.	.	32	YES
33	TERA	11	14.02 N	0.75 E	299.9	.	.	33	YES
34	TESSAOUA	13	13.75 N	7.98 E	209.1	.	.	34	NO
35	TILLABERY	22	14.20 N	1.45 E	209.1	TV	EH74	35	NO
36	TOUKOUNOUS	10	14.52 N	3.30 E	.	.	.	36	NO

AVERAGE PRECIPITATION (MM)

OBS.	JAN.	FEB.	MAR.	APR.	MAY	JUN.	JUL.	AUG.	SEP.	OCT.	NOV.	DEC.	ANNUAL
25	0	0	0	2	34	58	147	255	113	13	0	0	622
26	0	0	0	15	7	40	107	203	56	1	0	0	429
27	0	0	0	0	9	6	57	122	18	1	0	0	213
28	0	0	2	7	37	80	135	219	93	20	1	0	594
29	0	0	4	1	41	51	161	198	70	11	0	0	537
30	0	1	2	14	36	102	164	233	101	20	1	0	672
31	1	0	0	3	17	48	110	140	53	13	0	0	385
32	0	0	0	0	8	14	65	99	35	2	0	0	221
33	0	0	0	6	36	46	117	179	98	14	0	0	498
34	0	0	1	0	34	74	170	191	88	6	0	0	564
35	0	1	2	2	22	58	134	192	79	10	0	0	500
36	0	0	0	2	14	46	144	182	43	8	1	0	440

AVERAGE TEMPERATURE (DEGREES CENTIGRADE)

OBS.	JAN.	FEB.	MAR.	APR.	MAY	JUN.	JUL.	AUG.	SEP.	OCT.	NOV.	DEC.	ANNUAL
25	22.2	24.6	28.9	31.2	31.8	30.1	27.7	26.1	27.1	27.6	25.4	22.1	27.1
26	.	.	.	.	.	.	.	.	.	.	.	.	.
27	21.2	23.1	27.3	29.4	31.6	31.3	30.1	28.3	29.0	28.5	25.1	21.2	27.2
28	24.3	26.8	30.8	33.6	33.7	31.5	29.0	27.3	28.4	30.3	28.0	24.7	29.0
29	.	.	.	.	.	.	.	.	.	.	.	.	.
30	.	.	.	.	.	.	.	.	.	.	.	.	.
31	23.6	26.1	29.7	32.5	33.5	31.9	29.0	27.1	28.6	29.7	27.1	23.6	28.5
32	.	.	.	.	.	.	.	.	.	.	.	.	.
33	.	.	.	.	.	.	.	.	.	.	.	.	.
34	.	.	.	.	.	.	.	.	.	.	.	.	.
35	25.0	27.6	30.9	33.3	34.4	32.5	29.7	27.8	29.4	30.8	28.5	25.4	29.6
36	.	.	.	.	.	.	.	.	.	.	.	.	.

CLIMATOLOGICAL DATA FOR NIGER

GEOGRAPHY

OBS.	LOCATION	YEARS	LATITUDE	LONGITUDE	ELEVATION -M-	LIFEZONE	CCTX	ID	ONMAP
37	ZINDER	35	13.80 N	9.00 E	477.0	TV	EH75	37	NO

AVERAGE PRECIPITATION (MM)

OBS.	JAN.	FEB.	MAR.	APR.	MAY	JUN.	JUL.	AUG.	SEP.	OCT.	NOV.	DEC.	ANNUAL
37	0	0	0	1	25	51	159	217	70	7	0	0	530

AVERAGE TEMPERATURE (DEGREES CENTIGRADE)

OBS.	JAN.	FEB.	MAR.	APR.	MAY	JUN.	JUL.	AUG.	SEP.	OCT.	NOV.	DEC.	ANNUAL
37	22.7	25.5	29.5	32.3	32.7	31.2	27.8	26.4	28.0	29.6	26.8	22.9	27.9

NIGERIA*

Agriculture

Of 92,377,000 ha in total area, Nigeria was estimated in 1978 to have 23,990,000 ha of arable permanent cropland, 20,850,000 ha in permanent pasture, 31,069,000 ha in forest and woodland, and 15,168,000 ha in other terrain categories. In 1978, approximately 20,000 ha were irrigated. Of the 1979 population, 54.2% or 15,354,000 individuals of the economically active population were engaged in agriculture. The total population in 1979 was 74,595,000. According to FAO statistics for 1979, there were 250,000 horses, 700,000 asses, 12,000,000 cattle, 17,000 camels, 1,100,000 pigs, 8,500,000 sheep, 24,500,000 goats, 2,200,000 chickens, and 212,000 ducks.[2] Agricultural production is summarized in Table 1. Conventional and energetic yield data on most of these will be found in Appendixes I and II.

Climate

Data from Wernstedt[3] are presented in the Climatological Data Table (see Introduction for explanation). Based on the means for data points cited in Wernstedt, the mean annual precipitation is 18.5 dm (mean of 99 sites) and the mean annual temperature 26.5°C (mean of 25 sites). The averages would suggest that much of the country could be classified as Tropical Dry Forest Life Zone.[4]

Productivity

In the vegetation map, various generalized vegetation types of Nigeria are depicted. Estimated productivities, assuming reasonable management are

Vegetation type	MT/ha/yr
Montane vegetation	12—24
Short-grass savanna	3—6
Sudan savanna	3—9
Tall-grass savanna	6—12
High rain forest	18—36
Fresh-water swamp	12—36
Mangrove (salt-water swamp)	12—24

Economics

Food consumption per capita was 2308 cal/day, energy consumption 107 kg coal equivalent per capita in 1977. In 1970, the caloric supply was 89.0% of requirements. The total population growth rate was 2.5% in 1970. Population density was 72 persons per square kilometer in 1970, with 16.3% of the total population urban. Life expectancy at birth of those born in 1970 was 43 years. In 1977, with GNP at $510 per capita, agriculture accounted for 33.6% of GDP, manufacturing for 8.8%, other industry for 34.5%, and services for 23.1%. In 1977, food and beverage accounted for 5.7% of exports, nonfood agriculture for 0.4%, fuels, minerals, and metal for 93.3%, and other manufacturing for 0.4% of total merchandise exports. Food and beverage constituted 13.0% of imports, nonfood agriculture 0.3%, fuel and lubricants 1.8%, nonfuel minerals and metals 1.5%, machinery and equipment 47.6%, and other manufactured goods constituted 35.8% of total merchandise imports.[10]

Energetics

Nigeria has 17,400 million b.o.e. in oil, 0.5 billion b.o.e. in heavy oil reserves, 180 million ton coal (90 technically and economically recoverable), and some tar sand.

* Maps and table follow text.

Table 1
AGRICULTURAL PRODUCTION IN NIGERIA

Crop	Area planted (ha)	Yield (kg/ha)	Total production (MT)
Wheat	14,000	1,500	21,000
Rice paddy	528,000	1,894	1,000,000
Corn	1,665,000	901	1,500,000
Millet	5,000,000	620	3,100,000
Sorghum	6,000,000	631	3,785,000
Potatoes	3,000	14,000	35,000
Sweet potatoes	18,000	12,778	230,000
Cassava	1,150,000	10,000	11,500,000
Soybeans	195,000	385	75,000
Peanuts in shell	600,000	1,035	621,000
Sesame seed	235,000	311	73,000
Seed cotton	567,000	194	110,000
Cottonseed			74,000
Coconuts			90,000
Copra			10,000
Palm kernels			14,000
Palm oil			28,000
Tomatoes	35,000	8,000	280,000
Chilies and peppers, green	68,000	9,118	620,000
Sugar cane	16,000	51,282	800,000
Sugar (centrifugal, raw)			40,000
Sugar (noncentrifugal)			28,000
Oranges			7,000
Green coffee	6,000	750	5,000
Cocoa beans	7,000	500	4,000
Tobacco leaves	30,000	370	11,000
Cotton lint			43,000
Natural rubber			60,000

From FAO 1979 Production Yearbook, Vol. 33, Food and Agriculture Organization, Rome, 1980.

Although an oil exporter, Nigeria has a large population which could absorb most of their exportable surplus in the next decade unless significant new discoveries are made or alternative sources developed. A net oil-exporting developing country, Nigeria is expected to experience major fuelwood problems by the year 2000.[10]

Soils

The soils of Nigeria are shown on the map in seven groups: (1) sandy Latosols, (2) sandy Latosols with laterite, (3) Regosols, (4) Lithosols, (5) Grumusols, (6) Low-Humic Gley and Alluvial soils, and (7) Bog and Half-bog soils of swamps and marshes.

Sandy Latosols are acid, well-drained, deep, friable red soils having sandy loam surfaces and subsoils. They cover about 20% of Nigeria, mostly the undulating lowlands in the south-central part of the country. The fertility of these soils is very low, and the small mineral reserve is quickly exhausted when the soils are cleared and cropped. Tree and fruit crops are well adapted to these soils.

Most extensive of the soils in Nigeria are the sandy Latosols with laterite, which occur mostly on rolling uplands and hills and cover about half of Nigeria. They are acid, well-drained, commonly reddish-yellow soils, many of which are severely eroded. These Latosols have sandy loam surface textures and sandy clay or clay subsoils. Most contain hard laterite in the form of concretions or as an indurated ironpan, but the

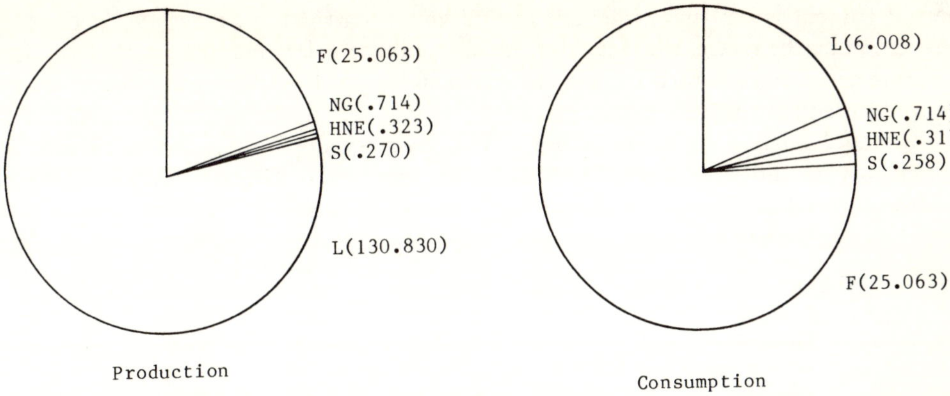

1978 Energy production and consumption in Nigeria; fuel types listed in million metric tons coal equivalents (MMTCE). 1 MMTCE = 27.78×10^{12} Btu. L = liquid fuel, S = solid fuel, HNE = hydro, nuclear, and electric fuel, F = firewood, NG = natural gas. (Data from United Nations, World Energy Supplies, 1973—1978, Statistical Papers Series J, 22, Department of International Economic and Social Affairs, Statistical Office, New York, 1979.)

laterite is generally deep enough so as not to interfere with root penetration of the crops commonly grown; in some places, however, it is at or near the surface. The fertility of these soils is moderately low, but the soils are responsive to fertilizers. Extensive areas of these soils are now in grassy and wooded savanna; they could be used more efficiently if grazing on them were controlled. The soils on the smoother areas are suitable for cultivated crops, although the production of perennial crops in the north is limited because of the pronounced dry season. Tree crops are best suited for the steepest land, and fruit crops for the less steep land. The severely eroded areas are almost worthless for agriculture.

Regosols cover about 15% of the country, mostly the sandy uplands along the northern border and in lowlands near the coast in the south. They are deep, well-drained, sandy soils that are droughty. In the north where rainfall is low, they are best suited for grazing, but in the south where rainfall is more plentiful, tree crops and fruit crops should do well.

Lithosols cover 10% of Nigeria, mostly on the Jos Plateau and the scattered hills and dissected uplands to the east. They are shallow, sandy, gravelly, and commonly stony soils; included with the Lithosols are numerous areas of bare rock. Forestry is the most practical use for these poor and generally eroded soils.

Grumusols cover less than 3% of Nigeria, mostly the nearly flat uplands near Gombe. They comprise dark gray to black clays that are calcareous in the subsoil. Though fertile, they are difficult to work because they are exceedingly sticky when wet and hard when dry. Their potential for crop production has not yet been fully realized.

Low-Humic Gley and Alluvial soils cover slightly more than 5% of Nigeria, mostly the Niger and Benue river troughs and a large stretch of upland plain near Lake Chad. They occur also along smaller streams, but cannot be shown at small scale. Both soils are dominantly acid and poorly drained; they range in texture from sands to clays. The Low-Humic Gley soils generally have a slightly heavier subsoil than surface soils, and the surface soil is moderately low in organic matter. The Alluvial soils have interbedded layers of sand and clay of varied thickness. Near Lake Chad, the Low-Humic Gley and Alluvial soils occur in close association with sandy Regosols, which occupy the higher lying areas, and dark wet clays in depressions together with Sononetz and Solonchak soils. The dark clays are somewhat like wet Grumusols, the Solonetz soils are alkaline in reaction, and the Solonchak soils have varied concentrations of salts. Where the salt content is not high, the soils are suitable for irrigation, especially in the dry season. Along the rivers, the wet clays, including those belonging to Low-Humic Gley soils, are especially adapted to rice; the better drained Alluvial soils on stream levees are well suited for garden crops.

Bog and Half-Bog soils, occurring in mangrove swamps along the coast and in marshes on the shores of Lake Chad, cover less than 3% of Nigeria. They are nonagricultural.

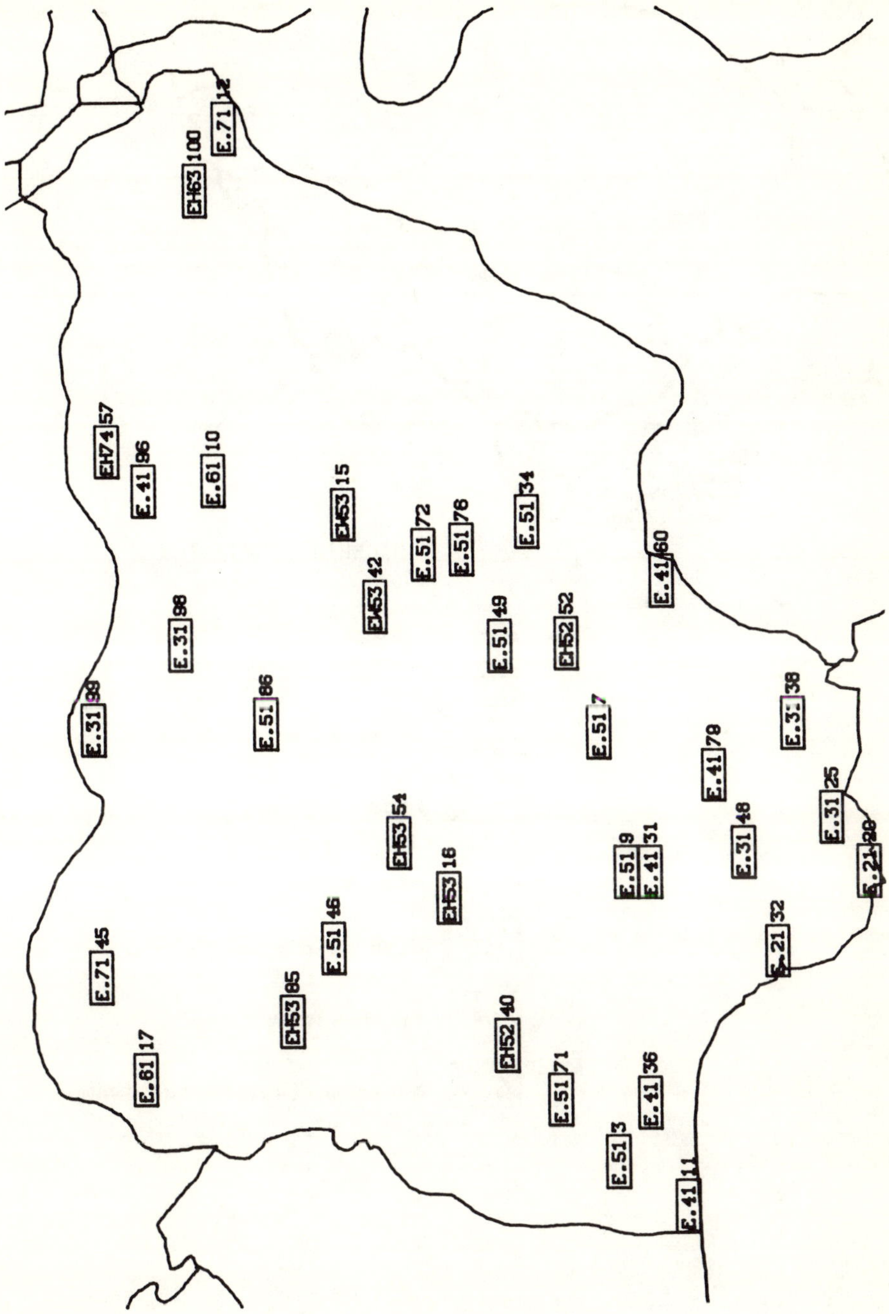

Selected weather stations for Nigeria.

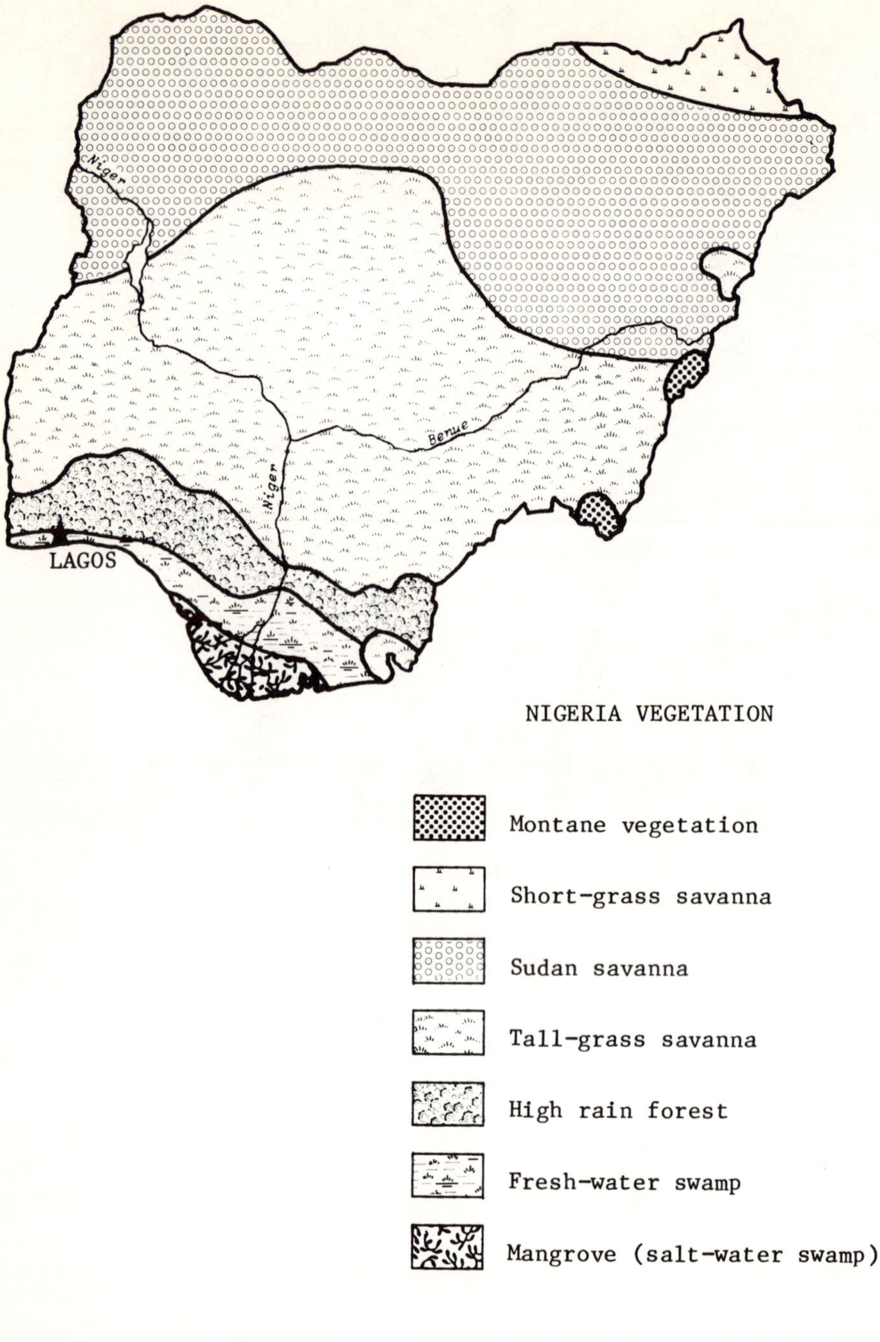
Niger
Niger
Benue
LAGOS
NIGERIA VEGETATION
Montane vegetation
Short-grass savanna
Sudan savanna
Tall-grass savanna
High rain forest
Fresh-water swamp
Mangrove (salt-water swamp)

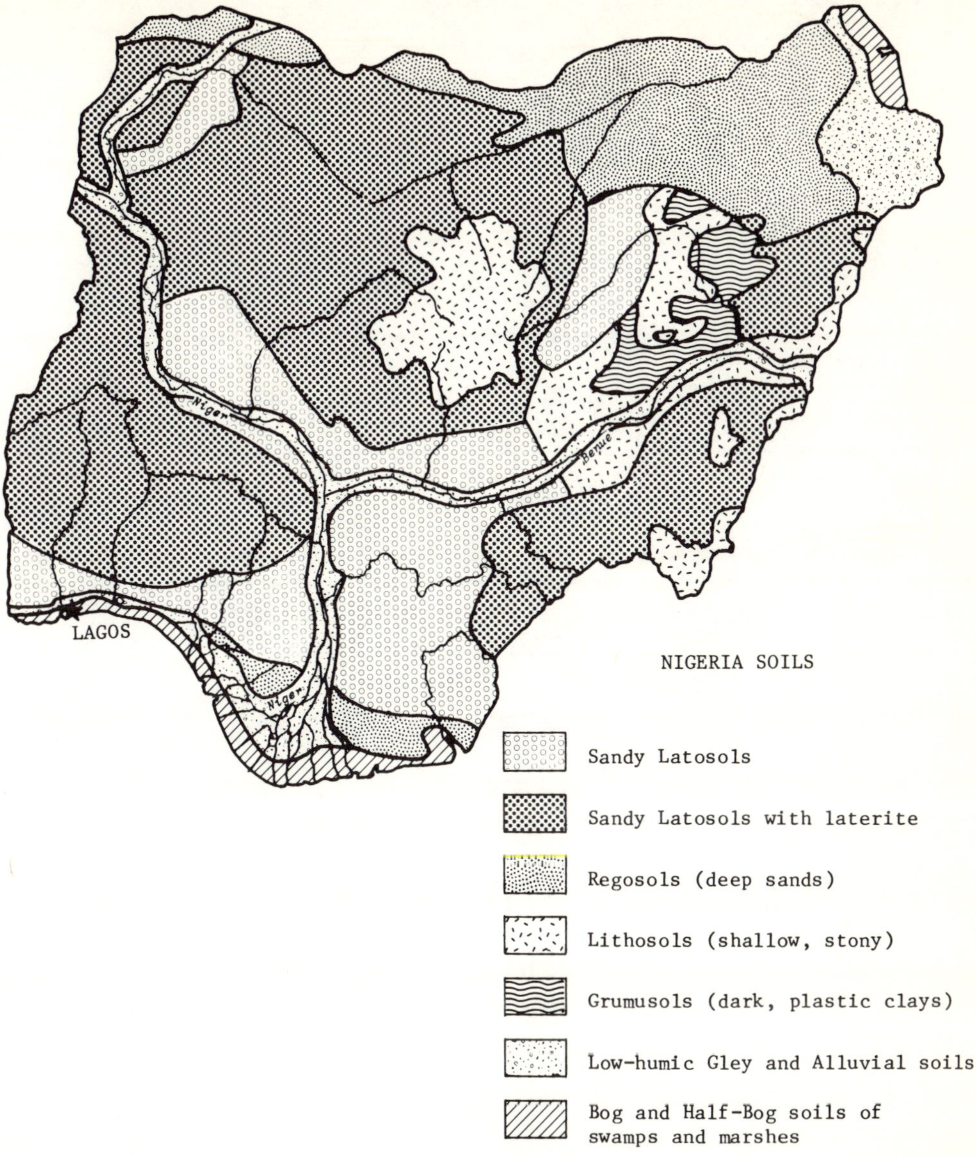

NIGERIA SOILS

CLIMATOLOGICAL DATA FOR NIGERIA

GEOGRAPHY

OBS.	LOCATION	YEARS	LATITUDE	LONGITUDE	ELEVATION -M-	LIFEZONE	CCTX	ID	ONMAP
1	ABA	25	5.12 N	7.37 E	61.0	.	.	1	NO
2	ABAK	16	4.97 N	7.80 E	30.5	.	.	2	NO
3	ABAKALAKI	44	6.32 N	8.10 E	106.7	.	.	90	NO
4	ABEOKUTA	50	7.18 N	3.33 E	67.1	.	.	3	YES
5	AFIKPO	47	5.88 N	7.93 E	118.9	.	.	4	NO
6	AGBOR	43	6.27 N	6.18 E	149.4	.	.	5	NO
7	AGEGE	42	6.65 N	3.30 E	30.5	.	.	89	NO
8	AHOADA	30	5.10 N	6.65 E	61.0	.	.	6	NO
9	ANKPA	25	7.37 N	7.63 E	274.3	.	.	7	YES
10	ASABA	27	6.20 N	6.75 E	91.4	.	.	8	NO
11	AUCHI	12	7.07 N	6.23 E	243.8	.	.	9	YES
12	AZARE	31	11.67 N	10.17 E	411.5	.	.	10	YES

AVERAGE PRECIPITATION (MM)

OBS.	JAN.	FEB.	MAR.	APR.	MAY	JUN.	JUL.	AUG.	SEP.	OCT.	NOV.	DEC.	ANNUAL
1	23	63	142	163	236	277	323	295	368	251	91	30	2266
2	13	79	150	170	231	287	335	317	427	356	112	28	2510
3	15	30	86	140	246	272	287	264	351	317	53	13	2075
4	13	28	89	140	150	188	170	84	140	132	46	13	1191
5	18	36	76	160	246	251	295	277	312	259	51	10	1991
6	15	43	119	168	211	274	279	193	325	218	53	15	1915
7	20	30	94	112	196	262	165	53	147	183	74	23	1359
8	28	76	127	173	264	317	345	300	483	317	112	25	2568
9	8	18	56	104	173	185	163	160	297	178	33	10	1384
10	13	23	61	112	203	221	234	208	302	206	36	8	1626
11	13	20	76	127	163	178	183	145	257	170	23	13	1367
12	0	0	0	8	61	86	193	287	119	18	5	0	777

AVERAGE TEMPERATURE (DEGREES CENTIGRADE)

OBS.	JAN.	FEB.	MAR.	APR.	MAY	JUN.	JUL.	AUG.	SEP.	OCT.	NOV.	DEC.	ANNUAL
1	.	.	.	.	.	.	.	.	.	.	.	.	.
2	.	.	.	.	.	.	.	.	.	.	.	.	.
3	.	.	.	.	.	.	.	.	.	.	.	.	.
4	.	.	.	.	.	.	.	.	.	.	.	.	.
5	.	.	.	.	.	.	.	.	.	.	.	.	.
6	.	.	.	.	.	.	.	.	.	.	.	.	.
7	.	.	.	.	.	.	.	.	.	.	.	.	.
8	.	.	.	.	.	.	.	.	.	.	.	.	.
9	.	.	.	.	.	.	.	.	.	.	.	.	.
10	.	.	.	.	.	.	.	.	.	.	.	.	.
11	.	.	.	.	.	.	.	.	.	.	.	.	.
12	.	.	.	.	.	.	.	.	.	.	.	.	.

CLIMATOLOGICAL DATA FOR NIGERIA

GEOGRAPHY

OBS.	LOCATION	YEARS	LATITUDE	LONGITUDE	ELEVATION -M-	LIFEZONE	CCTX	ID	ONMAP
13	BADAGRY	13	6.40 N	2.88 E	6.1	.	.	11	YES
14	BAMA	12	11.52 N	13.70 E	350.5	.	.	12	YES
15	BAMENDA	33	5.93 N	10.15 E	1722.1	.	.	13	NO
16	BANSO	24	6.20 N	10.67 E	1737.4	.	.	14	NO
17	BAUCHI	36	10.23 N	9.83 E	608.7	TD	EW53	15	YES
18	BENIN CITY	52	6.32 N	5.63 E	78.0	TM	EH32	91	NO
19	BIDA	30	9.07 N	5.98 E	184.4	TD	EH53	16	YES
20	BIRNIN KEBBI	39	12.47 N	4.18 E	274.3	.	.	17	YES
21	BONNY	43	4.47 N	7.17 E	2.1	.	.	18	NO
22	BRASS	23	4.32 N	6.23 E	3.0	.	.	20	YES
23	BUEA	28	4.15 N	9.23 E	914.4	.	.	21	NO
24	CALABAR	57	4.97 N	8.33 E	51.8	TM	EH21	22	NO

AVERAGE PRECIPITATION (MM)

OBS.	JAN.	FEB.	MAR.	APR.	MAY	JUN.	JUL.	AUG.	SEP.	OCT.	NOV.	DEC.	ANNUAL
13	23	51	112	109	284	371	188	36	97	188	61	18	1537
14	0	0	0	8	46	99	180	254	132	23	0	0	742
15	28	58	163	203	229	335	434	368	493	272	79	33	2695
16	8	41	117	130	168	193	277	274	330	236	53	8	1834
17	0	0	8	38	91	147	257	348	183	41	0	0	1113
18	20	36	99	163	211	307	323	211	312	244	74	15	2014
19	3	8	28	74	150	185	206	208	257	112	8	0	1110
20	0	0	5	13	69	109	185	267	155	18	0	0	820
21	86	127	203	254	417	765	665	427	561	462	302	99	4369
22	66	107	173	267	391	668	518	297	518	442	203	86	3736
23	30	63	160	185	244	264	472	538	505	312	107	28	2911
24	38	76	157	218	312	411	455	419	422	328	190	48	3076

AVERAGE TEMPERATURE (DEGREES CENTIGRADE)

OBS.	JAN.	FEB.	MAR.	APR.	MAY	JUN.	JUL.	AUG.	SEP.	OCT.	NOV.	DEC.	ANNUAL
13	.	.	.	.	.	.	.	.	.	.	.	.	.
14	.	.	.	.	.	.	.	.	.	.	.	.	.
15	.	.	.	.	.	.	.	.	.	.	.	.	.
16	.	.	.	.	.	.	.	.	.	.	.	.	.
17	23.0	24.9	27.7	29.7	28.6	26.4	24.6	23.8	24.3	25.7	25.1	23.6	25.7
18	26.3	27.4	27.8	27.6	26.8	25.3	24.6	24.3	25.0	25.6	26.8	26.6	26.2
19	27.7	29.6	30.6	30.9	28.7	27.1	26.1	25.6	26.0	26.8	28.2	27.3	27.9
20	.	.	.	.	.	.	.	.	.	.	.	.	.
21	.	.	.	.	.	.	.	.	.	.	.	.	.
22	.	.	.	.	.	.	.	.	.	.	.	.	.
23	.	.	.	.	.	.	.	.	.	.	.	.	.
24	26.6	27.4	27.7	27.4	27.1	26.5	25.1	25.2	25.7	26.0	26.5	26.7	26.5

CLIMATOLOGICAL DATA FOR NIGERIA

GEOGRAPHY

OBS.	LOCATION	YEARS	LATITUDE	LONGITUDE	ELEVATION -M-	LIFEZONE	CCTX	ID	ONMAP
25	DAURA	10	13.03 N	8.22 E	457.2	.	.	23	NO
26	DEBUNDSCHA	32	4.12 N	8.97 E	9.1	.	.	24	NO
27	DEGEMA	51	4.75 N	6.77 E	3.0	.	.	25	YES
28	EBUTE METTA	46	6.48 N	3.37 E	3.0	.	.	26	NO
29	EKET	11	4.63 N	7.92 E	24.4	.	.	27	NO
30	EKONA	12	4.22 N	9.35 E	365.8	.	.	28	NO
31	ENUGU	40	6.45 N	7.48 E	227.1	TD	EH42	29	NO
32	EPE	28	6.58 N	3.98 E	6.1	.	.	30	NO
33	EWU	12	6.80 N	6.23 E	182.9	.	.	31	YES
34	FORCADOS	50	5.37 N	5.43 E	3.0	.	.	32	YES
35	GUSAU	16	12.15 N	6.67 E	472.4	.	.	33	NO
36	HADEJIA	40	12.45 N	10.07 E	365.8	.	.	92	NO

AVERAGE PRECIPITATION (MM)

OBS.	JAN.	FEB.	MAR.	APR.	MAY	JUN.	JUL.	AUG.	SEP.	OCT.	NOV.	DEC.	ANNUAL
25	0	0	0	3	36	74	163	267	117	15	0	0	673
26	272	320	546	511	757	1232	1460	1372	1587	1204	658	384	10302
27	38	71	142	193	246	312	330	277	378	262	135	38	2423
28	23	43	109	132	251	406	262	63	145	180	56	25	1697
29	30	69	147	236	254	437	556	439	480	394	152	51	3246
30	23	51	163	193	224	211	409	442	419	302	104	20	2560
31	18	28	71	152	259	274	198	175	320	249	53	15	1814
32	18	38	99	117	221	338	279	84	218	175	69	20	1676
33	8	25	112	119	236	239	198	170	315	231	41	3	1697
34	48	79	170	254	384	577	688	328	587	417	152	56	3739
35	0	3	8	15	86	124	211	290	211	28	0	0	975
36	0	0	0	8	28	66	173	254	91	5	0	0	625

AVERAGE TEMPERATURE (DEGREES CENTIGRADE)

OBS.	JAN.	FEB.	MAR.	APR.	MAY	JUN.	JUL.	AUG.	SEP.	OCT.	NOV.	DEC.	ANNUAL
25	.	.	.	.	.	.	.	.	.	.	.	.	.
26	.	.	.	.	.	.	.	.	.	.	.	.	.
27	.	.	.	.	.	.	.	.	.	.	.	.	.
28	.	.	.	.	.	.	.	.	.	.	.	.	.
29	.	.	.	.	.	.	.	.	.	.	.	.	.
30	.	.	.	.	.	.	.	.	.	.	.	.	.
31	27.2	28.2	28.8	28.3	26.9	25.7	25.1	25.0	25.3	25.9	27.2	27.1	26.7
32	.	.	.	.	.	.	.	.	.	.	.	.	.
33	.	.	.	.	.	.	.	.	.	.	.	.	.
34	.	.	.	.	.	.	.	.	.	.	.	.	.
35	.	.	.	.	.	.	.	.	.	.	.	.	.
36	.	.	.	.	.	.	.	.	.	.	.	.	.

CLIMATOLOGICAL DATA FOR NIGERIA

GEOGRAPHY

OBS.	LOCATION	YEARS	LATITUDE	LONGITUDE	ELEVATION -M-	LIFEZONE	CCTX	ID	ONMAP
37	IBADAN	53	7.43 N	3.90 E	227.1	TD	EH52	88	NO
38	IBI	24	8.17 N	9.75 E	121.9	.	.	34	YES
39	IDAH	22	7.10 N	6.72 E	91.4	.	.	35	NO
40	IJEBU ODE	29	6.82 N	3.92 E	67.1	.	.	36	YES
41	IKEJA	14	6.58 N	3.33 E	38.1	TD	EH52	37	NO
42	IKOT EKPENE	50	5.18 N	7.72 E	61.0	.	.	38	YES
43	IKOYI	19	6.43 N	3.43 E	3.0	.	.	39	NO
44	ILORIN	42	8.43 N	4.50 E	366.7	TD	EH52	40	YES
45	ITU	16	5.20 N	7.98 E	45.7	.	.	41	NO
46	JOS	36	9.87 N	8.90 E	1289.3	SM	EW53	42	YES
47	KADUNA	19	10.60 N	7.45 E	645.0	TD	EW53	43	NO
48	KAFANCHAN	21	9.60 N	8.30 E	762.0	.	.	44	NO

AVERAGE PRECIPITATION (MM)

OBS.	JAN.	FEB.	MAR.	APR.	MAY	JUN.	JUL.	AUG.	SEP.	OCT.	NOV.	DEC.	ANNUAL
37	10	23	91	135	150	188	155	84	178	157	46	10	1227
38	3	5	33	89	168	190	147	150	224	142	10	0	1161
39	10	15	38	99	165	193	178	173	239	155	30	5	1300
40	18	36	84	130	208	257	251	104	201	226	79	13	1605
41	28	41	99	99	203	300	180	56	180	190	63	25	1466
42	38	58	130	193	246	282	335	307	351	262	102	33	2337
43	33	48	107	142	267	417	231	46	142	206	84	20	1742
44	10	18	63	102	170	193	142	132	251	165	28	13	1288
45	36	56	173	178	244	292	378	259	335	312	150	20	2433
46	3	8	28	84	203	226	325	279	216	43	3	3	1420
47	0	3	15	56	168	180	206	310	272	94	5	0	1308
48	5	8	36	84	155	206	259	361	333	145	10	0	1600

AVERAGE TEMPERATURE (DEGREES CENTIGRADE)

OBS.	JAN.	FEB.	MAR.	APR.	MAY	JUN.	JUL.	AUG.	SEP.	OCT.	NOV.	DEC.	ANNUAL
37	26.8	27.8	28.5	28.0	26.8	25.7	24.5	24.4	25.2	25.7	26.5	26.5	26.4
38	.	.	.	.	.	.	.	.	.	.	.	.	.
39	.	.	.	.	.	.	.	.	.	.	.	.	.
40	.	.	.	.	.	.	.	.	.	.	.	.	.
41	26.7	27.7	27.9	27.2	26.7	25.4	24.5	24.3	25.1	25.7	26.9	26.7	26.2
42	.	.	.	.	.	.	.	.	.	.	.	.	.
43	.	.	.	.	.	.	.	.	.	.	.	.	.
44	26.9	27.7	29.4	29.2	27.9	26.3	25.2	24.7	25.3	26.0	26.9	26.6	26.8
45	.	.	.	.	.	.	.	.	.	.	.	.	.
46	20.8	22.4	24.2	25.2	24.0	22.3	20.8	20.3	21.3	22.3	22.1	21.0	22.7
47	23.2	24.9	26.9	23.6	27.4	25.0	23.6	23.3	23.9	25.0	23.9	22.9	24.9
48	.	.	.	.	.	.	.	.	.	.	.	.	.

CLIMATOLOGICAL DATA FOR NIGERIA

GEOGRAPHY

OBS.	LOCATION	YEARS	LATITUDE	LONGITUDE	ELEVATION -M-	LIFEZONE	CCTX	ID	ONMAP
49	KANO	53	12.05 N	8.53 E	472.1	TV	EH64	93	NO
50	KASARAWA	25	12.95 N	5.20 E	350.8	.	.	45	YES
51	KATSINA	34	13.02 N	7.68 E	516.6	.	.	94	NO
52	KONTAGORA	33	10.37 N	5.48 E	396.2	.	.	46	YES
53	KUMBA	30	4.63 N	9.42 E	259.1	.	.	47	NO
54	KWALE	25	5.75 N	6.42 E	91.4	.	.	48	YES
55	LAFIA	26	8.48 N	8.50 E	182.9	.	.	49	YES
56	LAGOS	65	6.45 N	3.40 E	3.0	.	.	50	NO
57	LOKOJA	42	7.80 N	6.73 E	41.8	TD	EH52	51	NO
58	MAIDUGURI	42	11.85 N	13.08 E	354.2	TV	EH74	95	NO
59	MAKURDI	30	7.73 N	8.53 E	112.2	TD	EH52	52	YES
60	MAMFE	32	5.77 N	9.30 E	115.8	.	.	53	NO

AVERAGE PRECIPITATION (MM)

OBS.	JAN.	FEB.	MAR.	APR.	MAY	JUN.	JUL.	AUG.	SEP.	OCT.	NOV.	DEC.	ANNUAL
49	0	0	3	8	69	114	208	312	140	13	0	0	866
50	0	0	0	10	48	71	170	269	137	15	0	0	721
51	0	0	0	5	58	84	183	267	130	10	0	0	737
52	3	3	10	38	119	155	216	267	277	109	5	0	1201
53	25	84	185	224	259	239	295	307	302	320	140	25	2405
54	25	43	109	173	249	317	351	226	394	279	71	18	2256
55	3	5	25	71	157	163	201	249	236	142	8	0	1260
56	28	46	99	145	277	450	267	66	142	208	69	25	1821
57	5	5	48	94	155	160	173	165	211	137	13	3	1168
58	0	0	0	8	41	74	180	221	107	23	0	0	653
59	5	10	33	109	206	196	180	218	257	152	10	0	1377
60	33	79	160	206	325	437	513	465	564	452	152	38	3424

AVERAGE TEMPERATURE (DEGREES CENTIGRADE)

OBS.	JAN.	FEB.	MAR.	APR.	MAY	JUN.	JUL.	AUG.	SEP.	OCT.	NOV.	DEC.	ANNUAL
49	21.6	23.7	27.1	30.3	30.5	29.0	26.2	25.2	25.9	27.1	25.0	22.2	26.2
50	.	.	.	.	.	.	.	.	.	.	.	.	.
51	.	.	.	.	.	.	.	.	.	.	.	.	.
52	.	.	.	.	.	.	.	.	.	.	.	.	.
53	.	.	.	.	.	.	.	.	.	.	.	.	.
54	.	.	.	.	.	.	.	.	.	.	.	.	.
55	.	.	.	.	.	.	.	.	.	.	.	.	.
56	.	.	.	.	.	.	.	.	.	.	.	.	.
57	26.8	28.8	30.3	30.6	28.6	27.0	26.3	25.7	26.0	26.6	27.3	26.4	27.6
58	22.1	24.1	27.5	30.9	31.5	30.1	27.2	25.8	26.7	27.7	25.2	22.7	26.8
59	26.9	28.6	29.9	29.9	27.7	26.3	25.8	25.7	25.8	26.4	27.3	26.6	27.3
60	.	.	.	.	.	.	.	.	.	.	.	.	.

CLIMATOLOGICAL DATA FOR NIGERIA

GEOGRAPHY

OBS.	LOCATION	YEARS	LATITUDE	LONGITUDE	ELEVATION -M-	LIFEZONE	CCTX	ID	ONMAP
61	MINNA	42	9.62 N	6.53 E	258.5	TD	EH53	54	YES
62	MISSELLELE	11	4.12 N	9.45 E	76.2	.	.	55	NO
63	NARAGUTA	15	9.95 N	8.88 E	1158.2	.	.	56	NO
64	NGURU	16	12.85 N	10.47 E	335.3	TV	EH74	57	YES
65	NKWELLE	27	6.23 N	6.83 E	106.7	.	.	58	NO
66	OBIARUKU	21	5.85 N	6.15 E	61.0	.	.	59	NO
67	OBUDU	15	6.65 N	9.15 E	396.2	.	.	60	YES
68	OFFA	43	8.15 N	4.72 E	365.8	.	.	61	NO
69	OGBOMSHO	36	8.07 N	4.23 E	365.8	.	.	62	NO
70	OGOJA	32	6.65 N	8.70 E	228.6	.	.	105	NO
71	OGOJA	32	6.65 N	8.70 E	228.6	.	.	96	YES
72	OKENE	16	7.57 N	6.22 E	396.2	.	.	63	NO

AVERAGE PRECIPITATION (MM)

OBS.	JAN.	FEB.	MAR.	APR.	MAY	JUN.	JUL.	AUG.	SEP.	OCT.	NOV.	DEC.	ANNUAL
61	0	5	18	58	147	183	211	269	297	150	10	3	1351
62	13	48	119	173	185	246	577	478	315	251	104	13	2522
63	0	0	15	69	188	193	315	272	224	58	3	3	1339
64	0	0	0	3	28	61	124	236	104	10	3	0	569
65	13	30	69	152	216	231	284	211	315	257	58	10	1847
66	30	48	109	170	259	300	351	239	399	287	66	15	2273
67	18	15	69	135	251	277	196	201	333	330	63	10	1897
68	8	20	76	112	165	165	119	104	218	206	41	13	1247
69	10	23	76	112	160	160	137	107	213	180	48	13	1240
70	13	20	58	124	229	282	208	259	292	300	58	13	1857
71	13	20	58	124	229	282	208	259	292	300	58	13	1857
72	8	15	63	119	160	178	152	130	226	157	15	23	1247

AVERAGE TEMPERATURE (DEGREES CENTIGRADE)

OBS.	JAN.	FEB.	MAR.	APR.	MAY	JUN.	JUL.	AUG.	SEP.	OCT.	NOV.	DEC.	ANNUAL
61	27.1	28.9	29.8	30.3	28.3	25.8	25.1	24.7	25.1	26.0	26.6	26.7	27.1
62	.	.	.	.	.	.	.	.	.	.	.	.	.
63	.	.	.	.	.	.	.	.	.	.	.	.	.
64	21.6	23.8	27.1	30.2	30.6	30.7	28.2	26.6	27.2	27.6	25.2	22.4	26.7
65	.	.	.	.	.	.	.	.	.	.	.	.	.
66	.	.	.	.	.	.	.	.	.	.	.	.	.
67	.	.	.	.	.	.	.	.	.	.	.	.	.
68	.	.	.	.	.	.	.	.	.	.	.	.	.
69	.	.	.	.	.	.	.	.	.	.	.	.	.
70	.	.	.	.	.	.	.	.	.	.	.	.	.
71	.	.	.	.	.	.	.	.	.	.	.	.	.
72	.	.	.	.	.	.	.	.	.	.	.	.	.

CLIMATOLOGICAL DATA FOR NIGERIA

GEOGRAPHY

OBS.	LOCATION	YEARS	LATITUDE	LONGITUDE	ELEVATION -M-	LIFEZONE	CCTX	ID	ONMAP
73	OLOKEMEJI	35	7.42 N	3.53 E	100.6	.	.	64	NO
74	ONDO	49	7.10 N	4.83 E	287.1	TD	EW42	65	NO
75	ONITSHA	46	6.17 N	6.78 E	91.4	.	.	66	NO
76	OPOBO	50	4.57 N	7.55 E	3.0	.	.	67	NO
77	OSHOGBO	29	7.77 N	4.55 E	304.8	TD	EH52	68	NO
78	OWERRI	56	5.50 N	7.03 E	70.1	.	.	69	NO
79	OWO	22	7.20 N	5.58 E	304.8	.	.	70	NO
80	OYO	52	7.83 N	3.95 E	289.6	.	.	71	YES
81	PANKSHIN	26	9.33 N	9.42 E	1341.1	.	.	72	YES
82	PORT HARCOURT	44	4.85 N	7.02 E	19.5	TM	EH31	73	NO
83	POTISKUM	21	11.72 N	11.05 E	413.6	TV	EH64	74	NO
84	SAMARU	24	11.18 N	7.63 E	685.8	.	.	97	NO

AVERAGE PRECIPITATION (MM)

OBS.	JAN.	FEB.	MAR.	APR.	MAY	JUN.	JUL.	AUG.	SEP.	OCT.	NOV.	DEC.	ANNUAL
73	13	28	117	145	163	183	150	71	150	155	46	10	1229
74	10	36	109	157	170	213	236	132	236	196	74	18	1587
75	20	30	66	155	213	251	279	241	317	241	51	15	1882
76	48	84	160	198	366	554	617	511	554	366	190	48	3696
77	8	25	79	107	137	147	130	99	190	188	38	10	1158
78	20	46	124	201	274	307	356	315	429	277	74	20	2443
79	8	33	94	124	163	165	196	157	249	135	43	23	1389
80	5	25	81	124	157	160	119	84	193	157	48	10	1166
81	0	3	15	53	114	140	201	269	229	71	8	0	1102
82	30	66	130	190	244	320	328	312	409	277	140	46	2492
83	0	0	0	10	51	91	216	259	137	23	0	0	787
84	0	3	8	28	137	145	218	292	226	38	0	0	1095

AVERAGE TEMPERATURE (DEGREES CENTIGRADE)

OBS.	JAN.	FEB.	MAR.	APR.	MAY	JUN.	JUL.	AUG.	SEP.	OCT.	NOV.	DEC.	ANNUAL
73	.	.	.	.	.	.	.	.	.	.	.	.	.
74	26.1	26.9	27.6	27.1	26.2	25.3	24.0	23.7	24.5	25.2	26.5	26.5	25.8
75	.	.	.	.	.	.	.	.	.	.	.	.	.
76	.	.	.	.	.	.	.	.	.	.	.	.	.
77	26.3	27.3	28.4	28.0	27.0	26.0	24.4	23.9	25.1	25.9	26.6	26.3	26.3
78	.	.	.	.	.	.	.	.	.	.	.	.	.
79	.	.	.	.	.	.	.	.	.	.	.	.	.
80	.	.	.	.	.	.	.	.	.	.	.	.	.
81	.	.	.	.	.	.	.	.	.	.	.	.	.
82	27.1	27.8	28.0	27.7	27.1	26.6	25.4	25.5	25.8	26.2	26.8	27.0	26.7
83	21.4	23.8	27.3	30.4	31.2	29.4	26.6	25.2	26.7	26.8	24.6	22.0	26.3
84	.	.	.	.	.	.	.	.	.	.	.	.	.

CLIMATOLOGICAL DATA FOR NIGERIA

GEOGRAPHY

OBS.	LOCATION	YEARS	LATITUDE	LONGITUDE	ELEVATION -M-	LIFEZONE	CCTX	ID	ONMAP
85	SAMARU	24	11.18 N	7.63 E	685.8	.	.	107	NO
86	SAPALE	52	5.90 N	5.70 E	3.0	.	.	75	NO
87	SHENDAM	26	8.90 N	9.47 E	243.8	.	.	76	YES
88	SOKOTO	42	13.02 N	5.27 E	350.5	TV	EH74	77	NO
89	TOMBEL	13	4.75 N	9.65 E	365.8	.	.	78	NO
90	UFUMA	16	6.08 N	7.20 E	152.4	.	.	79	YES
91	UMUDIKE	24	5.47 N	7.55 E	121.9	.	.	108	NO
92	UMUDIKE	24	5.47 N	7.55 E	121.9	.	.	98	YES
93	UYO	17	5.03 N	7.93 E	61.0	.	.	109	NO
94	UYO	17	5.03 N	7.93 E	61.0	.	.	99	YES
95	VICTORIA	36	4.00 N	9.22 E	3.0	TW	EW12	80	NO
96	VOM	12	9.73 N	8.82 E	1219.2	.	.	81	NO

AVERAGE PRECIPITATION (MM)

OBS.	JAN.	FEB.	MAR.	APR.	MAY	JUN.	JUL.	AUG.	SEP.	OCT.	NOV.	DEC.	ANNUAL
85	0	3	8	28	137	145	218	292	226	38	0	0	1095
86	25	48	119	201	259	345	394	249	386	277	86	18	2408
87	0	3	15	71	147	185	190	226	251	127	5	0	1222
88	0	0	0	13	53	89	165	251	147	15	0	0	734
89	53	124	221	201	279	511	467	444	582	475	130	58	3546
90	18	38	91	165	257	249	292	239	302	267	58	8	1984
91	23	58	114	190	269	279	325	251	323	251	81	15	2182
92	23	58	114	190	269	279	325	251	323	251	81	15	2182
93	38	84	142	190	274	267	312	277	378	307	155	33	2484
94	38	84	142	190	274	267	312	277	378	307	155	33	2484
95	41	71	155	201	335	625	991	805	437	267	104	46	4077
96	3	5	25	97	206	196	274	295	213	56	8	0	1377

AVERAGE TEMPERATURE (DEGREES CENTIGRADE)

OBS.	JAN.	FEB.	MAR.	APR.	MAY	JUN.	JUL.	AUG.	SEP.	OCT.	NOV.	DEC.	ANNUAL
85	.	.	.	.	.	.	.	.	.	.	.	.	.
86	.	.	.	.	.	.	.	.	.	.	.	.	.
87	.	.	.	.	.	.	.	.	.	.	.	.	.
88	24.2	26.2	29.7	32.5	32.6	30.6	27.6	26.2	26.9	28.5	27.2	24.6	28.1
89	.	.	.	.	.	.	.	.	.	.	.	.	.
90	.	.	.	.	.	.	.	.	.	.	.	.	.
91	.	.	.	.	.	.	.	.	.	.	.	.	.
92	.	.	.	.	.	.	.	.	.	.	.	.	.
93	.	.	.	.	.	.	.	.	.	.	.	.	.
94	.	.	.	.	.	.	.	.	.	.	.	.	.
95	25.7	26.1	26.5	26.3	26.0	25.2	24.2	24.2	24.6	25.1	25.6	25.7	25.4
96	.	.	.	.	.	.	.	.	.	.	.	.	.

CLIMATOLOGICAL DATA FOR NIGERIA

GEOGRAPHY

OBS.	LOCATION	YEARS	LATITUDE	LONGITUDE	ELEVATION -M-	LIFEZONE	CCTX	ID	ONMAP
97	WAMBA	30	8.92 N	8.62 E	304.8	.	.	82	NO
98	WARRI	51	5.52 N	5.73 E	6.1	TM	EH22	83	NO
99	YABA	24	6.50 N	3.37 E	.	.	.	84	NO
100	YELWA	20	10.83 N	4.75 E	243.8	TD	EH53	85	YES
101	YOLA	42	9.22 N	12.47 E	175.3	TV	EH63	110	NO
102	YOLA	42	9.22 N	12.47 E	175.3	TV	EH63	100	YES
103	ZARIA	33	11.10 N	7.73 E	640.1	.	.	86	YES
104	ZUNGERU	19	9.80 N	6.15 E	137.2	.	.	87	NO

AVERAGE PRECIPITATION (MM)

OBS.	JAN.	FEB.	MAR.	APR.	MAY	JUN.	JUL.	AUG.	SEP.	OCT.	NOV.	DEC.	ANNUAL
97	0	8	30	81	152	190	338	378	335	132	18	3	1666
98	33	58	145	229	272	384	455	295	447	317	112	36	2781
99	15	43	94	135	229	348	244	74	135	155	63	25	1560
100	0	3	5	23	114	122	180	246	229	89	0	0	1008
101	0	0	10	43	127	157	168	185	211	79	5	0	986
102	0	0	10	43	127	157	168	185	211	79	5	0	986
103	0	3	8	43	122	147	221	300	224	38	3	0	1107
104	0	0	10	63	132	198	190	226	317	112	8	0	1257

AVERAGE TEMPERATURE (DEGREES CENTIGRADE)

OBS.	JAN.	FEB.	MAR.	APR.	MAY	JUN.	JUL.	AUG.	SEP.	OCT.	NOV.	DEC.	ANNUAL
97	.	.	.	.	.	.	.	.	.	.	.	.	.
98	26.6	27.4	28.0	27.9	27.1	26.2	25.2	25.6	25.5	26.1	27.0	26.8	26.6
99	.	.	.	.	.	.	.	.	.	.	.	.	.
100	24.8	27.9	30.7	32.3	30.6	27.7	26.6	25.6	26.1	27.2	26.3	24.6	27.6
101	26.7	28.8	31.5	32.2	30.3	27.7	26.6	26.3	26.3	27.5	27.9	26.9	28.3
102	26.7	28.8	31.5	32.2	30.3	27.7	26.6	26.3	26.3	27.5	27.9	26.9	28.3
103	.	.	.	.	.	.	.	.	.	.	.	.	.
104	.	.	.	.	.	.	.	.	.	.	.	.	.

PAKISTAN*

Agriculture

Of 80,394,000 ha in total area, Pakistan was estimated in 1978 to have 19,990,000 ha of arable or permanent cropland, 5,000,000 ha in permanent pasture, 2,860,000 ha in forest and woodland, and 50,022,000 ha in other terrain categories. In 1978, approximately 14,000,000 ha were irrigated. Of the 1979 population, 70.3% or 469,000 individuals of the economically active population were engaged in agriculture. The total population in 1979 was 1,802,000. According to FAO statistics for 1979, there were 480,000 horses, 66,000 mules, 2,371,000 asses, 14,992,000 cattle, 11,306,000 buffalo, 830,000 camels, 96,000 pigs, 24,185,000 sheep, 27,804,000 goats, 48,872,000 chickens, and 1,079,000 ducks.[10] Agricultural production is summarized in Table 1. Conventional and energetic yield data on most of these will be found in Appendixes I and II.

Climate

Data from Wernstedt[3] are presented in the Climatological Data Table (see Introduction for explanation). Based on the means for data points cited in Wernstedt, the mean annual precipitation is 3.3 dm (mean of 123 sites) and the mean annual temperature 23.0°C (mean of 51 sites). The averages would suggest that much of the country could be classified as Subtropical Thorn Woodland Life Zone.[4]

Productivity

Lacking the State Department vegetation maps for Pakistan, I will make my productivity estimates for the units of the soil map, generously provided by Dr. Ken Ackerson of the Soil Conservation Service.

Soil type	MT/ha/yr
Alluvial soils of humid regions	6—18
Alluvial soils of arid regions, solonetz, and solonchak	0—6
Saline alluvial soils	0—6
Latosolic soils	3—12
Red desert and reddish brown soils	3—6
Regosols and desertic soils	0—3
Lithosols	0—6

Spring[13] calculated the area and feedstock requirements for production of 1% of Pakistan's energy in ethanol. Relative contributions of liquid fuel and fuelwood to energy production and consumption in Pakistan are shown in the pie chart.

Economics

Food consumption per capita was 2281 cal/day, energy consumption 170 kg coal equivalent per capita in 1977. In 1970, the caloric supply was 94.0% of requirements. The total population growth rate was 2.8% in 1970. Population density was 75 persons per square kilometer in 1970, with 24.9% of the total population urban. Life expectancy at birth of those born in 1970 was 48 years. In 1977, with GNP at $200 per capita, agriculture accounted for 32.6% of GDP, manufacturing for 15.4%, other industry for 7.9%, and services for 44.0%. In 1977, food and beverage constituted 29.3% of exports, nonfood agriculture for 6.0%, fuels, minerals, and metal for 4.9%, machines for 2.3%, and other manufacturing for 57.5% of total merchandise exports. Food and

* Maps and table follow text.

Table 1
AGRICULTURAL PRODUCTION IN PAKISTAN

Crop	Area planted (ha)	Yield (kg/ha)	Total production (MT)
Wheat	6,696,000	1,485	9,944,000
Rice paddy	1,975,000	2,508	3,953,000
Barley	177,000	730	129,000
Corn	656,000	1,290	846,000
Millet	629,000	493	310,000
Sorghum	477,000	587	280,000
Potatoes	30,000	10,667	320,000
Sweet potatoes	18,000	8,901	162,000
Dry beans	125,000	496	62,000
Chick peas	1,224,000	439	538,000
Lentils	106,000	368	39,000
Peanuts in shell	40,000	1,425	57,000
Castor beans	20,000	750	15,000
Rapeseed	420,000	578	243,000
Sesame seed	34,000	397	14,000
Linseed	13,000	530	7,000
Seed cotton	1,891,000	1,032	1,951,000
Cottonseed			1,301,000
Hempseed			3,200
Chilies, peppers, green	52,000	1,577	82,000
Dry onions	31,000	10,645	330,000
Garlic	4,000	7,805	32,000
Grapes	3,000	11,923	31,000
Dates			215,000
Sugar cane	738,000	37,639	27,763,000
Sugar beets	8,000	37,975	300,000
Sugar (centrifugal, raw)			662,000
Sugar (noncentrifugal)			1,700,000
Apples			92,000
Pears			38,000
Peaches and nectarines			11,000
Plums			31,000
Oranges			540,000
Tang. mand. clement satsuma			210,000
Lemons and limes			32,000
Grapefruit and pomelo			2,000
Apricots			31,000
Mangoes			600,000
Bananas			125,000
Tobacco leaves	53,000	1,361	72,000
Hemp fiber and tow	10,000	696	7,000
Jute and substitutes	1,000	773	1,000
Cotton lint			650,000

From FAO 1979 Yearbook, Vol. 33, Food and Agriculture Organization, Rome, 1980.

beverage constituted 17.0% of imports, nonfood agriculture 4.4%, fuel and lubricants 15.8%, nonfuel minerals and metals 3.2%, machinery and equipment 28.0%, and other manufactured goods constituted 31.6% of total merchandise imports.[10]

Energetics

Pakistan has 200 million b.o.e. in oil, 1375 million tons coal, and a hydroelectric potential of 20,000 MW. Pakistan has the potential to become self sufficient within this decade. Her attitude towards private oil exploration activities has softened. Pakistan already generates some nuclear power. Importing less than 50% (26 to 50%) of its commercial energy demands as oil, Pakistan is expected to be experiencing fuelwood problems by the year 2000.[10]

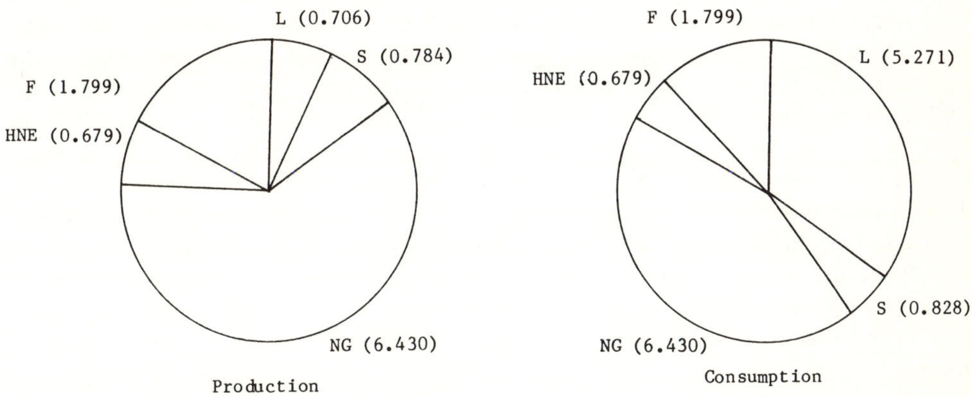

1978 Energy production and consumption in Pakistan; fuel types listed in million metric tons coal equivalents (MMTCE). 1 MMTCE = 27.78×10^{12} Btu. L = liquid fuel, S = solid fuel, HNE = hydro, nuclear, and electric fuel, F = firewood, NG = natural gas. (Data from United Nations, World Energy Supplies, 1973—1978, Statistical Papers Series J, 22, Department of International Economic and Social Affairs, Statistical Office, New York, 1979.)

Soils

The principal soils of Pakistan are shown on the Soils Map in seven groups: (1) Alluvial soils of humid regions, (2) Alluvial soils of arid regions, Solonetz, and Solonchak soils, (3) wet saline Alluvial soils of marshes and swamps, (4) Latosolic soils, (5) Red Desert and Reddish Brown soils, (6) Regosols and desertic soils, and (7) Lithosols.

The most productive soils of the country are the Alluvial soils of the humid regions, which cover about three fourths of East Pakistan. These are neutral to slightly acid silts, sandy loams, and clays. The silty soils, common along the Ganges, have a particularly high organic content. The sandy loams occur along the Brahmaputra and its tributaries; they are rich in phosphate but deficient in some of the other essential plant nutrients. Clays occur in the lower lying areas away from streams; many of them are wet. In old river channels, however, some of the wet soils have a peaty surface layer, and others have peat interlayered with silts and clays; they are unsuitable for crops, unless drained.

Alluvial soils of arid regions have associated with them Solonetz and Solonchak soils on the flood plains of the Indus River and its tributaries in West Pakistan. The Alluvial soils are the calcareous, nonsaline to slightly saline, better drained soils of the flood plains. Under irrigation, they constitute the best cropland in this part of the country. The Solonchak soils are the highly saline soils, which in many places have a high water table. The Solonchak soils are so saline that economic crop production on them is not feasible; it could be made so on most of the soils by leaching the salts and providing drainage. The Solonetz soils are too alkaline for economic crop production; their reclamation by leaching would be difficult because of the slowly permeable clayey subsoils. Textures of the surface layers of the soils in this association range from loamy sand to silt loam; gravel is intermixed in some places. Textures at lower depths are commonly sandy. Although the soils are well supplied with most plant nutrients, both nitrogen and available phosphorus tend to be low; under otherwise good management, plants will respond to applications of these elements.

Wet saline Alluvial soils occur in the coastal salt marshes of West Pakistan and in the mangrove swamps of Bangladesh. These soils become periodically inundated and are mostly nonagricultural. Some could be reclaimed for rice, but would require protection from inundation as well as other ameliorative measures.

Latosolic soils occur in Bangladesh. They are reddish, highly leached, generally acid soils that are low in plant nutrients. Most productive for agriculture are those on the high alluvial plains near Dacca and to the east where they appear to consist of deep, friable, permeable clays, characteristic of Red Latosols. Least productive and difficult to work when dry are the latosolic soils on the plains in the western part of Bangladesh. These soils, impoverished of plant nutrients, are droughty because of high concentrations of pea-size concretions throughout the subsoil, and become hard when dry. Latosolic soils in the Chittagong hills are generally unsuitable for cultivation because of their location on steep slopes. These soils are moderately deep sandy clays, eroded in many places, and on flats are underlain by laterite gravel. On some of the hilly areas, they support shifting agriculture. They are best suited, however, to forest and tree crops.

Red Desert and Reddish Brown soils occur on alluvial fans in the central and western parts of West Pakistan. Each of these soils has gravel on the surface, and, near the base of mountains, rocks and boulders. Red Desert soils, which are common south of 32°N. latitude, are sandy loams to loamy sandy with limy reddish sandy clay loam subsoils. Reddish Brown soils, which receive more rain than the Reddish Desert soils, occur mainly in the basins north of about 32°N. latitude. They have a grayish-brown sandy loam surface containing some organic matter and a brown clayey subsoil. The Red Desert and Reddish Brown soils support some grazing. Where water is available, they are fairly productive for crops.

Regosols and desertic soils occur extensively in West Pakistan, east of the Indus River and in dry basins in the extreme west. They are sandy soils with low water-retention capacities. Too dry for crops, they support sparse grazing.

Lithosols cover nearly half of West Pakistan, from the Hindu Kush mountains in the north to the Arabian Sea in the south. Lithosols are shallow, stony soils, barren of vegetation. Some agriculture can be carried on in intermountain valleys, on included Alluvial and other soils that cannot be shown at the small scale of the map. Associated with the Lithosols are extensive areas of bare rock and some snowcapped peaks.

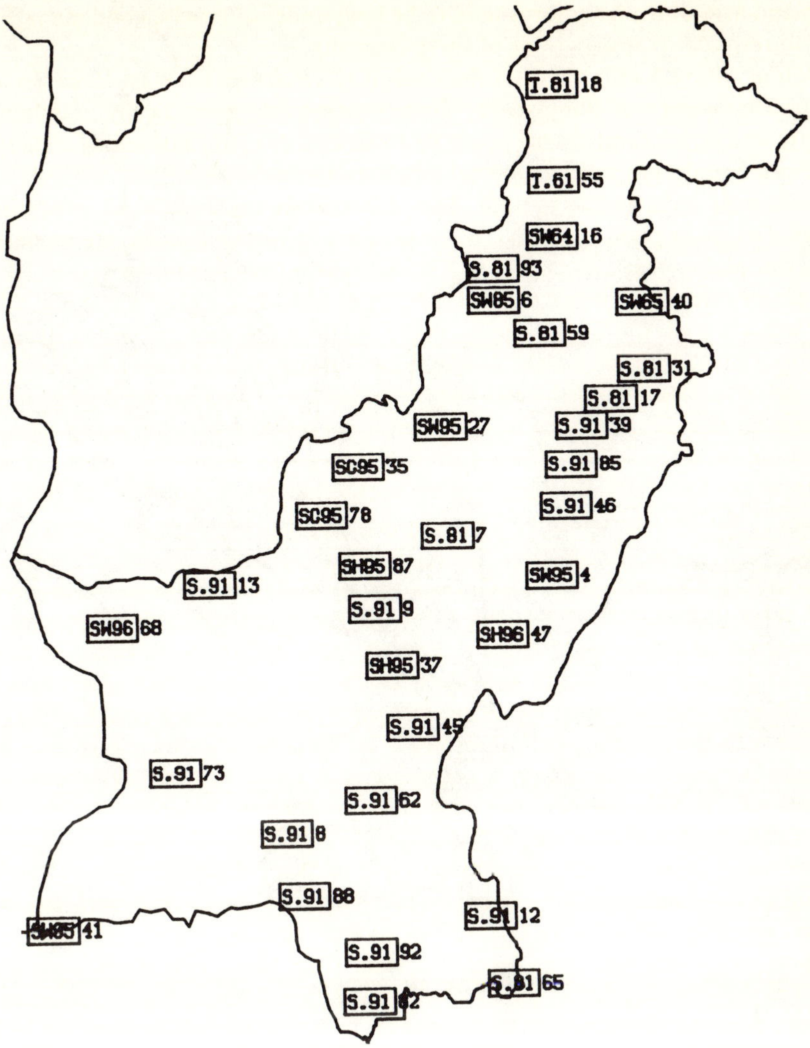

Selected weather stations for Pakistan.

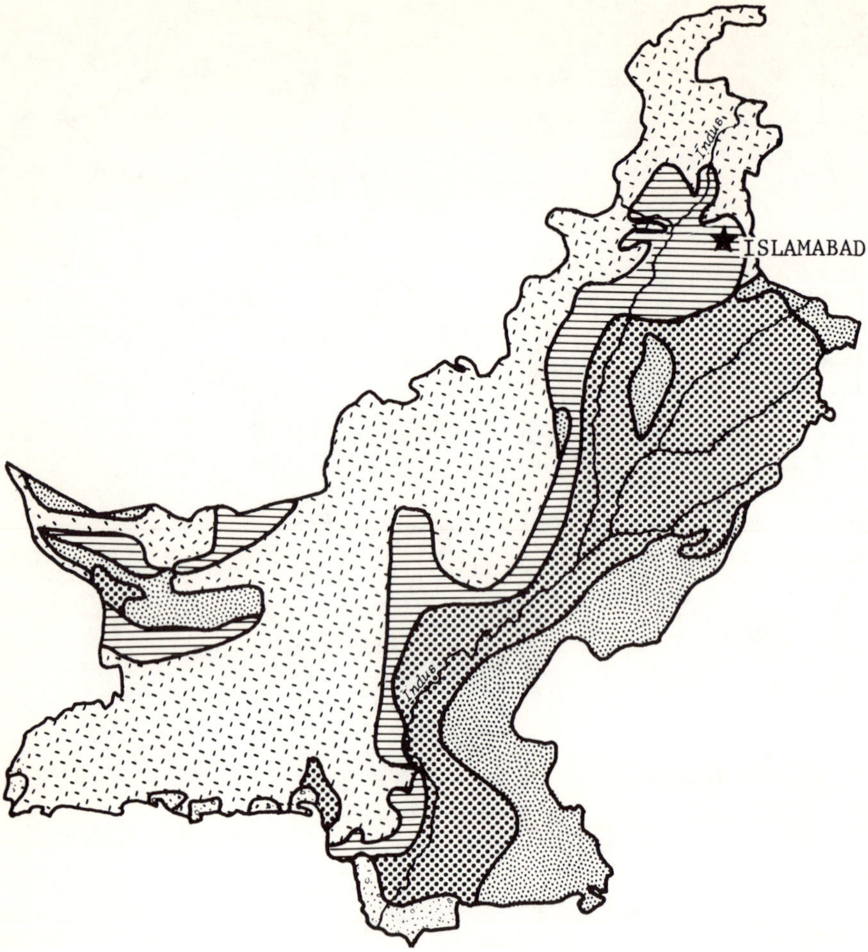

PAKISTAN SOILS

Alluvial soils of humid regions

Alluvial soils of arid regions, Solonetz, and Solonchak soils

Wet saline Alluvial soils of marshes and swamps

Latosolic soils

Red Desert and Reddish Brown soils

Regosols and desertic soils

Lithosols

CLIMATOLOGICAL DATA FOR PAKISTAN

GEOGRAPHY

OBS.	LOCATION	YEARS	LATITUDE	LONGITUDE	ELEVATION -M-	LIFEZONE	CCTX	ID	ONMAP
1	ABBOTTABAD	75	34.20 N	73.25 E	.	.	.	101	NO
2	ALIPUR	73	29.37 N	70.98 E	.	.	.	1	NO
3	ATTOCK	39	33.87 N	72.33 E	.	.	.	2	NO
4	BADIN	30	24.65 N	68.83 E	.	TX	SH95	3	NO
5	BAHAWALPUR	30	29.40 N	71.68 E	.	TX	SW95	4	YES
6	BANDA DAUD SHAH	53	33.27 N	71.20 E	.	.	.	5	NO
7	BANNU	30	32.98 N	70.60 E	.	TT	SW85	6	YES
8	BARKHAN	29	29.92 N	69.57 E	.	.	.	7	YES
9	BELA	29	26.20 N	66.33 E	.	.	.	8	YES
10	BELLPAT	51	29.00 N	68.08 E	.	.	.	9	YES
11	BHAKKAR	75	31.67 N	71.13 E	.	.	.	10	NO
12	BHERRA	75	32.47 N	72.97 E	.	.	.	11	NO

AVERAGE PRECIPITATION (MM)

OBS.	JAN.	FEB.	MAR.	APR.	MAY	JUN.	JUL.	AUG.	SEP.	OCT.	NOV.	DEC.	ANNUAL
1	87	104	114	89	56	79	228	250	94	35	18	46	1200
2	6	8	6	6	4	8	45	36	10	1	1	4	134
3	51	48	59	39	20	20	83	106	51	7	6	22	512
4	3	5	3	1	3	9	107	62	38	1	2	2	236
5	6	6	11	6	5	6	50	38	9	2	0	4	142
6	33	33	48	41	29	44	109	115	48	14	6	14	535
7	22	30	47	32	14	21	70	56	18	3	3	11	327
8	13	17	19	20	28	36	91	67	24	11	4	10	337
9	11	24	15	14	18	12	71	35	16	3	2	9	231
10	5	8	7	2	3	5	24	21	7	1	2	3	87
11	10	12	16	12	9	13	57	48	16	1	2	5	201
12	24	24	27	20	18	26	93	109	38	4	5	8	396

AVERAGE TEMPERATURE (DEGREES CENTIGRADE)

OBS.	JAN.	FEB.	MAR.	APR.	MAY	JUN.	JUL.	AUG.	SEP.	OCT.	NOV.	DEC.	ANNUAL
1	.	.	.	.	.	.	.	.	.	.	.	.	.
2	.	.	.	.	.	.	.	.	.	.	.	.	.
3	.	.	.	.	.	.	.	.	.	.	.	.	.
4	17.2	20.7	25.8	30.2	33.1	33.2	31.2	29.8	29.8	28.9	24.2	19.1	26.9
5	13.3	16.5	21.8	27.9	33.2	35.7	34.3	32.9	31.3	26.5	20.3	15.2	25.7
6	.	.	.	.	.	.	.	.	.	.	.	.	.
7	11.8	14.4	19.0	24.7	30.7	33.9	33.3	32.2	30.2	24.9	18.6	13.6	23.9
8	.	.	.	.	.	.	.	.	.	.	.	.	.
9	.	.	.	.	.	.	.	.	.	.	.	.	.
10	.	.	.	.	.	.	.	.	.	.	.	.	.
11	.	.	.	.	.	.	.	.	.	.	.	.	.
12	.	.	.	.	.	.	.	.	.	.	.	.	.

CLIMATOLOGICAL DATA FOR PAKISTAN

GEOGRAPHY

OBS.	LOCATION	YEARS	LATITUDE	LONGITUDE	ELEVATION -M-	LIFEZONE	CCTX	ID	ONMAP
13	CHACHRO	27	25.15 N	70.30 E	.	.	.	12	YES
14	CHAGAI	14	29.33 N	64.73 E	.	.	.	13	YES
15	CHAKWAL	75	32.93 N	72.88 E	.	.	.	14	NO
16	CHAMAN	30	30.92 N	66.37 E	.	SX	SW95	15	NO
17	CHERAT	30	33.82 N	71.88 E	.	SD	SW64	16	YES
18	CHINIOT	75	31.67 N	73.00 E	.	.	.	17	YES
19	CHITRAL	31	35.87 N	71.97 E	.	.	.	18	YES
20	CHUNIAN	75	30.95 N	74.02 E	.	.	.	19	NO
21	DALBANDIN	30	28.88 N	64.42 E	.	SO	SW95	20	NO
22	DERA GAZI-KHAN	77	30.08 N	70.73 E	.	.	.	21	NO
23	DIPLO	27	24.45 N	69.58 E	.	.	.	22	NO
24	FATEHJANG	59	33.55 N	72.72 E	.	.	.	23	NO

AVERAGE PRECIPITATION (MM)

OBS.	JAN.	FEB.	MAR.	APR.	MAY	JUN.	JUL.	AUG.	SEP.	OCT.	NOV.	DEC.	ANNUAL
13	2	3	1	2	7	12	106	78	44	6	2	1	264
14	23	19	12	11	3	0	2	3	0	3	4	14	94
15	30	33	35	28	24	34	101	130	57	8	6	14	499
16	71	49	42	18	5	0	10	1	0	0	4	35	234
17	73	89	15	69	40	21	113	137	38	19	13	38	767
18	13	17	17	12	13	30	86	75	33	3	2	7	309
19	22	37	87	92	33	9	4	6	8	20	6	21	345
20	17	14	12	8	10	30	87	104	44	4	1	5	335
21	25	18	12	5	2	1	7	0	0	0	1	13	84
22	9	9	11	10	9	11	38	34	11	2	2	6	151
23	2	3	1	2	9	11	119	62	44	7	1	2	261
24	45	49	54	41	26	33	129	173	61	13	5	24	653

AVERAGE TEMPERATURE (DEGREES CENTIGRADE)

OBS.	JAN.	FEB.	MAR.	APR.	MAY	JUN.	JUL.	AUG.	SEP.	OCT.	NOV.	DEC.	ANNUAL
13	.	.	.	.	.	.	.	.	.	.	.	.	.
14	.	.	.	.	.	.	.	.	.	.	.	.	.
15	.	.	.	.	.	.	.	.	.	.	.	.	.
16	6.8	9.6	14.1	20.0	25.6	29.6	31.4	29.9	25.6	19.7	13.7	9.1	19.6
17	6.6	8.4	12.7	18.2	24.8	28.8	26.7	25.1	24.3	19.9	14.4	9.1	18.3
18	.	.	.	.	.	.	.	.	.	.	.	.	.
19	.	.	.	.	.	.	.	.	.	.	.	.	.
20	.	.	.	.	.	.	.	.	.	.	.	.	.
21	9.1	12.4	17.3	23.1	28.4	32.0	33.4	31.6	26.8	21.0	14.8	10.1	21.7
22	.	.	.	.	.	.	.	.	.	.	.	.	.
23	.	.	.	.	.	.	.	.	.	.	.	.	.
24	.	.	.	.	.	.	.	.	.	.	.	.	.

CLIMATOLOGICAL DATA FOR PAKISTAN

GEOGRAPHY

OBS.	LOCATION	YEARS	LATITUDE	LONGITUDE	ELEVATION -M-	LIFEZONE	CCTX	ID	ONMAP
25	FAZILKA	57	29.30 N	70.52 E	.	.	.	24	NO
26	FEROZEPORO	53	30.92 N	74.63 E	.	.	.	25	NO
27	FORT ABBAS	24	29.20 N	72.87 E	.	TX	SH96	26	NO
28	FORT SANDEMAN	30	31.33 N	69.45 E	.	ST	SW95	27	YES
29	GANDAWA	29	28.60 N	67.50 E	.	.	.	28	NO
30	GUJARKHAN	59	33.32 N	73.38 E	.	.	.	29	NO
31	GUJRANWALA	75	32.10 N	74.18 E	.	.	.	30	NO
32	HAFIZABAD	75	32.05 N	73.70 E	.	.	.	31	YES
33	HANGU	53	33.53 N	71.07 E	.	.	.	32	NO
34	HARIPUR (HAZARA)	75	33.98 N	72.93 E	.	SD	SW65	33	NO
35	HARNAI	30	30.08 N	68.00 E	.	.	.	34	NO
36	HINDUBAGH	18	30.82 N	67.75 E	.	WT	SC95	35	YES

AVERAGE PRECIPITATION (MM)

OBS.	JAN.	FEB.	MAR.	APR.	MAY	JUN.	JUL.	AUG.	SEP.	OCT.	NOV.	DEC.	ANNUAL
25	14	14	11	9	9	28	80	70	32	4	1	6	277
26	25	24	19	12	9	40	126	118	48	6	2	10	439
27	11	9	6	7	6	14	72	41	22	2	2	4	196
28	24	30	41	27	22	13	52	47	4	1	3	14	279
29	5	11	7	5	2	7	38	17	6	1	1	7	106
30	47	50	44	33	27	45	165	190	71	11	6	23	712
31	33	31	29	18	20	46	155	166	65	8	5	15	589
32	23	25	21	17	15	34	114	116	42	6	3	11	425
33	36	41	66	55	38	48	117	138	59	19	10	17	643
34	57	63	73	56	29	40	148	166	74	19	11	33	769
35	23	34	36	19	13	19	79	68	24	6	5	20	343
36	54	45	43	17	16	1	30	19	2	1	7	29	264

AVERAGE TEMPERATURE (DEGREES CENTIGRADE)

OBS.	JAN.	FEB.	MAR.	APR.	MAY	JUN.	JUL.	AUG.	SEP.	OCT.	NOV.	DEC.	ANNUAL
25	.	.	.	.	.	.	.	.	.	.	.	.	.
26	.	.	.	.	.	.	.	.	.	.	.	.	.
27	13.2	16.3	22.1	28.8	34.6	36.5	34.6	33.1	31.9	27.4	21.3	14.9	26.2
28	5.9	9.1	14.0	20.0	25.5	29.9	30.3	29.4	26.5	20.1	13.9	8.7	19.4
29	.	.	.	.	.	.	.	.	.	.	.	.	.
30	.	.	.	.	.	.	.	.	.	.	.	.	.
31	.	.	.	.	.	.	.	.	.	.	.	.	.
32	.	.	.	.	.	.	.	.	.	.	.	.	.
33	.	.	.	.	.	.	.	.	.	.	.	.	.
34	10.1	12.9	17.7	22.1	27.6	32.7	31.1	29.3	27.9	23.3	17.2	12.8	22.0
35	.	.	.	.	.	.	.	.	.	.	.	.	.
36	3.7	6.7	11.4	17.0	21.7	25.6	27.4	26.8	23.2	16.9	11.3	6.5	16.5

CLIMATOLOGICAL DATA FOR PAKISTAN

GEOGRAPHY

OBS.	LOCATION	YEARS	LATITUDE	LONGITUDE	ELEVATION -M-	LIFEZONE	CCTX	ID	ONMAP
37	HYDERABAD	30	25.37 N	68.37 E	.	TX	SH96	36	NO
38	JACOBABAD	30	28.28 N	68.43 E	.	TO	SH95	37	YES
39	JAMESABAD	27	25.27 N	69.30 E	.	.	.	38	NO
40	JHANG	75	31.32 N	72.37 E	.	.	.	39	YES
41	JHELUM	13	32.93 N	73.73 E	.	TV	SW65	40	YES
42	JIWANI	16	25.05 N	61.75 E	.	TX	SW95	41	YES
43	KALAT	30	29.03 N	66.58 E	.	WX	SC95	42	NO
44	KARACHI CITY	62	24.87 N	67.05 E	.	TX	SW95	43	NO
45	KASUR	75	31.12 N	74.50 E	.	.	.	44	NO
46	KHAIRPUR-MIRA	12	27.50 N	68.83 E	.	.	.	45	YES
47	KHANEWALA	22	30.28 N	72.00 E	.	.	.	46	YES
48	KHANPUR	30	28.65 N	70.65 E	.	TO	SH96	47	YES

AVERAGE PRECIPITATION (MM)

OBS.	JAN.	FEB.	MAR.	APR.	MAY	JUN.	JUL.	AUG.	SEP.	OCT.	NOV.	DEC.	ANNUAL
37	4	5	1	2	4	6	68	44	15	3	1	3	155
38	7	9	8	2	4	6	27	22	1	0	1	3	88
39	2	4	3	2	5	10	99	66	20	5	2	1	219
40	11	12	18	12	10	23	76	58	24	2	2	7	253
41	55	34	66	20	19	39	248	241	106	25	14	24	891
42	68	18	10	9	1	0	9	2	0	0	10	23	149
43	55	47	37	15	6	3	31	14	2	0	3	19	233
44	11	12	8	4	2	18	82	36	15	0	4	4	196
45	20	19	17	11	9	38	116	101	48	6	2	9	396
46	5	8	3	2	0	24	46	28	0	0	0	2	119
47	7	12	7	3	8	14	55	39	11	2	1	7	165
48	4	3	4	3	0	5	46	31	6	0	0	4	120

AVERAGE TEMPERATURE (DEGREES CENTIGRADE)

OBS.	JAN.	FEB.	MAR.	APR.	MAY	JUN.	JUL.	AUG.	SEP.	OCT.	NOV.	DEC.	ANNUAL
37	17.1	20.6	25.9	30.8	34.1	34.2	32.5	36.3	30.9	29.3	24.2	19.1	27.5
38	14.7	18.3	23.9	29.9	35.0	37.1	35.3	33.6	32.2	28.1	22.0	16.6	27.2
39	.	.	.	.	.	.	.	.	.	.	.	.	.
40	.	.	.	.	.	.	.	.	.	.	.	.	.
41	11.9	14.6	19.7	25.7	30.9	33.6	31.3	29.7	29.2	24.7	17.9	13.1	23.5
42	18.8	20.3	23.3	26.6	29.7	30.9	29.7	28.4	27.9	26.9	23.9	20.3	25.6
43	2.8	5.1	9.1	13.9	18.7	22.3	24.3	23.1	18.9	13.3	8.3	4.6	13.7
44	19.1	20.8	24.5	27.4	30.3	30.7	29.8	28.8	27.9	27.8	24.7	20.4	26.0
45	.	.	.	.	.	.	.	.	.	.	.	.	.
46	.	.	.	.	.	.	.	.	.	.	.	.	.
47	.	.	.	.	.	.	.	.	.	.	.	.	.
48	13.7	16.9	22.7	28.7	33.8	36.0	34.7	33.3	31.9	27.3	20.9	10.2	26.3

CLIMATOLOGICAL DATA FOR PAKISTAN

GEOGRAPHY

OBS.	LOCATION	YEARS	LATITUDE	LONGITUDE	ELEVATION -M-	LIFEZONE	CCTX	ID	ONMAP
49	KHUSHAB	30	32.30 N	72.35 E	.	TT	SW85	48	NO
50	KOHAT	75	33.62 N	71.50 E	.	.	.	49	NO
51	LAHORE	30	31.58 N	74.30 E	.	TT	SW85	103	NO
52	LANDIKOTAL	13	34.10 N	71.15 E	.	ST	TW85	50	NO
53	LARKANA	27	27.53 N	68.30 E	.	.	.	51	NO
54	LEIAH	75	30.98 N	70.97 E	.	.	.	52	NO
55	LORALAI	20	30.37 N	68.60 E	.	SX	SC94	53	NO
56	LYALLPUR	30	31.42 N	73.08 E	.	TT	SW85	102	NO
57	MAILSI	75	29.77 N	72.25 E	.	.	.	54	NO
58	MALAKAND	38	34.57 N	71.95 E	.	.	.	55	YES
59	MAND	30	26.08 N	62.07 E	.	.	.	56	NO
60	MARDAN	75	34.23 N	72.08 E	.	.	.	57	NO

AVERAGE PRECIPITATION (MM)

OBS.	JAN.	FEB.	MAR.	APR.	MAY	JUN.	JUL.	AUG.	SEP.	OCT.	NOV.	DEC.	ANNUAL
49	20	25	30	21	19	30	104	88	30	6	3	10	388
50	31	36	59	39	32	25	77	78	39	14	11	14	455
51	31	23	24	16	8	39	122	123	80	10	4	11	490
52	37	34	83	76	37	10	34	33	8	14	17	28	412
53	3	7	5	3	3	3	25	27	11	1	1	4	90
54	10	9	14	13	7	11	50	52	19	2	1	6	193
55	26	20	40	23	20	9	35	40	5	3	7	17	244
56	16	18	23	14	9	29	97	98	29	5	2	8	346
57	6	8	5	5	8	13	33	42	16	1	1	4	142
58	75	101	119	73	24	22	128	186	56	19	8	53	863
59	38	38	17	15	11	5	5	6	2	4	1	18	160
60	43	46	58	36	16	17	81	120	43	8	10	21	499

AVERAGE TEMPERATURE (DEGREES CENTIGRADE)

OBS.	JAN.	FEB.	MAR.	APR.	MAY	JUN.	JUL.	AUG.	SEP.	OCT.	NOV.	DEC.	ANNUAL
49	12.2	15.1	20.3	26.7	32.2	34.7	32.9	32.1	30.8	26.0	19.2	13.8	24.7
50	.	.	.	.	.	.	.	.	.	.	.	.	.
51	12.2	15.2	20.4	26.6	31.9	33.9	32.1	31.2	30.0	25.4	15.9	13.8	24.3
52	6.4	8.3	12.6	18.3	24.2	29.3	29.1	27.6	25.4	19.8	12.5	8.2	18.4
53	.	.	.	.	.	.	.	.	.	.	.	.	.
54	.	.	.	.	.	.	.	.	.	.	.	.	.
55	5.6	8.3	13.4	18.7	23.7	27.0	27.3	26.0	23.4	18.4	12.8	7.9	17.7
56	12.1	15.0	19.9	26.2	31.9	34.3	32.8	31.8	30.5	25.8	19.3	14.1	24.5
57	.	.	.	.	.	.	.	.	.	.	.	.	.
58	.	.	.	.	.	.	.	.	.	.	.	.	.
59	.	.	.	.	.	.	.	.	.	.	.	.	.
60	.	.	.	.	.	.	.	.	.	.	.	.	.

CLIMATOLOGICAL DATA FOR PAKISTAN

GEOGRAPHY

OBS.	LOCATION	YEARS	LATITUDE	LONGITUDE	ELEVATION -M-	LIFEZONE	CCTX	ID	ONMAP
61	MASTUNG	30	29.73 N	66.93 E	.	.	.	58	NO
62	MIANWALI	69	32.53 N	71.55 E	.	.	.	59	YES
63	MIRPUR KHAS	27	25.55 N	69.08 E	.	.	.	60	NO
64	MONTGOMERY	30	30.67 N	73.10 E	.	TT	SW95	61	NO
65	MORO	27	26.60 N	67.98 E	.	.	.	62	YES
66	MULTAN	28	30.18 N	71.48 E	.	TX	SW95	63	NO
67	MURREE	30	33.90 N	73.40 E	.	WM	SC44	64	NO
68	NAGAR PARKAR	45	24.33 N	70.73 E	.	.	.	65	YES
69	NAROWAL	19	32.07 N	74.90 E	.	.	.	66	NO
70	NAWABSHAH	27	26.25 N	68.42 E	.	TX	SH95	67	NO
71	NOKKINDI	28	28.80 N	62.77 E	.	TO	SW96	68	YES
72	NURPUR	58	31.90 N	71.92 E	.	.	.	69	NO

AVERAGE PRECIPITATION (MM)

OBS.	JAN.	FEB.	MAR.	APR.	MAY	JUN.	JUL.	AUG.	SEP.	OCT.	NOV.	DEC.	ANNUAL
61	35	40	28	19	6	4	7	4	4	3	6	21	177
62	19	21	25	20	16	25	67	67	27	4	5	8	303
63	2	4	3	4	4	10	103	60	15	1	1	2	209
64	12	12	17	6	7	23	74	75	25	2	2	7	261
65	2	6	6	2	4	7	46	52	15	0	1	2	140
66	8	10	14	7	6	12	46	31	20	1	2	5	162
67	120	112	155	102	61	107	362	358	134	53	21	54	1640
68	1	3	2	1	6	17	145	117	55	7	2	2	358
69	27	36	28	15	17	49	184	209	54	6	8	18	648
70	2	5	4	2	3	11	49	50	15	1	1	2	143
71	15	8	6	4	1	0	1	0	0	0	1	4	39
72	12	17	21	11	10	23	67	63	30	2	4	5	264

AVERAGE TEMPERATURE (DEGREES CENTIGRADE)

OBS.	JAN.	FEB.	MAR.	APR.	MAY	JUN.	JUL.	AUG.	SEP.	OCT.	NOV.	DEC.	ANNUAL
61	.	.	.	.	.	.	.	.	.	.	.	.	.
62	.	.	.	.	.	.	.	.	.	.	.	.	.
63	.	.	.	.	.	.	.	.	.	.	.	.	.
64	13.3	15.6	20.9	27.2	32.4	34.6	33.2	32.2	30.9	26.2	19.7	14.4	24.9
65	.	.	.	.	.	.	.	.	.	.	.	.	.
66	12.3	15.8	22.1	27.4	32.2	36.0	34.2	33.3	31.6	26.1	19.1	14.0	25.3
67	2.9	4.3	8.1	13.4	18.4	21.1	19.7	18.7	17.5	14.2	10.1	5.8	12.9
68	.	.	.	.	.	.	.	.	.	.	.	.	.
69	.	.	.	.	.	.	.	.	.	.	.	.	.
70	14.9	18.1	24.1	29.1	33.7	35.6	33.7	33.0	31.1	27.5	21.6	16.7	26.6
71	11.6	15.1	19.9	25.9	31.4	34.6	35.7	34.2	29.9	24.6	18.3	13.3	24.6
72	.	.	.	.	.	.	.	.	.	.	.	.	.

CLIMATOLOGICAL DATA FOR PAKISTAN

GEOGRAPHY

OBS.	LOCATION	YEARS	LATITUDE	LONGITUDE	ELEVATION -M-	LIFEZONE	CCTX	ID	ONMAP
73	NUSKI	18	29.55 N	66.02 E	.	SX	SW95	70	NO
74	NUTTAL	51	28.72 N	68.17 E	.	.	.	71	NO
75	PAKPATTAN	75	30.33 N	73.45 E	.	.	.	72	NO
76	PANJGUR	30	26.97 N	64.10 E	.	.	.	73	YES
77	PARACHINAR	30	33.90 N	70.10 E	.	WD	SC64	74	NO
78	PASNI	29	25.27 N	63.47 E	.	TX	SW94	75	NO
79	PESHAWA6	30	34.02 N	71.55 E	.	ST	TW84	76	NO
80	PHALIA	75	32.47 N	73.60 E	.	.	.	77	NO
81	QUETTA	30	30.20 N	67.00 E	.	WX	SC95	78	YES
82	RAJANPUR	75	28.50 N	70.17 E	.	.	.	79	NO
83	RAWALPINDI	30	33.60 N	73.07 E	.	SD	SW65	80	NO
84	SARGODHA	37	32.08 N	72.67 E	.	TT	SW85	81	NO

AVERAGE PRECIPITATION (MM)

OBS.	JAN.	FEB.	MAR.	APR.	MAY	JUN.	JUL.	AUG.	SEP.	OCT.	NOV.	DEC.	ANNUAL
73	40	29	24	7	3	0	4	1	0	0	2	16	127
74	4	7	7	1	3	5	20	18	4	1	3	5	79
75	11	11	8	7	8	22	67	54	31	1	1	5	226
76	21	24	15	10	3	3	20	9	1	0	1	14	122
77	82	76	136	104	64	47	122	112	54	18	13	31	858
78	43	33	8	7	2	6	16	3	1	0	2	12	132
79	39	41	65	42	14	7	34	41	14	10	10	15	331
80	29	30	29	24	19	39	118	149	56	7	5	11	515
81	52	53	43	18	10	4	21	9	0	3	5	25	244
82	6	8	7	6	8	5	21	25	7	1	1	3	98
83	63	63	81	42	23	55	233	258	85	21	12	23	960
84	14	23	24	21	16	29	95	96	39	5	2	9	373

AVERAGE TEMPERATURE (DEGREES CENTIGRADE)

OBS.	JAN.	FEB.	MAR.	APR.	MAY	JUN.	JUL.	AUG.	SEP.	OCT.	NOV.	DEC.	ANNUAL
73	10.8	13.6	18.3	24.6	29.4	33.2	34.7	33.2	28.8	23.3	16.8	11.7	23.2
74	.	.	.	.	.	.	.	.	.	.	.	.	.
75	.	.	.	.	.	.	.	.	.	.	.	.	.
76	.	.	.	.	.	.	.	.	.	.	.	.	.
77	4.1	6.1	10.1	15.3	20.5	24.6	24.9	23.8	21.7	16.9	14.4	7.1	15.6
78	18.6	19.8	23.3	26.6	29.6	30.5	29.7	28.2	27.5	26.6	23.4	19.8	25.3
79	10.7	13.2	17.4	22.9	29.1	32.9	32.6	30.9	28.9	23.6	17.4	12.5	22.7
80	.	.	.	.	.	.	.	.	.	.	.	.	.
81	3.6	6.3	10.6	15.5	20.4	24.2	26.6	25.2	20.2	14.2	8.9	5.3	15.1
82	.	.	.	.	.	.	.	.	.	.	.	.	.
83	9.8	12.5	17.1	23.0	29.1	32.2	30.6	28.9	27.9	23.1	16.4	11.4	21.8
84	11.8	14.6	19.7	25.4	29.7	34.1	32.6	31.4	30.1	25.3	17.7	13.3	23.8

CLIMATOLOGICAL DATA FOR PAKISTAN

GEOGRAPHY

OBS.	LOCATION	YEARS	LATITUDE	LONGITUDE	ELEVATION -M-	LIFEZONE	CCTX	ID	ONMAP
85	SHAHBANDAR	50	24.13 N	67.93 E	.	.	.	82	YES
86	SHAHDADKOT	27	30.82 N	72.13 E	.	.	.	83	NO
87	SHIKARPUR	60	27.97 N	68.70 E	.	.	.	84	NO
88	SHOTKOT	75	30.82 N	72.13 E	.	.	.	85	YES
89	SIALKOT	30	32.50 N	74.52 E	.	TV	SW65	86	NO
90	SIBI	30	29.55 N	67.88 E	.	TX	SH95	87	YES
91	SONMIANI	29	25.42 N	66.67 E	.	.	.	88	YES
92	SUKKUR	30	27.70 N	68.87 E	.	TO	SH95	89	NO
93	TANDO ADAM	27	25.73 N	68.68 E	.	.	.	90	NO
94	TANK	61	32.23 N	70.48 E	.	.	.	91	NO
95	TATTA	62	24.73 N	67.97 E	.	.	.	92	YES
96	THALL	51	33.40 N	70.60 E	.	.	.	93	YES

AVERAGE PRECIPITATION (MM)

OBS.	JAN.	FEB.	MAR.	APR.	MAY	JUN.	JUL.	AUG.	SEP.	OCT.	NOV.	DEC.	ANNUAL
85	4	7	4	1	2	15	120	53	17	1	1	2	227
86	4	10	7	4	2	5	25	30	17	2	1	5	112
87	5	8	6	3	2	3	18	26	8	1	2	4	84
88	9	11	12	12	7	20	59	56	21	2	2	5	215
89	46	42	41	20	13	56	15	248	97	21	8	20	883
90	17	18	18	6	4	7	38	20	5	1	5	6	143
91	8	16	7	4	1	9	44	18	7	0	1	7	122
92	5	7	6	2	2	3	33	23	4	2	1	1	90
93	1	5	2	2	3	9	71	43	13	0	1	5	155
94	14	19	27	20	11	16	64	59	15	4	3	8	261
95	5	8	4	3	3	19	102	49	13	0	2	2	211
96	22	26	46	34	17	23	80	83	27	10	5	12	385

AVERAGE TEMPERATURE (DEGREES CENTIGRADE)

OBS.	JAN.	FEB.	MAR.	APR.	MAY	JUN.	JUL.	AUG.	SEP.	OCT.	NOV.	DEC.	ANNUAL
85	.	.	.	.	.	.	.	.	.	.	.	.	.
86	.	.	.	.	.	.	.	.	.	.	.	.	.
87	.	.	.	.	.	.	.	.	.	.	.	.	.
88	.	.	.	.	.	.	.	.	.	.	.	.	.
89	12.0	14.9	19.9	26.4	32.1	33.9	31.1	29.7	29.3	24.9	18.3	13.4	23.8
90	13.8	17.2	22.9	29.8	35.3	38.1	36.6	35.2	27.7	27.8	21.3	15.8	27.2
91	.	.	.	.	.	.	.	.	.	.	.	.	.
92	15.1	18.3	23.6	29.4	34.1	35.5	34.2	32.9	31.8	28.3	22.3	17.2	26.9
93	.	.	.	.	.	.	.	.	.	.	.	.	.
94	.	.	.	.	.	.	.	.	.	.	.	.	.
95	.	.	.	.	.	.	.	.	.	.	.	.	.
96	.	.	.	.	.	.	.	.	.	.	.	.	.

CLIMATOLOGICAL DATA FOR PAKISTAN

GEOGRAPHY

OBS.	LOCATION	YEARS	LATITUDE	LONGITUDE	ELEVATION -M-	LIFEZONE	CCTX	ID	ONMAP
97	TOBATEK SINGH	41	30.90 N	72.50 E	.	.	.	94	NO
98	TURBAT	17	25.98 N	63.07 E	.	TX	SH95	95	NO
99	UMARKOT	18	25.37 N	69.73 E	.	TX	SH95	96	NO
100	WADH	10	27.35 N	66.50 E	.	.	.	97	NO
101	WAZIRABAD	75	32.38 N	74.17 E	.	.	.	98	NO
102	ZAFARWAL	55	32.33 N	74.88 E	.	.	.	99	NO

AVERAGE PRECIPITATION (MM)

OBS.	JAN.	FEB.	MAR.	APR.	MAY	JUN.	JUL.	AUG.	SEP.	OCT.	NOV.	DEC.	ANNUAL
97	7	9	12	13	8	23	54	63	25	5	2	7	230
98	41	21	18	6	10	6	25	9	2	1	2	14	155
99	1	3	1	1	1	10	50	76	23	5	1	1	172
100	58	25	25	9	6	12	18	8	0	0	0	1	162
101	31	30	28	17	17	42	138	164	63	5	5	14	555
102	51	41	35	20	19	67	216	234	79	8	5	16	792

AVERAGE TEMPERATURE (DEGREES CENTIGRADE)

OBS.	JAN.	FEB.	MAR.	APR.	MAY	JUN.	JUL.	AUG.	SEP.	OCT.	NOV.	DEC.	ANNUAL
97	.	.	.	.	.	.	.	.	.	.	.	.	.
98	18.1	20.3	25.3	29.9	34.8	36.4	34.1	33.4	32.1	29.4	24.5	19.8	28.3
99	16.6	20.1	25.9	30.3	33.7	34.1	32.0	24.8	30.2	28.3	23.1	18.1	26.9
100	.	.	.	.	.	.	.	.	.	.	.	.	.
101	.	.	.	.	.	.	.	.	.	.	.	.	.
102	.	.	.	.	.	.	.	.	.	.	.	.	.

PANAMA*

Agriculture

Of 7,708,000 ha in total area, Panama was estimated in 1978 to have 566,000 ha of arable or permanent cropland, 1,161,000 ha in permanent pasture, 4,156,000 ha in forest and woodland and 1,716,000 ha in other terrain categories. In 1978, approximately 26,000 ha were irrigated. Of the 1979 population, 35.2% or 224,000 individuals of the economically active population were engaged in agriculture. The total population in 1979 was 1,899,000. According to FAO statistics for 1979, there were 165,000 horses, 5000 mules, 1,423,000 cattle, 205,000 pigs, 6000 goats, 4,900,000 chickens, 124,000 ducks, and 37,000 turkeys.[2] Agricultural production is summarized in Table 1. Conventional and energetic yield data on most of these will be found in Appendixes I and II.

Climate

Data from Wernstedt[3] are presented in the Climatological Data Table (see Introduction for explanation). Based on the means for data points cited in Wernstedt, the mean annual precipitation is 26.5 dm (mean of 61 sites) and the mean annual temperature 26.8°C (mean of 5 sites). The averages would suggest that much of the country could be classified as Tropical Moist Forest Life Zone.[4]

Productivity

The Life Zone Map, based on Holdridge and Budowski,[34] shows the Holdridge Life Zones of Panama. The annual productivities in these Life Zones are estimated at:

Vegetation type	MT/ha/yr
Tropical Moist Forest	18—36
Tropical Moist Forest (Transition)	12—36
Tropical Dry Forest	12—18
Tropical Dry Forest (Transition)	12—18
Subtropical Moist Forest	12—18
Subtropical Dry Forest	6—12
Lower Montane Wet Forest	18—24
Montane Wet Forest	12—18

Recently, I toured Panama in a survey of annual biomass productivity and standing biomass.[29] Dr. C. Curtis, University of Delaware, and I worked with the recent Atlas Nacional De Panama to calculate productivity and standing biomass. We assigned productivity values to the units in the Atlas of Panama. For the Holdridge Life Zones of that Atlas (not included here) we project the following values:

Tropical Dry Forest	12—18 MT/ha/yr
Tropical Moist Forest	18—36 MT/ha/yr
Tropical Wet Forest	18—36 MT/ha/yr
Premontane Dry Forest	6—12 MT/ha/yr
Premontane Moist Forest	12—18 MT/ha/yr
Premontane Wet Forest	18—24 MT/ha/yr
Premontane Rain Forest	18—24 MT/ha/yr
Lower Montane Moist Forest	12—18 MT/ha/yr
Lower Montane Wet Forest	12—18 MT/ha/yr
Lower Montane Rain Forest	12—18 MT/ha/yr
Montane Wet Forest	6—12 MT/ha/yr
Montane Rain Forest	6—12 MT/ha/yr

* Maps and table follow text.

Table 1
AGRICULTURAL PRODUCTION IN PANAMA

Crop	Area planted (ha)	Yield (kg/ha)	Total production (MT)
Rice paddy	110,000	1,818	200,000
Corn	68,000	956	65,000
Potatoes	1,000	11,000	11,000
Cassava	5,000	8,511	40,000
Dry beans	10,000	312	3,000
Sesame seed		500	
Coconuts			24,000
Copra			1,000
Cabbages		9,897	2,000
Tomatoes	1,000	20,833	25,000
Cucumbers and gherkins		7,797	
Chilies and peppers, green		6,667	1,000
Dry onions		12,250	2,000
Carrots		9,074	1,000
Watermelons		6,356	1,000
Cantaloupes and other melons		7,946	1,000
Sugar cane	44,000	59,408	2,624,000
Sugar (centrifugal, raw)			225,000
Sugar (noncentrifugal)			2,000
Oranges			65,000
Avocados			2,000
Mangoes			26,000
Pineapples			8,000
Bananas			1,000,000
Plantains			100,000
Green coffee	24,000	238	6,000
Cocoa beans	4,000	250	1,000
Tobacco leaves	1,000	1,373	2,000

From FAO 1979 Production Yearbook, Vol. 33, Food and Agriculture Organization, Rome, 1980.

For another map from the Atlas, we projected the following values:

Tropical Evergreen Forest	18—36 MT/ha/yr
Subtropical Evergreen Forest	18—24 MT/ha/yr
Upland Evergreen Forest	12—18 MT/ha/yr
Tropical Semideciduous Forest	18—36 MT/ha/yr
Tropical Deciduous Forest	12—36 MT/ha/yr
Forests and Floodplains	10—30 MT/ha/yr
Cultivated areas, savannas, and secondary vegetation	10 MT/ha/yr

During my Panama stay, I worked with a University of Georgia team[35] who actually measured the standing biomass in the deciduous forest (tropical moist) at 263 to 368 MT/ha, the cloud forest (premontane wet) at 272 MT/ha, the fresh water swamp forest (tropical gallery) at 1176 MT/ha, and the salt water swamp forest (mangrove) at 280 MT/ha/yr. If measured biomass higher than the 1176 for the swamp forest has been reported, I have not seen the report. I estimate the annual productivity of forest types of Darien as follows:

Thorn and littoral	3—12 MT/ha/yr
Deciduous seasonal	18—36 MT/ha/yr
Monsoon	18—36 MT/ha/yr

Salt water swamp	12—18 MT/ha/yr
Brackish water swamp	18—36 MT/ha/yr
Fresh water swamp	18—36 MT/ha/yr
Fresh water swamp (with palms)	18—36 MT/ha/yr
Evergreen seasonal	18—36 MT/ha/yr
Premontane rain	18—36 MT/ha/yr
Cloud	18—36 MT/ha/yr
Mossy or elfin	6—18 MT/ha/yr

Economics

Food consumption per capita was 2341 cal/day, energy consumption 981 kg coal equivalent per capita in 1977. In 1970, the caloric supply was 116% of requirements. The total population growth rate was 2.9% in 1970. Population density was 20 persons per square kilometer in 1980, with 48.1% of the total population urban. Life expectancy at birth of those born in 1970 was 66 years. In 1977, with GNP at $1200 per capita, agriculture accounted for 19.2% of GDP, manufacturing for 15.9%, other industry for 7.9%, and services for 57.0%. In 1977, food and beverage constituted 62.7% of exports, nonfood agriculture for 0.2%, fuels, minerals, and metal for 28.0%, machines for 1.6%, and other manufacturing for 7.6% of total merchandise exports. Food and beverage constituted 10.1% of imports, nonfood agriculture 0.4%, fuel and lubricants 32.8%, nonfuel minerals and metals 0.7%, machinery and equipment 19.3%, and other manufactured goods constituted 36.7% of total merchandise imports.[10]

Energetics

According to AID, in 1975 ca. 71% of Panama's energy was produced from oil, all imported, with 13% from hydro, and 15% from wood and bagasse. Overall energy demand is increasing at the rate of 6.8%/yr. Geothermal sources are being investigated near the Panamanian border. The price of electric energy is more expensive than in other Latin American countries (except Uruguay) with regular gas at $0.264/ℓ, super at 0.291, diesel at 0.145, kerosene at 0.145, with industrial energy at 0.052 per kWh, residential at 0.0614, and commercial at 0.0564. According to our contacts, petrol-generated energy in Yaviza is 0.17 per kWh as compared with a projected wood-generated cost of 0.52, exactly ten times the 1976 industrial rate overall in Panama.

Micro-hydros (5 to 50 kW) are being financed under AID loans. MIDA (Ministry of Agricultural Development) has a mandate to begin alcohol production from sugar cane and other agricultural products and wastes and use it for fuel in transportation and other sectors.

IRHE has purchased five windmills. Municipal sewage and refuse is the responsibility of other GOP organizations. Thus IRHE has focused on five sources — micro-hydros, biomass, agricultural and animal wastes, and solar.

Following an aggregate of nearly 5 years in Central America, mostly Panama, with emphasis on Darien, I feel rather safe in the following conclusion:

I assume Panama needs 15,000,000 bbl of oil per year and I assume average Panama forests could easily produce an annual 25 m^3/ha fire wood. I have elsewhere assumed that 1 m^3 of wood is energetically equivalent to 2 bbl of oil, suggesting a return of 50 bbl of oil per hectare per year. Based on these assumptions, Panama could satisfy its energy requirements renewably with 300,000 ha (ca. 4% of Panama) in fuelwood plantations producing the equivalent of 50 bbl/ha/yr. Pessimists about firewood production could scale yields to 20 bbl/ha/yr and still require only 10% of Panama to satisfy her energy requirement.

According to the World Bank, Panama has a hydroelectric potential of 2400 MW. There has been little systematic evaluation of geological information to identify coal occurrences or potential coal-bearing regions; basic exploration work is required to establish an inventory of commercial coal prospects. Importing more than 75% of its commercial energy demands as oil, Panama is not expected to be experiencing fuel-wood problems by the year 2000.[10]

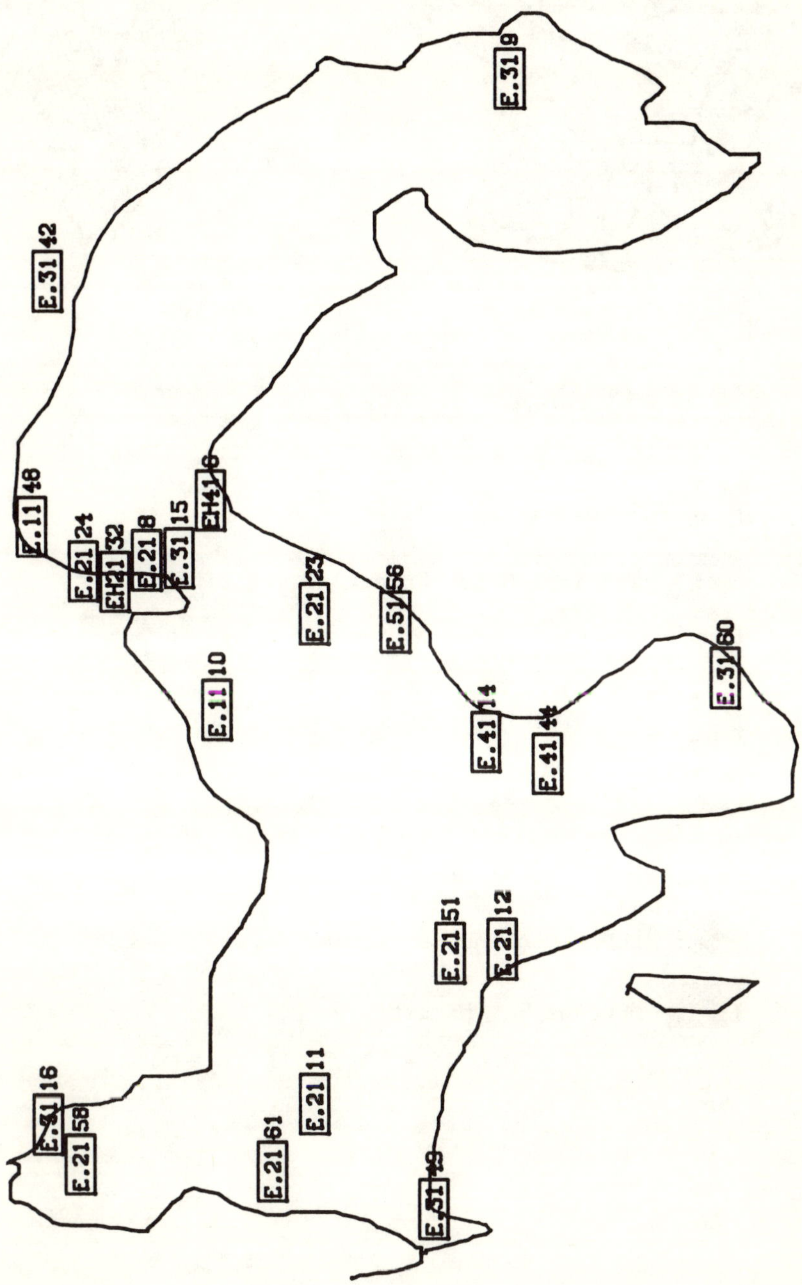

Selected weather stations for Panama.

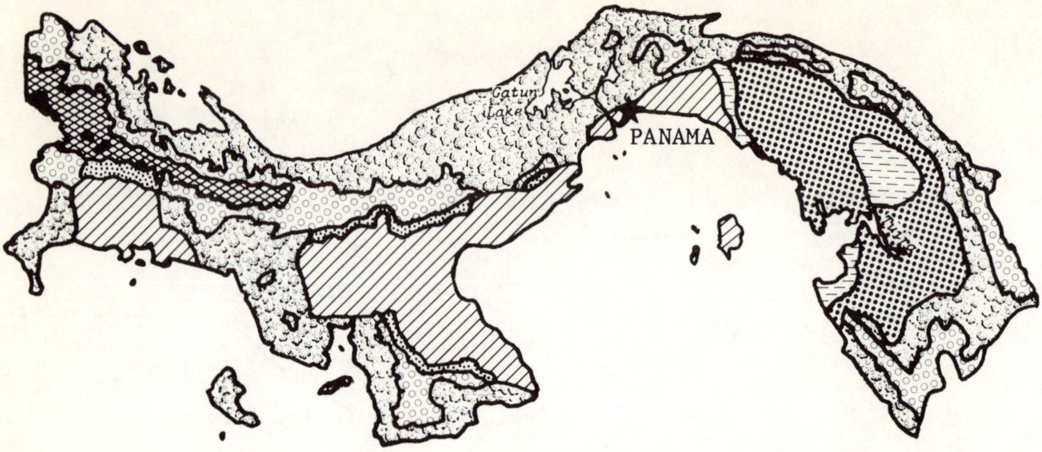

PANAMA VEGETATION

- Tropical Moist Forest
- Tropical Moist Forest (Transition)
- Tropical Dry Forest
- Tropical Dry Forest (Transition)
- Subtropical Moist Forest
- Subtropical Dry Forest
- Lower Montane Wet Forest
- Montane Wet Forest

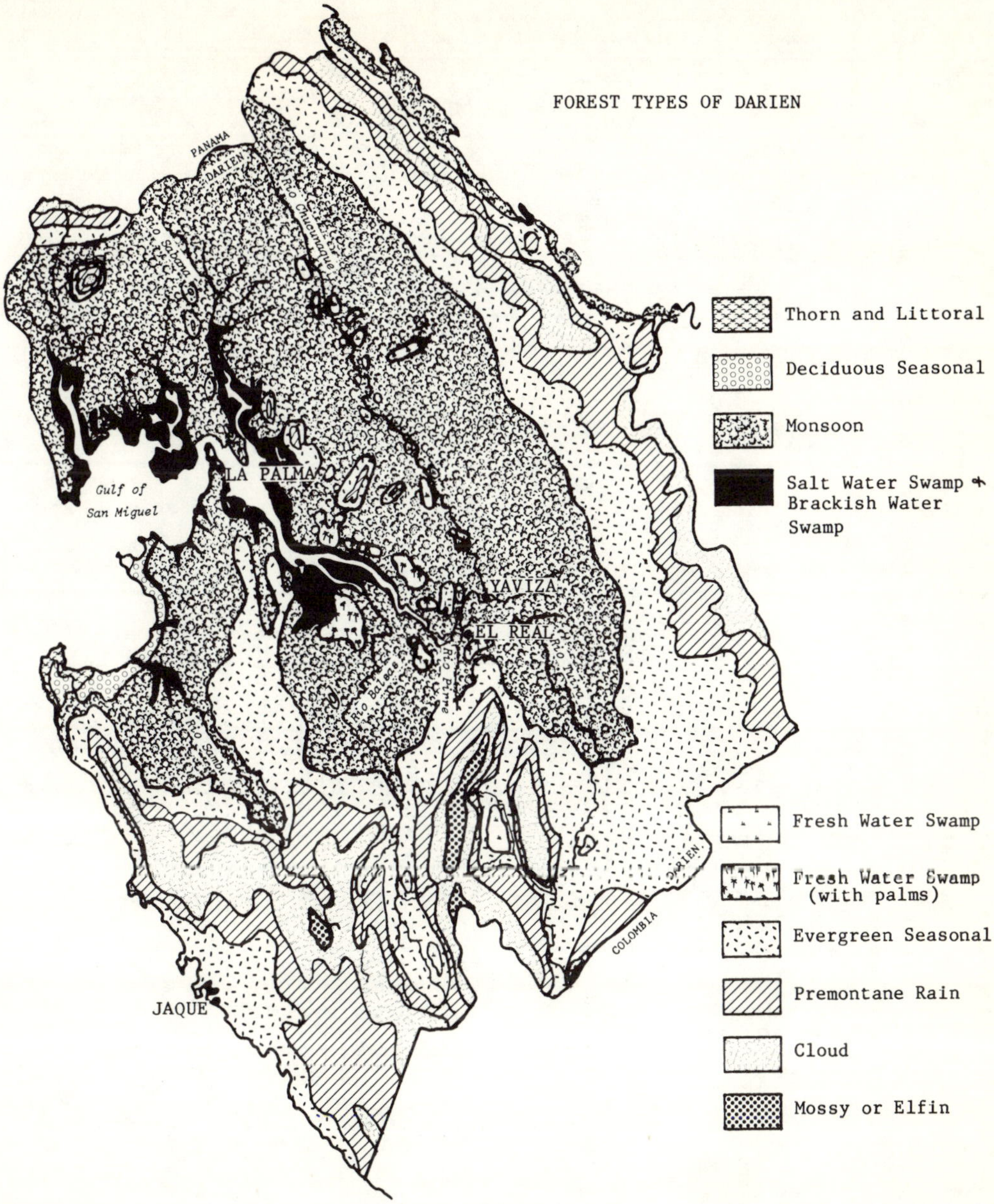
FOREST TYPES OF DARIEN
PANAMA
DARIEN
Gulf of
San Miguel
LA PALMA
YAVIZA
EL REAL
JAQUE
DARIEN
COLOMBIA
Thorn and Littoral
Deciduous Seasonal
Monsoon
Salt Water Swamp &
Brackish Water
Swamp
Fresh Water Swamp
Fresh Water Swamp
(with palms)
Evergreen Seasonal
Premontane Rain
Cloud
Mossy or Elfin

CLIMATOLOGICAL DATA FOR PANAMA

GEOGRAPHY

OBS.	LOCATION	YEARS	LATITUDE	LONGITUDE	ELEVATION -M-	LIFEZONE	CCTX	ID	ONMAP
1	AGUA CLARA	24	9.37 N	79.72 W	457.2	.	.	1	NO
2	AGUA SUCIA	15	9.25 N	79.72 W	304.8	.	.	2	NO
3	ALANJE	10	8.30 N	82.53 W	.	.	.	3	NO
4	ALBROOK AFB	18	8.98 N	79.57 W	12.5	TD	EH41	4	NO
5	BALBOA DOCKS	67	8.97 N	79.57 W	9.1	.	.	5	NO
6	BALBOA HEIGHTS	68	8.97 N	79.55 W	30.5	TD	EH41	6	YES
7	BAMBITO	10	8.83 N	82.62 W	.	.	.	7	NO
8	BARRIO COLORADO	41	9.17 N	79.83 W	64.0	.	.	8	YES
9	BOCA DE CUPE	10	8.05 N	77.58 W	.	.	.	9	YES
10	BOCA DE TOABRE	10	8.95 N	80.53 W	.	.	.	10	YES
11	CALDERA	10	8.65 N	82.38 W	.	.	.	11	YES
12	CAMARON	10	8.07 N	81.65 W	.	.	.	12	YES

AVERAGE PRECIPITATION (MM)

OBS.	JAN.	FEB.	MAR.	APR.	MAY	JUN.	JUL.	AUG.	SEP.	OCT.	NOV.	DEC.	ANNUAL
1	89	40	49	126	402	329	327	219	315	499	568	398	3362
2	52	29	19	89	319	279	285	255	282	380	391	237	2617
3	19	75	17	142	226	267	264	328	279	389	282	133	2421
4	45	14	17	48	223	199	203	204	201	296	263	178	1892
5	29	15	16	76	203	200	185	194	193	263	255	135	1764
6	39	17	15	69	204	206	188	196	199	275	263	126	1787
7	22	20	24	84	247	429	209	227	296	457	239	76	2330
8	58	33	30	87	277	277	294	317	263	352	461	269	2718
9	42	53	54	158	328	344	246	222	202	196	170	81	2096
10	286	144	125	390	396	318	351	399	257	346	411	656	4078
11	17	75	104	229	307	440	415	425	642	866	332	66	3918
12	14	42	33	164	270	482	403	474	534	812	595	122	3945

AVERAGE TEMPERATURE (DEGREES CENTIGRADE)

OBS.	JAN.	FEB.	MAR.	APR.	MAY	JUN.	JUL.	AUG.	SEP.	OCT.	NOV.	DEC.	ANNUAL
1	.	.	.	.	.	.	.	.	.	.	.	.	.
2	.	.	.	.	.	.	.	.	.	.	.	.	.
3	.	.	.	.	.	.	.	.	.	.	.	.	.
4	26.1	27.6	27.9	28.6	27.2	26.8	27.1	27.4	27.2	26.7	26.6	27.0	27.2
5	.	.	.	.	.	.	.	.	.	.	.	.	.
6	26.7	26.9	27.6	27.9	27.2	26.7	26.9	26.8	26.6	26.2	26.2	26.6	26.8
7	.	.	.	.	.	.	.	.	.	.	.	.	.
8	.	.	.	.	.	.	.	.	.	.	.	.	.
9	.	.	.	.	.	.	.	.	.	.	.	.	.
10	.	.	.	.	.	.	.	.	.	.	.	.	.
11	.	.	.	.	.	.	.	.	.	.	.	.	.
12	.	.	.	.	.	.	.	.	.	.	.	.	.

CLIMATOLOGICAL DATA FOR PANAMA

GEOGRAPHY

OBS.	LOCATION	YEARS	LATITUDE	LONGITUDE	ELEVATION -M-	LIFEZONE	CCTX	ID	ONMAP
13	CANDELARIA	32	9.38 N	79.53 W	100.6	.	.	13	NO
14	CANELOS (LOS)	10	8.12 N	80.68 W	.	.	.	14	YES
15	CANO	43	9.07 N	79.82 W	33.5	.	.	15	YES
16	CHANGUINOLA	47	9.48 N	82.48 W	4.9	.	.	16	YES
17	CHASE	10	9.45 N	82.83 W	.	.	.	17	NO
18	CHIFIBRILLO	44	9.20 N	79.62 W	61.0	.	.	18	NO
19	CHIGUIRI ARRIBA	10	8.68 N	80.22 W	.	.	.	19	NO
20	CHOCO	23	9.27 N	79.52 W	121.9	.	.	20	NO
21	CHORRO (EL)	19	8.97 N	79.98 W	42.7	.	.	21	NO
22	CIENTO	17	9.30 N	79.73 W	38.1	.	.	22	NO
23	CIRI GRANDE	15	8.65 N	80.08 W	1121.7	.	.	23	YES
24	COCO SOLO	31	9.37 N	79.88 W	4.0	.	.	24	YES

AVERAGE PRECIPITATION (MM)

OBS.	JAN.	FEB.	MAR.	APR.	MAY	JUN.	JUL.	AUG.	SEP.	OCT.	NOV.	DEC.	ANNUAL
13	109	63	42	143	355	351	350	374	320	350	430	328	3214
14	1	5	23	75	117	138	159	268	262	368	361	37	1813
15	49	26	21	77	260	239	242	253	244	343	370	200	2324
16	239	158	145	183	215	200	250	235	128	140	257	341	2493
17	153	127	125	138	219	199	249	176	71	137	263	277	2135
18	30	17	10	72	270	296	317	298	283	375	331	135	2434
19	89	44	31	165	295	435	393	461	334	441	410	308	3406
20	43	21	12	65	313	307	353	340	335	385	399	200	2773
21	71	27	30	83	284	223	215	269	314	321	374	209	2419
22	89	53	44	109	363	575	348	370	332	490	576	365	3714
23	95	46	33	100	365	374	321	363	399	475	353	224	3149
24	78	35	36	104	316	353	396	414	298	432	637	341	3440

AVERAGE TEMPERATURE (DEGREES CENTIGRADE)

OBS.	JAN.	FEB.	MAR.	APR.	MAY	JUN.	JUL.	AUG.	SEP.	OCT.	NOV.	DEC.	ANNUAL
13	.	.	.	.	.	.	.	.	.	.	.	.	.
14	.	.	.	.	.	.	.	.	.	.	.	.	.
15	.	.	.	.	.	.	.	.	.	.	.	.	.
16	.	.	.	.	.	.	.	.	.	.	.	.	.
17	.	.	.	.	.	.	.	.	.	.	.	.	.
18	.	.	.	.	.	.	.	.	.	.	.	.	.
19	.	.	.	.	.	.	.	.	.	.	.	.	.
20	.	.	.	.	.	.	.	.	.	.	.	.	.
21	.	.	.	.	.	.	.	.	.	.	.	.	.
22	.	.	.	.	.	.	.	.	.	.	.	.	.
23	.	.	.	.	.	.	.	.	.	.	.	.	.
24	.	.	.	.	.	.	.	.	.	.	.	.	.

CLIMATOLOGICAL DATA FOR PANAMA

GEOGRAPHY

OBS.	LOCATION	YEARS	LATITUDE	LONGITUDE	ELEVATION -M-	LIFEZONE	CCTX	ID	ONMAP
25	COCOLI	37	8.98 N	79.62 W	21.3	.	.	25	NO
26	CRISTOBAL	96	9.35 N	79.92 W	12.2	TM	EH21	26	NO
27	DIVISA	14	8.13 N	80.67 W	22.9	.	.	27	NO
28	ESCANDALOSA	10	9.43 N	79.58 W	458.7	.	.	28	NO
29	FERNANDEZ	15	8.93 N	79.95 W	253.0	.	.	29	NO
30	FRIJOLES	45	9.18 N	79.80 W	30.5	.	.	30	NO
31	GAMBOA	82	9.12 N	79.70 W	33.5	.	.	31	NO
32	GATUN	62	9.27 N	79.93 W	29.0	TM	EH21	32	YES
33	HARINO (EL)	10	8.62 N	80.57 W	.	.	.	33	NO
34	HORNITO	10	8.73 N	82.23 W	.	.	.	34	NO
35	INDIO RIO	44	9.22 N	79.53 W	88.4	.	.	35	NO
36	ISLA GRANDE	36	9.63 N	79.57 W	73.2	.	.	36	NO

AVERAGE PRECIPITATION (MM)

OBS.	JAN.	FEB.	MAR.	APR.	MAY	JUN.	JUL.	AUG.	SEP.	OCT.	NOV.	DEC.	ANNUAL
25	38	9	10	62	257	257	237	248	241	309	301	156	2125
26	84	39	38	103	317	332	393	391	317	317	575	308	3215
27	10	4	1	34	222	199	155	173	220	280	248	72	1617
28	174	64	75	129	304	349	381	344	297	278	524	280	3198
29	69	36	26	71	271	238	227	244	248	358	312	214	2313
30	51	32	24	85	275	279	303	302	283	379	463	221	2697
31	37	18	14	77	264	246	249	267	249	310	310	158	2199
32	84	47	45	121	328	294	324	346	284	415	549	313	3149
33	80	28	22	55	171	275	136	202	243	344	202	78	1836
34	320	207	220	264	219	400	280	340	295	385	387	406	3722
35	28	17	9	63	286	332	372	356	351	391	361	133	2701
36	60	40	23	48	285	278	290	283	237	299	457	284	2586

AVERAGE TEMPERATURE (DEGREES CENTIGRADE)

OBS.	JAN.	FEB.	MAR.	APR.	MAY	JUN.	JUL.	AUG.	SEP.	OCT.	NOV.	DEC.	ANNUAL
25	.	.	.	.	.	.	.	.	.	.	.	.	.
26	26.8	26.8	27.1	27.4	27.3	27.1	27.0	26.9	27.0	26.8	26.4	26.7	26.9
27	.	.	.	.	.	.	.	.	.	.	.	.	.
28	.	.	.	.	.	.	.	.	.	.	.	.	.
29	.	.	.	.	.	.	.	.	.	.	.	.	.
30	.	.	.	.	.	.	.	.	.	.	.	.	.
31	.	.	.	.	.	.	.	.	.	.	.	.	.
32	26.7	26.7	27.1	27.5	27.4	27.1	26.9	26.9	27.2	27.5	28.0	26.6	27.1
33	.	.	.	.	.	.	.	.	.	.	.	.	.
34	.	.	.	.	.	.	.	.	.	.	.	.	.
35	.	.	.	.	.	.	.	.	.	.	.	.	.
36	.	.	.	.	.	.	.	.	.	.	.	.	.

CLIMATOLOGICAL DATA FOR PANAMA

GEOGRAPHY

OBS.	LOCATION	YEARS	LATITUDE	LONGITUDE	ELEVATION -M-	LIFEZONE	CCTX	ID	ONMAP
37	JEFE (EL)	16	9.23 N	79.40 W	975.4	.	.	37	NO
38	LIMPIO	15	9.35 N	79.45 W	548.6	.	.	38	NO
39	MADDEN DAM	67	9.20 N	79.62 W	76.2	TM	EH21	39	NO
40	MADRONAL	26	9.23 N	79.62 W	61.0	.	.	40	NO
41	MONTE LIRIO	58	9.23 N	79.85 W	85.3	.	.	41	NO
42	NARGANA	14	9.48 N	78.52 W	3.0	.	.	42	YES
43	NUEVA CALIFORNIA	10	8.78 N	82.65 W	.	.	.	43	NO
44	OCU	10	7.93 N	80.78 W	.	.	.	44	YES
45	PEDRO MIGUEL	59	9.02 N	79.62 W	33.5	.	.	45	NO
46	PEDRO MIGUEL R.	38	9.05 N	79.62 W	30.5	.	.	46	NO
47	PELUCA	30	9.38 N	79.57 W	106.7	.	.	47	NO
48	PORTOBELO	55	9.53 N	79.67 W	4.6	.	.	48	YES

AVERAGE PRECIPITATION (MM)

OBS.	JAN.	FEB.	MAR.	APR.	MAY	JUN.	JUL.	AUG.	SEP.	OCT.	NOV.	DEC.	ANNUAL
37	85	52	30	80	437	408	373	405	465	599	392	250	3575
38	103	66	56	108	315	311	309	309	317	406	335	288	2923
39	26	13	12	75	280	292	308	297	281	355	574	146	2659
40	34	14	9	58	304	278	301	280	275	350	393	183	2479
41	61	46	41	102	284	290	294	325	292	402	502	288	2926
42	85	42	18	87	295	243	290	274	180	223	452	176	2365
43	1	36	32	98	286	378	266	305	439	491	305	29	2666
44	1	0	0	84	291	305	156	189	195	390	237	57	1905
45	30	11	10	81	249	236	233	231	225	301	290	148	2045
46	31	10	12	69	250	250	246	242	241	286	299	153	2090
47	98	55	38	141	339	324	324	336	296	331	428	330	3039
48	129	74	65	165	448	425	464	494	342	402	637	411	4057

AVERAGE TEMPERATURE (DEGREES CENTIGRADE)

OBS.	JAN.	FEB.	MAR.	APR.	MAY	JUN.	JUL.	AUG.	SEP.	OCT.	NOV.	DEC.	ANNUAL
37	.	.	.	.	.	.	.	.	.	.	.	.	.
38	.	.	.	.	.	.	.	.	.	.	.	.	.
39	25.6	25.8	26.4	27.1	26.9	26.5	26.4	26.3	26.3	26.1	25.8	25.7	26.2
40	.	.	.	.	.	.	.	.	.	.	.	.	.
41	.	.	.	.	.	.	.	.	.	.	.	.	.
42	.	.	.	.	.	.	.	.	.	.	.	.	.
43	.	.	.	.	.	.	.	.	.	.	.	.	.
44	.	.	.	.	.	.	.	.	.	.	.	.	.
45	.	.	.	.	.	.	.	.	.	.	.	.	.
46	.	.	.	.	.	.	.	.	.	.	.	.	.
47	.	.	.	.	.	.	.	.	.	.	.	.	.
48	.	.	.	.	.	.	.	.	.	.	.	.	.

CLIMATOLOGICAL DATA FOR PANAMA

GEOGRAPHY

OBS.	LOCATION	YEARS	LATITUDE	LONGITUDE	ELEVATION -M-	LIFEZONE	CCTX	ID	ONMAP
49	PUERTO ARMUELLES	37	8.28 N	82.87 W	15.2	.	.	49	YES
50	RAICES (LAS)	35	9.10 N	79.98 W	33.5	.	.	50	NO
51	REMEDIOS	12	8.23 N	81.67 W	41.1	.	.	51	YES
52	RIO GRANDE	10	8.42 N	80.48 W	18.0	.	.	52	NO
53	SADDLE FIVE	26	9.20 N	79.60 W	85.3	.	.	53	NO
54	SALAMANCA	53	9.32 N	79.58 W	82.3	.	.	54	NO
55	SAN MIGUEL	21	9.45 N	79.50 W	515.1	.	.	55	NO
56	SANTA CLARA	24	8.40 N	80.12 W	36.6	.	.	56	YES
57	SANTA ROSA	41	8.20 N	80.67 W	24.4	.	.	57	NO
58	SEIYIC	10	9.38 N	82.67 W	.	.	.	58	YES
59	SUMMIT	42	9.07 N	79.65 W	82.3	.	.	59	NO
60	TONOSI	42	7.38 N	80.38 W	15.2	.	.	60	YES

AVERAGE PRECIPITATION (MM)

OBS.	JAN.	FEB.	MAR.	APR.	MAY	JUN.	JUL.	AUG.	SEP.	OCT.	NOV.	DEC.	ANNUAL
49	25	18	26	82	236	236	203	266	292	544	293	85	2306
50	59	31	24	83	261	234	220	241	244	345	336	210	2288
51	27	35	26	120	368	434	403	317	403	691	446	117	3386
52	7	1	4	33	151	174	127	143	216	303	228	57	1445
53	31	13	9	53	292	304	335	306	298	345	363	162	2511
54	35	20	11	73	263	299	296	291	305	363	362	163	2482
55	154	62	66	180	356	341	340	369	313	352	445	344	3322
56	11	3	1	15	154	141	110	139	127	186	172	77	1137
57	9	10	1	36	215	203	169	195	232	353	247	72	1742
58	228	122	237	245	181	265	199	193	194	284	328	305	2779
59	23	9	10	79	262	231	257	255	251	306	324	149	2156
60	26	2	4	46	224	237	225	223	231	418	358	130	2124

AVERAGE TEMPERATURE (DEGREES CENTIGRADE)

OBS.	JAN.	FEB.	MAR.	APR.	MAY	JUN.	JUL.	AUG.	SEP.	OCT.	NOV.	DEC.	ANNUAL
49	.	.	.	.	.	.	.	.	.	.	.	.	.
50	.	.	.	.	.	.	.	.	.	.	.	.	.
51	.	.	.	.	.	.	.	.	.	.	.	.	.
52	.	.	.	.	.	.	.	.	.	.	.	.	.
53	.	.	.	.	.	.	.	.	.	.	.	.	.
54	.	.	.	.	.	.	.	.	.	.	.	.	.
55	.	.	.	.	.	.	.	.	.	.	.	.	.
56	.	.	.	.	.	.	.	.	.	.	.	.	.
57	.	.	.	.	.	.	.	.	.	.	.	.	.
58	.	.	.	.	.	.	.	.	.	.	.	.	.
59	.	.	.	.	.	.	.	.	.	.	.	.	.
60	.	.	.	.	.	.	.	.	.	.	.	.	.

CLIMATOLOGICAL DATA FOR PANAMA

GEOGRAPHY

OBS.	LOCATION	YEARS	LATITUDE	LONGITUDE	ELEVATION -M-	LIFEZONE	CCTX	ID	ONMAP
61	VOLCAN (EL)	10	8.78 N	82.70 W	1219.2	.	.	61	YES

AVERAGE PRECIPITATION (MM)

OBS.	JAN.	FEB.	MAR.	APR.	MAY	JUN.	JUL.	AUG.	SEP.	OCT.	NOV.	DEC.	ANNUAL
61	26	27	45	168	468	480	340	391	533	632	340	89	3539

AVERAGE TEMPERATURE (DEGREES CENTIGRADE)

OBS.	JAN.	FEB.	MAR.	APR.	MAY	JUN.	JUL.	AUG.	SEP.	OCT.	NOV.	DEC.	ANNUAL
61	.	.	.	.	.	.	.	.	.	.	.	.	.

PARAGUAY*

Agriculture

Of 40,675,000 ha in total area, Paraguay was estimated in 1978 to have 1,155,000 ha of arable or permanent cropland, 15,100,000 ha in permanent pasture, 20,380,000 ha in forest and woodland, and 3,095,000 ha in other terrain categories. In 1978, approximately 55,000 ha were irrigated. Of the 1979 population, 49.3% or 472,000 individuals of the economically active population were engaged in agriculture. The total population in 1979 was 2,979,000. According to FAO statistics for 1979, there were 334,000 horses, 13,000 mules, 28,000 asses, 5,203,000 cattle, 1,273,000 pigs, 423,000 sheep, 126,000 goats, 12,471,000 chickens, 303,000 ducks, and 43,000 turkeys.[2] Agricultural production is summarized in Table 1. Conventional and energetic yield data on most of these will be found in Appendixes I and II.

Climate

Data from Wernstedt[3] are presented in the Climatological Data Table (see Introduction for explanation). Based on the means for data points cited in Wernstedt, the mean annual precipitation is 13.4 dm (mean of 19 sites) and the mean annual temperature 23.3°C (mean of 16 sites). The averages would suggest that much of the country could be classified as Subtropical Moist Forest Life Zone.[4]

Productivity

The vegetation map depicts vegetation types in Paraguay. Based on the Wernstedt means for Paraguay, I estimate the annual potential productivities, assuming reasonable inputs at:

Vegetation type	MT/ha/yr
Broadleaf evergreen forest	12—24
Thorny shrubs	3—6
Deciduous scrub forest	6—12
Thorny shrubs and grass	3—6
Grass	3—6
Cultivated vegetation	3—12

Economics

Food consumption per capita was 2824 cal/day, energy consumption 200 kg coal equivalant per captia in 1977. In 1970, the caloric supply was 121.0% of requirements. The total population growth rate was 2.6% in 1976. Population density was 6 persons per square kilometer in 1970, with 37.1% of the total population urban. Life expectancy at birth of those born in 1970 was 61 years. In 1977, with GNP at $750 per capita, agriculture accounted for 34.1% of GDP, manufacturing for 17.1%, other industry for 6.0%, and services for 42.8%. In 1976, food and beverage consituted 60.8% of exports, nonfood agriculture for 24.8%, fuels, minerals, and metal for 1.1%, and other manufacturing for 13.2% total merchandise exports. Food and beverage constituted 15.1% of imports, nonfood agriculture 0.3%, fuel and lubricants 24.9%, nonfuel minerals and metals 1.1%, machinery and equipment 34.0%, and other manufactured goods constituted 24.5% of total merchandise imports.[10]

* Maps and table follow text.

Table 1
AGRICULTURAL PRODUCTION IN PARAGUAY

Crop	Area planted (ha)	Yield (kg/ha)	Total production (MT)
Wheat	52,000	1,114	58,000
Rice paddy	30,000	1,890	57,000
Corn	377,000	1,553	585,000
Sorghum	7,000	1,234	9,000
Potatoes	1,000	8,828	9,000
Sweet potatoes	15,000	8,267	124,000
Cassava	120,000	12,450	1,494,000
Dry beans	79,000	731	58,000
Dry broad beans	16,000	938	15,000
Dry peas	4,000	875	4,000
Soybeans	360,000	1,524	549,000
Peanuts in shell	24,000	980	23,000
Castor beans	10,000	1,000	10,000
Seed cotton	360,000	639	230,000
Cottonseed			145,000
Tung oil			16,000
Palm kernels			16,000
Palm oil			4,500
Tomatoes	3,000	18,667	56,000
Dry onions	4,000	6,250	25,000
Garlic	1,000	2,143	2,000
Watermelons	13,000	4,923	64,000
Cantaloupes and other melons	6,000	4,000	22,000
Grapes	2,000	7,500	15,000
Wine			7,000
Sugar cane	48,000	35,417	1,700,000
Sugar (centrifugal, raw)			69,000
Apples			1,000
Plums			2,000
Oranges			247,000
Tang. mand. clement satsuma			36,000
Lemons and limes			17,000
Grapefruit and pomelo			51,000
Avocados			2,000
Mangoes			30,000
Pineapples			70,000
Bananas			255,000
Papayas			13,000
Green coffee	16,000	450	7,000
Tobacco leaves	16,000	1,226	19,000
Cotton lint			71,000

From FAO 1979 Production Yearbook, Vol. 33, Food and Agriculture Organization, Rome, 1980.

Energetics

Paraguay has a hydroelectric potential of 6000 MW. Importing more than 75% of its commerical energy demands as oil, Paraguay is not expected to be experiencing fuelwood problems by the year 2000.[10]

Soils

The distribution and classification of the soils of Paraguay are shown on the accompanying map in nine map units: (1) Alluvial soils, (2) Saline and saline-alkali soils, (3)

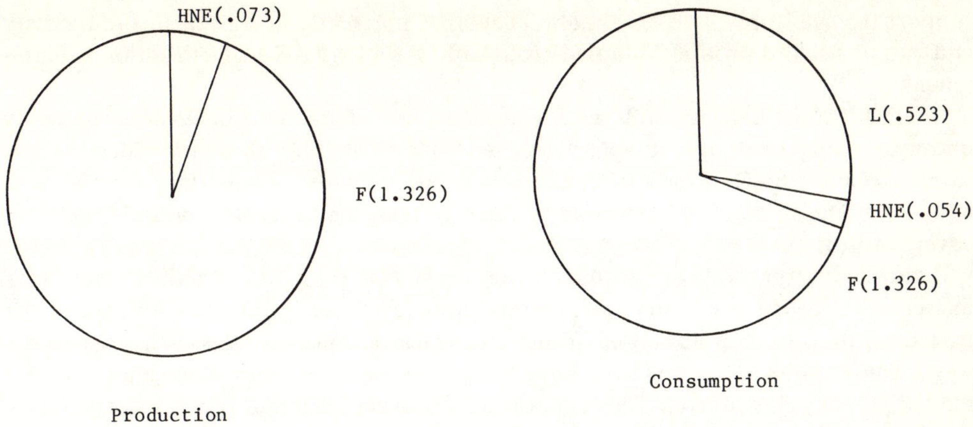

1978 Energy production and consumption in Paraguay; fuel types listed in million metric tons coal equivalents (MMTCE). 1 MMTCE = 27.78 × 10^{12} Btu. L = liquid fuel, S = solid fuel, HNE = hydro, nuclear, and electric fuel, F = firewood, NG = natural gas. (Data from United Nations, World Energy Supplies, 1973—1978, Statistical Papers Series J, 22, Department of International Economic and Social Affairs, Statistical Office, New York, 1979.)

Planosols, (4) Grumosols, (5) Latosols, (6) Reddish Brown Lateritic soils, (7) Reddish Chestnut soils, (8) Reddish Brown soils, and (9) Red-Yellow Podzolic soils.

The Alluvial soils, located most extensively along the Paraguay River and Parana River, are of little agriculture value. They have level to very gently undulating surfaces and are variously well to poorly drained and include areas of swamps and marshes. They support a natural vegetation of forest and grasses. Areas not subject to frequent inundation have some potential for use as cropland and pasture.

Saline and saline-alkali soils have dual limitations for agricultural use — impeded internal drainage and relatively high levels of salinity. In addition, they are situated in an environment characterized in part by a shortage of moisture during the winter months. Topographically the soils are in a wide, nearly level basin. They are suited for production of adapted pasture grasses; improvement of these soils for more intensive use in general would not be feasible.

Planosols are extensive soils in the central part of Paraguay. They have a dense, compact subsurface layer of clay, which impedes entry and percolation of water through the profile. These soils are wet during most of the rainy season occuring during October, November, and March. Planosols are flat with some minor depressions where accumulated water stagnates for long periods. These soils support a natural vegetation of tall grasses, clumps of palm, and some forest on better-drained sites along streams and on higher-lying ground. With adequate control of surface water, they can be expected to produce moderate yields of small grains and adapted pasture grasses.

Grumosols are on low-lying plains on either side of the Paraguay River. They are clay soils that shrink and crack deeply when dry and expand when wet. They are difficult to manage as crop land inasmuch as they can be worked only at a narrow range of moisture content. Associated soils are moderately deep and shallow clays on low hills and ridges of limestone. Grumosols support a natural vegetation of grassland with forest remnants on the better-drained and higher-lying hills and ridges. Grumosols support production of good yields of cultivated crops if carefully managed.

Latosols are on undulating plains in the eastern part of Paraguay. They are red clay soils with massive, porous subsoils and little readily apparent difference between surface and subsoils. In that part of the map unit that is generally aligned along the 56th meridian, the Latosols have sandy surfaces which grade into clays with depth. Latosols support tropical forest and savannah. They have inherently low levels of productivity and can be used to produce adapted crops only if a high level of fertilization is maintained.

Reddish Brown Lateritic soils are inextensively distributed in southeastern Paraguay on more strongly rolling and slightly higher-lying plains than those on which the Latosols are situated. They have more strongly expressed subsoils than the Latosols, have a high content of iron, and are more productive than the Latosols. Natural vegetative cover is forest.

Reddish Chestnut soils are the most extensive soils in Paraguay, dominating approximately a third of the country west of the Paraguay River. In the eastern part of the map unit, they are in a humid environment, which becomes progressively drier to the west. The Reddish Chestnut soils have a dark medium textured surface layer and a reddish, blocky clay subsoil. They support a natural vegetation of forest and savannah, the forest in the western part of the unit being less dense than that in the eastern part of the map unit. These soils are suited for production of crops such as cotton, in the east without supplemental irrigation.

Reddish Brown soils are in the extreme western, and driest, part of Paraguay. These soils are thinner and lighter in color that the Reddish Chestnut soils and they are somewhat sandier. Natural vegetation is open and lower than that to the east and the savannah has a higher proportion of thorny shrubs. Because of the droughtiness of the Reddish Brown soils, seasonal grazing is the most practical use made of these soils.

Red-yellow Podzolic soils are extensive on the rolling plains and hills east of the Paraguay River. Characteristically, they have dark red to brownish yellow medium textured surface soils with subsoils higher in clay content; they are low to moderately low in bases. Red-Yellow Podzolic soils are low in productivity and require a high level of management for acceptable yields. They support a natural vegetation of broadleaf forest with some interspersed areas of grassland.

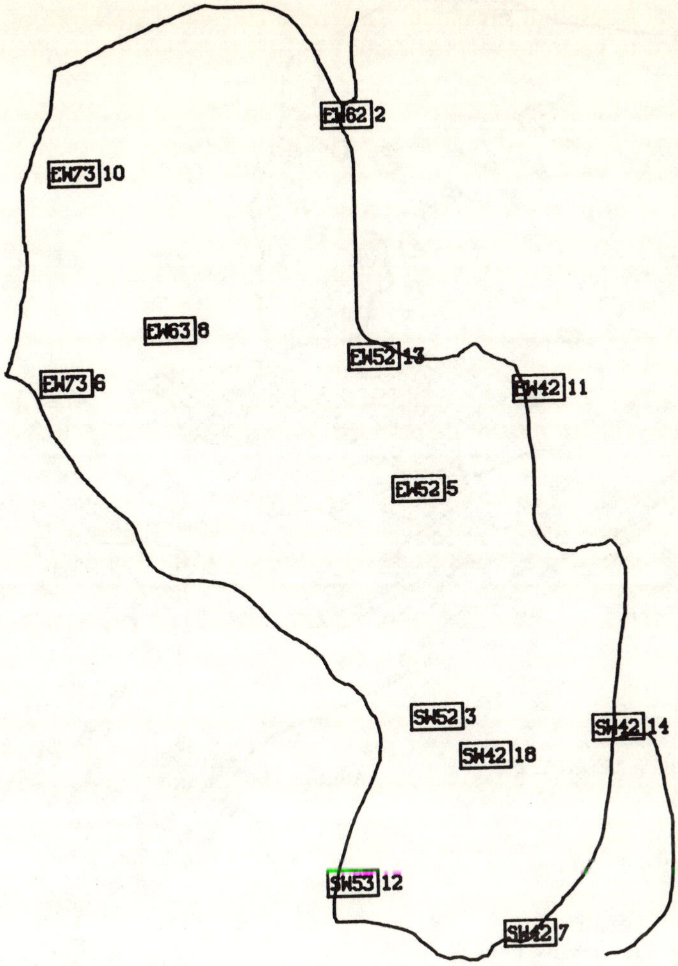

Selected weather stations for Paraguay.

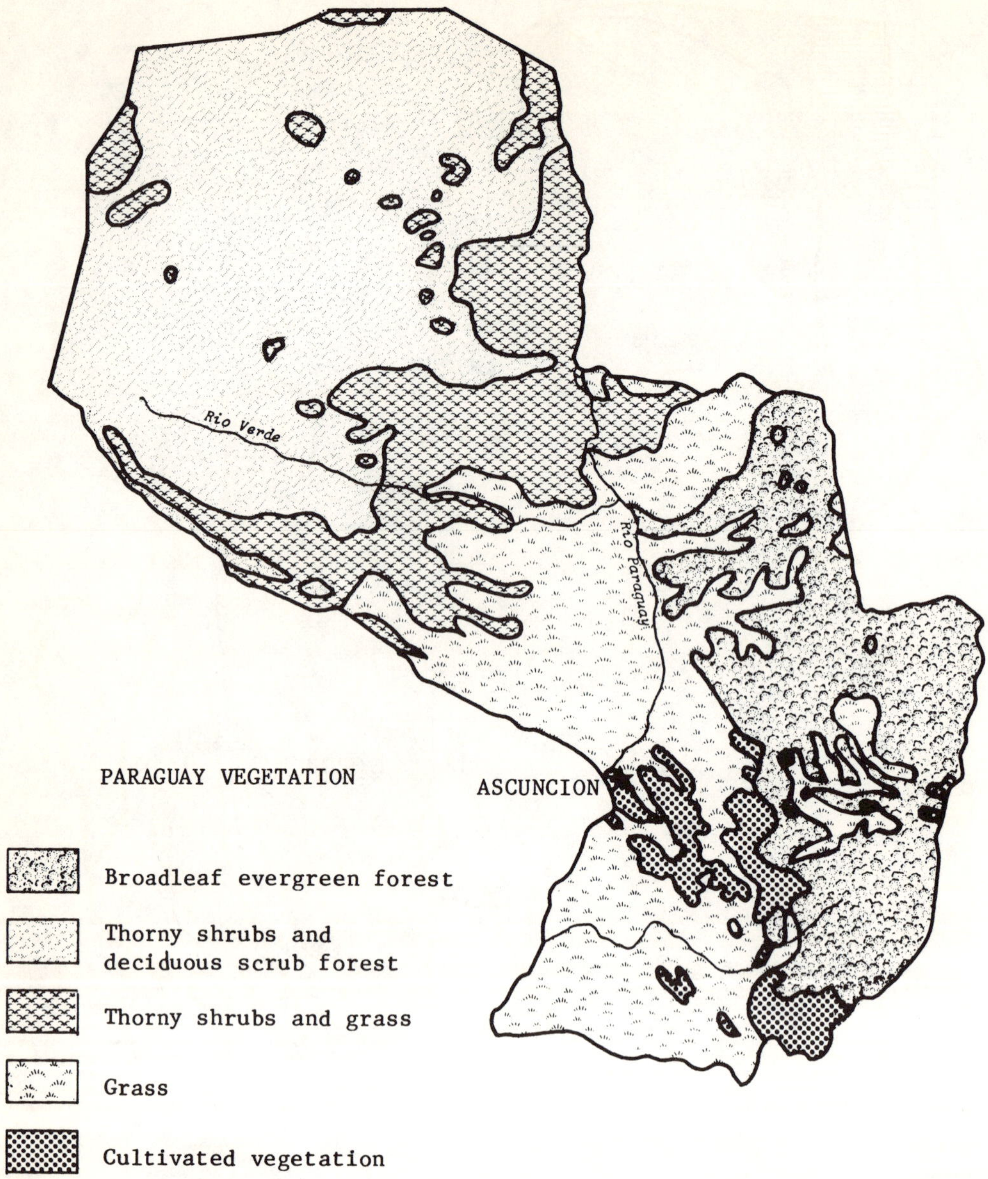

Rio Verde
Rio Paraguay
PARAGUAY VEGETATION
ASCUNCION
Broadleaf evergreen forest
Thorny shrubs and deciduous scrub forest
Thorny shrubs and grass
Grass
Cultivated vegetation

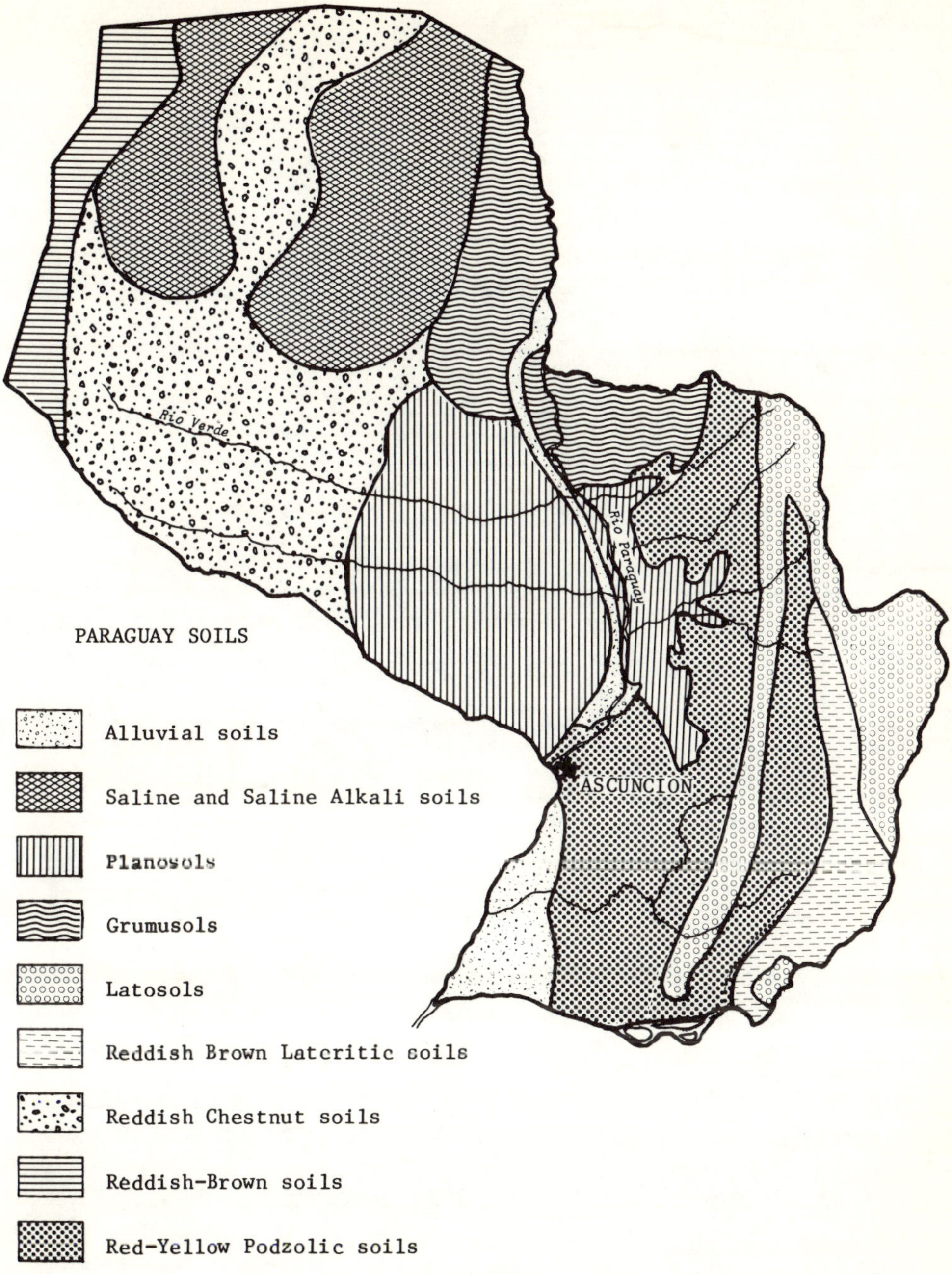
Rio Verde
Rio Paraguay
ASCUNCION
PARAGUAY SOILS
Alluvial soils
Saline and Saline Alkali soils
Planosols
Grumusols
Latosols
Reddish Brown Latcritic soils
Reddish Chestnut soils
Reddish-Brown soils
Red-Yellow Podzolic soils

CLIMATOLOGICAL DATA FOR PARAGUAY

GEOGRAPHY

OBS.	LOCATION	YEARS	LATITUDE	LONGITUDE	ELEVATION -M-	LIFEZONE	CCTX	ID	ONMAP
1	ASUNCION	21	25.28 S	57.65 W	118.9	TD	SW53	1	NO
2	BAHIA NEGRA	20	20.23 S	58.17 W	96.6	TV	EH62	2	YES
3	CAACUPE	10	25.40 S	57.10 W	228.0	SM	SW52	3	YES
4	CAPITAN MIRANDA	10	27.28 S	55.82 W	.	.	.	4	NO
5	CONCEPCION	10	23.42 S	57.30 W	84.7	TD	EW52	5	YES
6	DR PEDRO P PENA	10	22.45 S	62.05 W	79.9	TV	EW73	6	YES
7	ENCARNACION	20	27.33 S	55.83 W	79.9	SM	SW42	7	YES
8	ESTIGARRIBA MCAL	20	22.02 S	60.62 W	181.1	TV	EW63	8	YES
9	HORQUETA	10	23.40 S	56.83 W	124.1	TD	EW52	9	NO
10	NUEVA ASUNCION	10	20.67 S	61.83 W	314.9	TV	EW73	10	YES
11	PEDRO J CABELLERO	10	22.58 S	55.65 W	645.0	SM	EW42	11	YES
12	PILAR	20	26.85 S	58.32 W	54.9	SM	SW53	12	YES

AVERAGE PRECIPITATION (MM)

OBS.	JAN.	FEB.	MAR.	APR.	MAY	JUN.	JUL.	AUG.	SEP.	OCT.	NOV.	DEC.	ANNUAL
1	159	134	174	154	93	79	48	35	75	110	139	116	1316
2	137	135	109	72	54	41	23	20	47	70	124	104	938
3	155	172	162	165	98	72	52	42	92	132	147	177	1465
4	99	212	116	158	73	66	150	190	146	146	72	180	1608
5	144	154	134	120	120	59	48	63	68	111	141	148	1310
6	82	110	80	46	21	6	5	1	15	33	61	109	568
7	135	152	139	164	142	134	94	95	140	164	128	136	1624
8	114	107	101	68	41	27	17	12	17	65	99	97	764
9	111	131	109	136	126	72	77	37	88	163	140	107	1297
10	68	90	78	67	40	10	3	6	3	50	40	91	544
11	172	158	148	116	101	104	44	53	77	165	186	196	1520
12	151	128	186	158	97	68	41	38	72	119	118	110	1286

AVERAGE TEMPERATURE (DEGREES CENTIGRADE)

OBS.	JAN.	FEB.	MAR.	APR.	MAY	JUN.	JUL.	AUG.	SEP.	OCT.	NOV.	DEC.	ANNUAL
1	28.9	28.3	26.3	23.7	20.9	18.6	18.4	20.6	22.1	24.7	26.7	28.8	24.0
2	28.8	28.3	27.4	26.0	23.6	22.0	21.7	24.1	25.6	27.5	28.5	29.1	26.1
3	25.9	25.4	24.0	21.3	18.9	17.2	16.9	19.0	19.8	22.5	24.4	26.0	21.8
4	.	.	.	.	.	.	.	.	.	.	.	.	.
5	27.7	27.4	25.9	23.2	19.1	18.9	18.7	20.4	22.5	25.3	26.6	28.1	23.7
6	28.6	28.0	25.3	23.5	20.9	17.7	17.5	20.5	23.3	26.1	27.8	28.7	24.0
7	26.9	26.3	24.3	21.2	18.2	16.2	16.1	18.1	19.7	22.0	24.3	26.4	21.7
8	29.7	28.7	26.9	24.6	21.8	19.8	19.5	23.0	25.2	27.0	28.4	29.4	25.3
9	28.6	28.2	26.9	24.9	22.2	20.7	20.2	22.8	24.2	25.1	26.3	27.8	24.8
10	28.2	27.0	25.5	23.8	21.4	19.0	19.7	22.8	25.4	27.5	27.2	28.3	24.7
11	23.6	23.4	22.7	20.7	18.1	16.9	16.8	19.2	20.8	21.8	22.9	23.6	20.9
12	28.0	27.3	25.2	22.2	19.6	16.9	15.6	18.4	20.2	22.9	25.3	27.4	22.4

CLIMATOLOGICAL DATA FOR PARAGUAY

GEOGRAPHY

OBS.	LOCATION	YEARS	LATITUDE	LONGITUDE	ELEVATION -M-	LIFEZONE	CCTX	ID	ONMAP
13	PTO CASADO	20	22.28 S	57.87 W	86.6	TD	EW52	13	YES
14	PTO PTE FRANCO	17	25.53 S	54.58 W	125.0	SM	SW42	14	YES
15	PTO PTE STROESSNER	20	25.53 S	54.60 W	196.6	.	.	15	NO
16	SAN JUAN BAUTISTA	10	26.67 S	57.15 W	149.7	SM	SW42	16	NO
17	SAN LORENZO	10	25.37 S	57.57 W	119.8	SM	SW42	17	NO
18	VILLARRICA	20	25.75 S	56.43 W	154.8	SM	SW42	18	YES
19	YACY RETA (ISLA)	10	27.40 S	56.45 W	86.0	.	.	19	NO

AVERAGE PRECIPITATION (MM)

OBS.	JAN.	FEB.	MAR.	APR.	MAY	JUN.	JUL.	AUG.	SEP.	OCT.	NOV.	DEC.	ANNUAL
13	148	105	127	116	76	62	29	31	44	135	150	140	1164
14	133	151	157	152	143	88	74	84	132	190	152	148	1602
15	137	334	157	215	207	126	108	128	176	336	112	346	2380
16	159	156	139	144	80	92	81	82	98	183	149	152	1515
17	150	181	187	132	87	75	48	55	94	150	168	167	1543
18	165	153	164	141	118	106	73	69	99	175	137	139	1539
19	103	154	146	174	92	102	87	56	96	200	198	70	1478

AVERAGE TEMPERATURE (DEGREES CENTIGRADE)

OBS.	JAN.	FEB.	MAR.	APR.	MAY	JUN.	JUL.	AUG.	SEP.	OCT.	NOV.	DEC.	ANNUAL
13	28.8	28.3	26.8	24.7	22.2	20.2	20.2	22.5	24.4	26.1	27.4	28.4	25.0
14	26.6	26.2	24.8	21.7	19.0	17.3	16.7	18.6	20.3	22.3	24.3	26.1	22.0
15	.	.	.	.	.	.	.	.	.	.	.	.	.
16	26.6	25.9	24.3	21.0	18.2	16.3	16.9	18.0	20.0	22.5	24.9	26.8	21.8
17	27.5	26.8	25.1	22.0	19.3	18.1	17.3	19.2	20.6	23.7	25.7	27.4	22.7
18	27.1	26.7	24.9	21.8	19.1	17.2	17.2	19.3	20.4	22.7	24.9	26.9	22.3
19	.	.	.	.	.	.	.	.	.	.	.	.	.

PERU

Agriculture

Of 128,522,000 ha in total area, Peru was estimated in 1978 to have 3,430,000 ha of arable or permanent cropland, 27,120,000 ha in permanent pasture, 73,800,000 ha in forest and woodland, and 23,650,000 ha in other terrain categories. In 1978, approximately 1,180,000 ha were irrigated. Of the 1979 population, 38.1% or 1,939,000 individuals of the economically active population were engaged in agriculture. The total population in 1979 was 17,291,000. According to FAO statistics for 1979, there were 648,000 horses, 216,000 mules, 486,000 asses, 4,187,000 cattle, 2,200,000 pigs, 14,473,000 sheep, 2,000,000 goats, and 36,000,000 chickens.[2] Agricultural production is summarized in Table 1. Conventional and energetic yield data on most of these will be found in Appendixes I and II.

Climate

Data from Wernstedt[3] are presented in the Climatological Data Table (see Introduction for explanation). Based on the means for data points cited in Wernstedt, the mean annual precipitation is 6.7 dm (mean of 56 sites) and the mean annual temperature 18.4°C (mean of 42 sites). The averages would suggest that much of the country could be classified as Subtropical Dry Forest Life Zone.[4]

Productivity

The vegetation map depicts the distribution of the major vegetation types of Peru and their estimated annual potential productivities, assuming reasonable inputs:

Vegetation type	MT/ha/yr
Tropical rain forest	18—36
Mountain rain forest	12—24
Mountain tall grass and scrub	6—12
Mountain short grass and alpine wastes	0—6
Coastal desert and scrub	0—6

Economics

Food consumption per capita was 2274 cal/day, energy consumption 646 kg coal equivalent per captia in 1977. In 1970, the caloric supply was 99.0% of requirements. The total population growth rate was 2.9% in 1970. Population density was 11 persons per square kilometer in 1970, with 57.4% of the total population urban. Life expectancy at birth of those born in 1970 was 53 years. In 1977, with GNP at $720 per capita, agriculture accounted for 20.3% of GDP, manufacturing for 19.2%, other industry for 12.1%, and services for 48.3%. In 1970, food and beverage constituted 43.5% of exports, nonfood agriculture for 5.9%, fuels, minerals, and metal for 49.1%, machines for 6.1%, and other manufacturing for 1.4% of total merchandise exports. Food and beverage constituted 20.1% of imports, nonfood agriculture 3.8%, fuel and lubricants 2.3%, nonfuel minerals and metals 1.5%, machinery and equipment 34.5%, and other manufactured good constituted 37.8% of total merchandise imports.[10]

Energetics

Peru has 655 million b.o.e. in oil, 102 in associated gas, 216 in nonassociated gas, 1.5 billion b.o.e. in heavy oil reserves, 1122 million ton coal (105 technically and economically recoverable), and a hydroelectric potential of 12,500 MW. The World Bank is currently involved with Peru in enhanced recovery projects. A net oil-exporting developing country, Peru is not expected to experience major fuelwood problems by the year 2000.[10]

Table 1
AGRICULTURAL PRODUCTION IN PERU

Crop	Area planted (ha)	Yield (kg/ha)	Total production (MT)
Wheat	95,000	1,000	95,000
Rice paddy	122,000	4,463	545,000
Barley	185,000	946	175,000
Corn	360,000	1,667	600,000
Rye	1,000	1,000	1,000
Oats	1,000	909	1,000
Sorghum	19,000	3,158	60,000
Potatoes	255,000	6,667	1,700,000
Sweet potatoes	16,000	10,369	166,000
Cassava	38,000	10,958	416,000
Dry beans	78,000	790	62,000
Dry broad beans	23,000	933	21,000
Dry peas	20,000	752	15,000
Chick peas	4,000	641	3,000
Lentils	2,000	550	1,000
Soybeans	3,000	2,000	6,000
Peanuts in shell	5,000	1,404	7,000
Castor beans		2,000	
Sesame seed		938	
Linseed		636	
Seed cotton	146,000	1,781	260,000
Cottonseed			162,000
Olives			12,000
Coconuts			9,000
Palm oil			5,000
Cabbages	3,000	14,407	39,000
Artichokes		6,774	2,000
Tomatoes	6,000	14,036	79,000
Cauliflower	1,000	13,364	15,000
Pumpkins, squash, gourds	5,000	12,792	61,000
Cucumbers and gherkins	1,000	11,429	7,000
Chilies and peppers, green	3,000	5,538	14,000
Dry onions	8,000	21,250	170,000
Garlic	2,000	6,000	13,000
Green beans	12,000	2,371	28,000
Green peas	12,000	2,512	31,000
Carrots	2,000	17,727	39,000
Watermelons	3,000	11,481	31,000
Cantaloupes and other melons	1,000	13,784	10,000
Grapes	11,000	5,823	66,000
Wine			10,000
Sugar cane	53,000	126,415	6,700,000
Sugar (centrifugal, raw)			695,000
Sugar (noncentrifugal)			15,000
Apples			81,000
Pears			8,000
Peaches and nectarines			32,000
Plums			2,000
Oranges			166,000
Tang. mand. clement. satsuma			21,000
Lemons and limes			86,000
Grapefruit and pomelo			6,000
Avocados			75,000
Mangoes			72,000

Table 1 (continued)
AGRICULTURAL PRODUCTION IN PERU

Crop	Area planted (ha)	Yield (kg/ha)	Total production (MT)
Pineapples			58,000
Plantains			742,000
Papayas			69,000
Strawberries			4,000
Green coffee	130,000	554	72,000
Cocoa beans	6,000	667	4,000
Tea	4,000	750	3,000
Tobacco leaves	3,000	1,123	3,000
Jute and substitutes	4,000	1,143	4,000
Cotton lint			72,000
Natural rubber			2,000

From FAO 1979 Production Yearbook, Vol. 33, Food and Agriculture Organization, Rome, 1980.

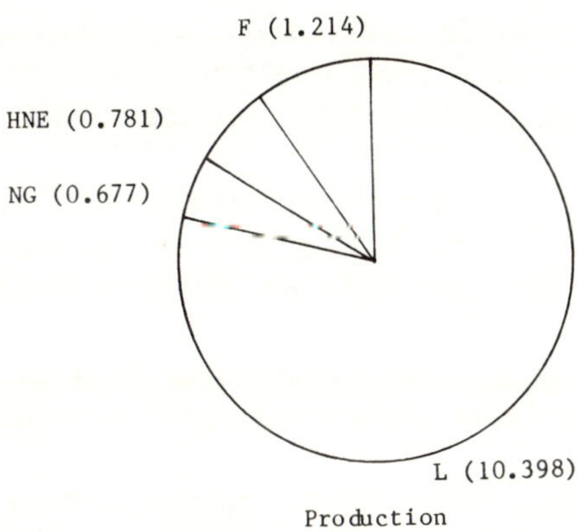

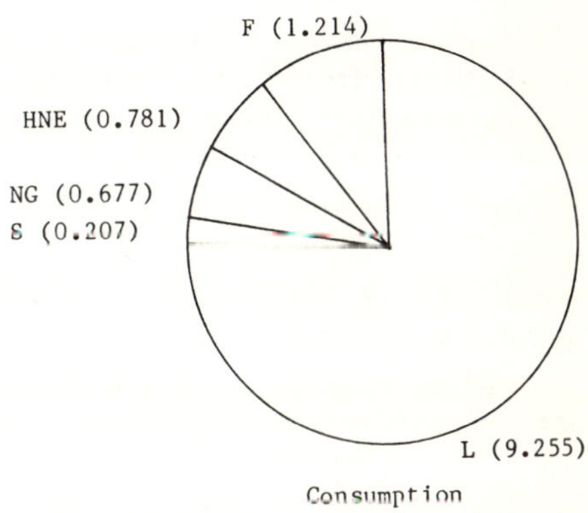

1978 Energy production and consumption in Peru; fuel types listed in million metric tons coal equivalents (MMTCE). 1 MMTCE = 27.78×10^{12} Btu. L = liquid fuel, S = solid fuel, HNE = hydro, nuclear, and electric fuel, F = firewood, NG = natural gas. (Data from United Nations, World Energy Supplies, 1973—1978, Statistical Papers Series J, 22, Department of International Economic and Social Affairs, Statistical Office, New York, 1979.)

Soils

Six soils associations in Peru are shown on the Soils Map: (1) Latosols, Alluvial soils, and Humic Gley soils of the upper Amazon Basin, (2) Latosols of the eastern Andean slopes, (3) Lithosols of the Andean Highlands, (4) Lithosols of the western Andean slopes, (5) Regosols and Desert soils, and (6) Alluvial soils. Except for the Alluvial soils, which are distributed throughout the country, the soils of Peru occur in northwest-southeast trending areas, closely related to the physiographic regions of the country.

Latosols, the most extensive soils in Peru, are reddish or yellowish and have thin surface soils containing little organic matter over subsoils ranging from sandy clay loam to clay. Because of high acidity and low natural fertility, Latosols can be used for economic crop production only under highly skilled management. At present, farms on these soils, typically small and isolated, grow mainly subsistence crops. Cacao and coffee are grown in some places on the less steep slopes between 650 and 2000 m elevation.

In the broad upper Amazon Basin, Latosols occupy fairly level to undulating areas some 5 to 20 m above the present river levels and are formed from very thick, water-laid materials. In this region, they are associated with well-drained Alluvial soils on the low terraces and flood plains and with the extensive areas of poorly drained, acid Humic Gley soils.

On the eastern slopes of the Andes, Latosols extend over both hills and mountains; here they are derived from a variety of sedimentary and igneous rocks.

In the Andean Highlands, the Lithosols are generally shallow and stony with large areas of bare rock; at the highest elevations they are covered with snow and ice during part or all of the year. Because of the cold climate, land use is generally limited to grazing. Below about 5000 m, however, subsistence crops are grown. Included with Lithosols are numerous areas of Alpine Meadow soils, all too small to show on the map. Many areas of these wet soils are used for grazing; some, reclaimed by drainage, are used for cultivated crops.

On the western Andean slopes, the soils are also Lithosols; because the region is more arid, vegetation is sparse. Except for a few scattered irrigated areas, the land is used only to graze goats, sheep, and llamas.

Regosols and Desert soils, which occur in the arid coast, have developed mostly from water-deposited material. Regosols occupy numerous *tablazos* (level to gently sloping lands formed from marine sediments) along the coast. These soils consist mostly of unconsolidated, stratified, and often mixed stone, gravel, sand, silt, and clay, and are salty, gypsiferous, or calcareous in some layers. Receiving little moisture (as fog or mist), the soils support only occasional desert shrubs. The associated Desert soils in the northwestern corner of the country, through developed under slightly more precipitation, also support only sparse vegetation. Some cultivated crops may be grown where water is available for irrigation.

Alluvial soils are, by far, the leading agricultural soils in the country. Well-drained, they are deficient in nitrogen, medium to low in available phosphorus, but generally well supplied with available potassium.

In the west, the Alluvial soils form irrigated oases in the coastal valleys of the otherwise arid region. They range from shallow coarse sand and gravel to deep fine-textured soils. In general, the soils are neutral to slightly alkaline in reaction; in many areas soluble salts, gypsum, and calcium carbonate are present.

Alluvial soils of the intermountain Andean valleys, where irrigation is essential for good production, are usually intensively cultivated. The soils occur in deep, narrow valleys; most of the valley floors have narrow strips of recent alluvium and a wider band of level to gently sloping natural terraces composed of colluvium and a local alluvium from the adjacent hillsides. The soils are calcareous in some places, and neutral or slightly alkaline in others. Between 2500 and 3500 m elevation, the crops include corn, potatoes, barley, wheat, alfalfa, apples, peaches, pears, and vegetables. In the deeper, frostfree valleys (at elevations of 1500 to 2500 m), citrus, mangoes, and sugar cane are also grown, usually with supplementary irrigation.

In the Titicaca Basin, the Alluvial soils have been formed largely from relatively recent water-laid material, and are neutral to moderately acid. These soils are moderately well to well-drained, with textures ranging from sandy loams to clay loams. The depth is variable; much of the well-drained soil is underlain by gravel at 0.3 m or so. In most areas, the water table stands fairly high, 1 to 5 m below the surface. For many years, much of this land has been intensively cropped, and a large area is in pasture. Included with these Alluvial soils are extensive areas of Humic Gley soils.

In the upper Amazon Basin, the well-drained Alluvial soils of the flood plains and low terraces, though inextensive, offer the best possibilities for agricultural development. These soils are deep, medium textured, fertile, and slightly acid to moderately alkaline in reaction. Isolated subsistence crop production indicates possibilities for a permanent agriculture. Major disadvantages of these soils are their inaccessibility, the narrowness of the soil areas along the main rivers, and the periodic danger of floods in the lower areas.

Alluvial soils are also included with the Latosols of the eastern Andean slopes, occurring on the narrow valley floors. Because of their position, these valleys receive less rainfall and have a more pronounced dry season than the eastern slopes as a whole. The valley along Rio Urubamba northwest and southeast of Cuzco is an example of this condition. The soils of these valleys are generally well-drained and medium textured. Slightly to moderately acid, with a fair content of organic matter, they can nurture most tropical and subtropical crops.

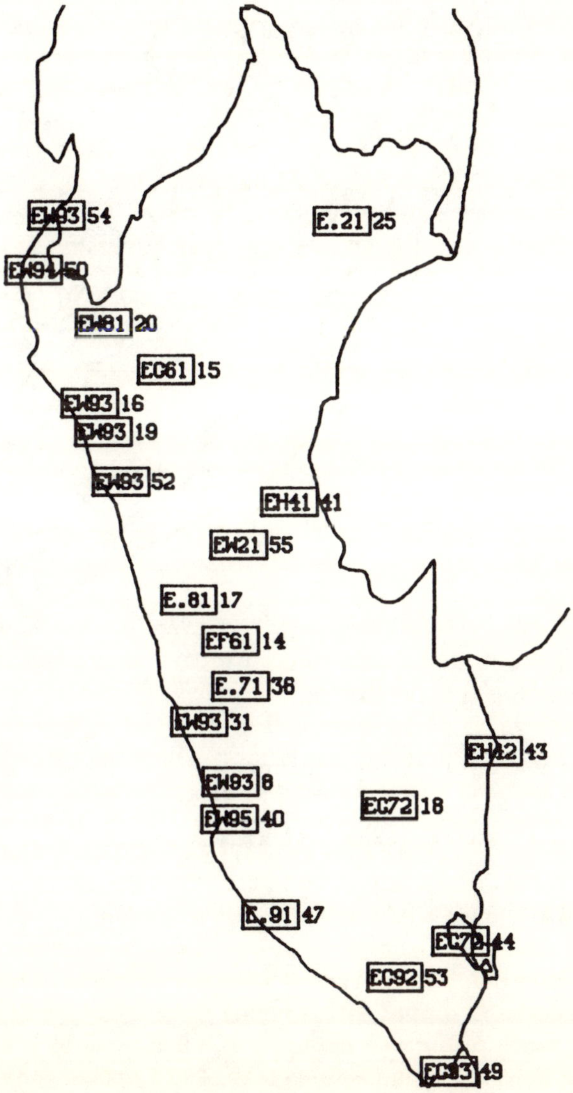

Selected weather stations for Peru.

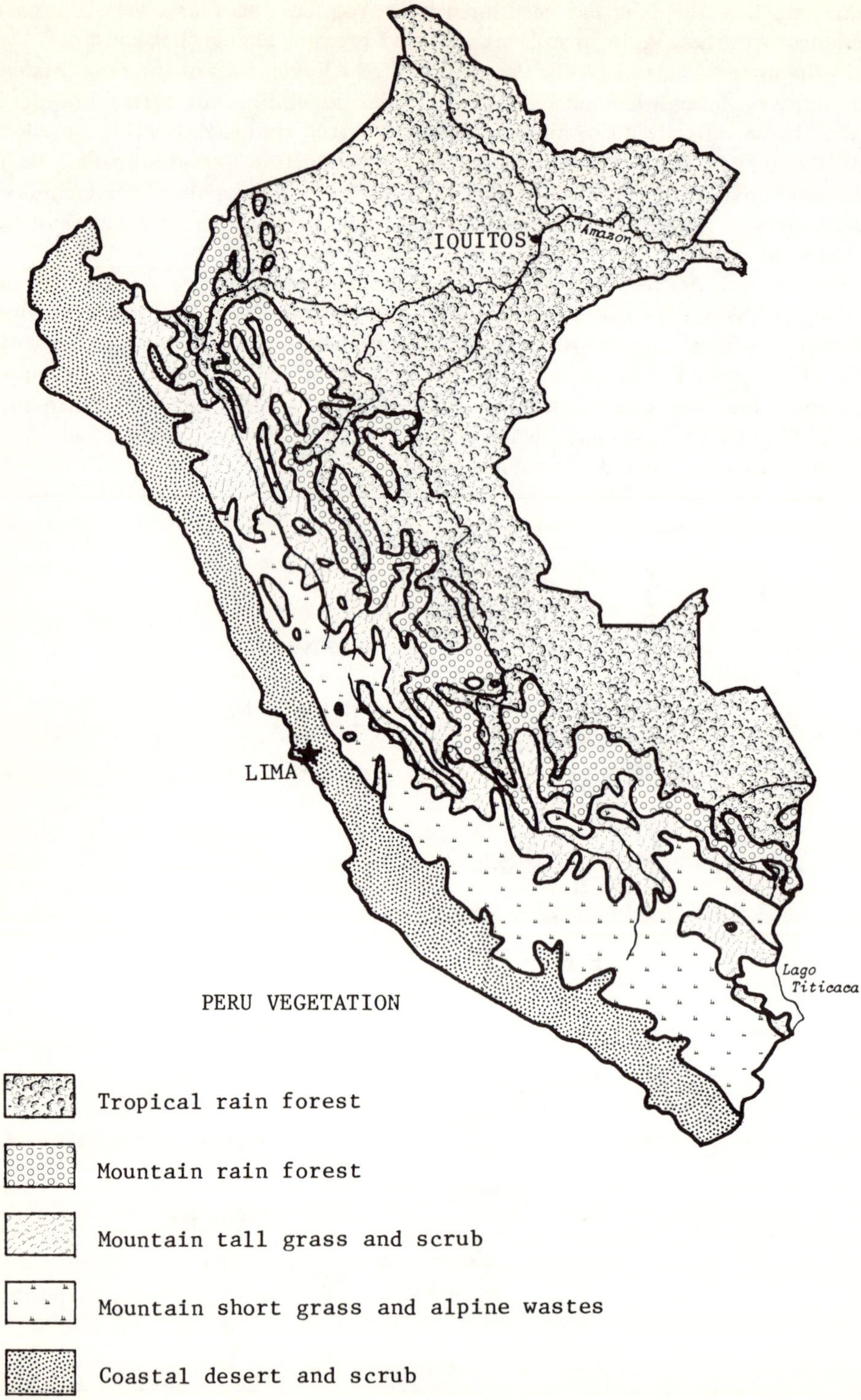
IQUITOS
Amazon
LIMA
Lago Titicaca
PERU VEGETATION
Tropical rain forest
Mountain rain forest
Mountain tall grass and scrub
Mountain short grass and alpine wastes
Coastal desert and scrub

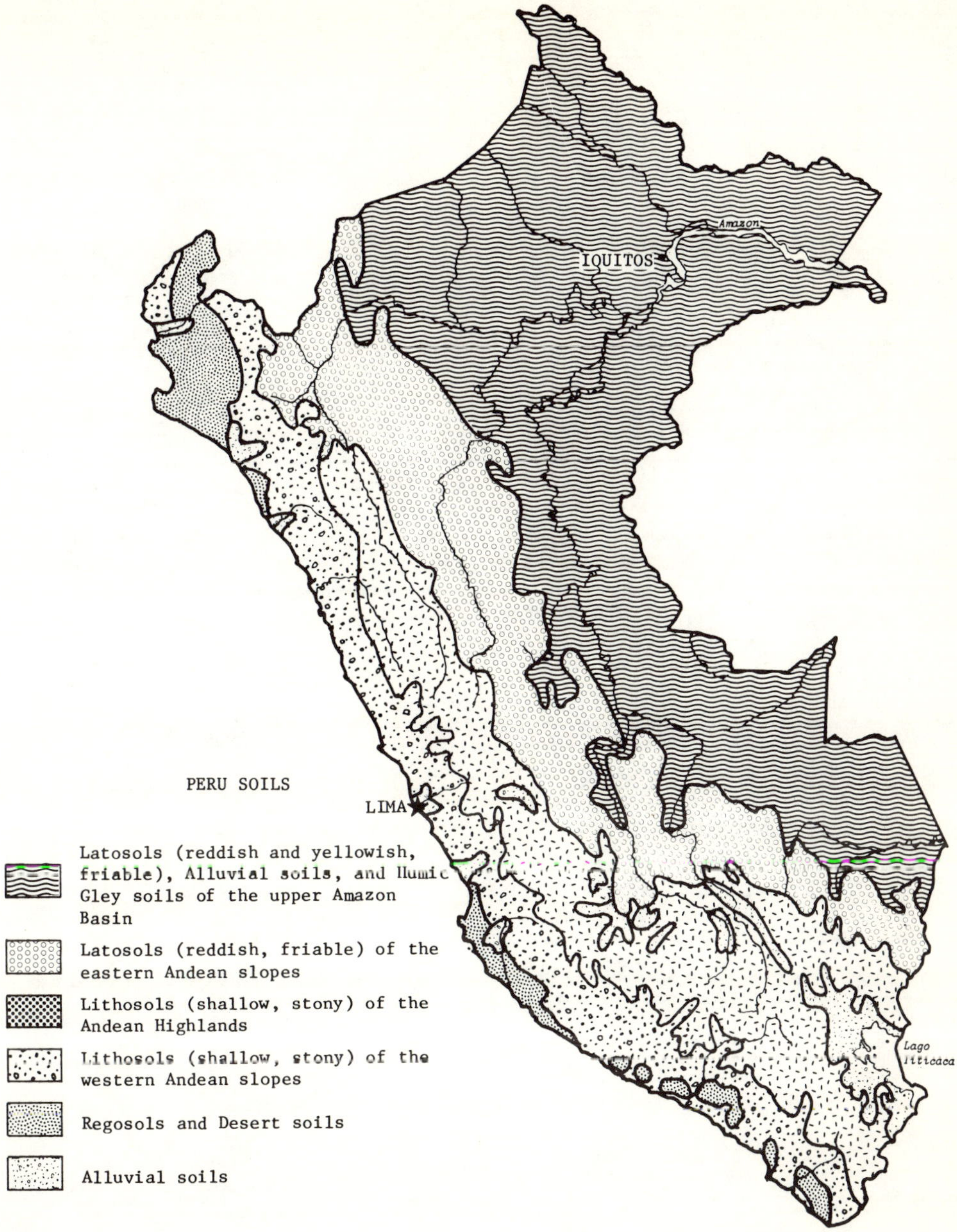
Amazon
IQUITOS
LIMA
Lago
Titicaca
PERU SOILS
Latosols (reddish and yellowish, friable), Alluvial soils, and Humic Gley soils of the upper Amazon Basin
Latosols (reddish, friable) of the eastern Andean slopes
Lithosols (shallow, stony) of the Andean Highlands
Lithosols (shallow, stony) of the western Andean slopes
Regosols and Desert soils
Alluvial soils

CLIMATOLOGICAL DATA FOR PERU

GEOGRAPHY

OBS.	LOCATION	YEARS	LATITUDE	LONGITUDE	ELEVATION -M-	LIFEZONE	CCTX	ID	ONMAP
1	ACCNOCOCHA	.	13.05 S	75.20 W	4518.7	BW	EF62	1	NO
2	ALTO (EL)	27	4.25 S	81.23 W	269.7	SO	EW94	2	NO
3	AREQUIPA	10	16.32 S	71.55 W	2522.2	WO	EC93	3	NO
4	BEBEDERO	16	6.80 S	79.32 W	321.9	SO	EW93	4	NO
5	BELLAVISTA	16	6.98 S	76.72 W	.	.	.	5	NO
6	CAJAMARCA	16	7.13 S	78.47 W	2620.4	WD	EC61	6	NO
7	CAMPO DE MARTE	14	12.07 S	77.03 W	136.9	SO	EW93	7	NO
8	CANETE	17	13.12 S	76.20 W	111.9	SO	EW93	8	YES
9	CARAZ	10	9.05 S	77.82 W	2287.2	.	.	9	NO
10	CARTAVIO	17	7.90 S	79.22 W	50.9	SO	EW93	10	NO
11	CASA GRANDE	27	7.68 S	79.18 W	157.9	SO	EW93	11	NO
12	CASAPALCA	.	11.65 S	76.23 W	4189.8	.	.	12	NO

AVERAGE PRECIPITATION (MM)

OBS.	JAN.	FEB.	MAR.	APR.	MAY	JUN.	JUL.	AUG.	SEP.	OCT.	NOV.	DEC.	ANNUAL
1	164	168	147	61	16	4	2	8	22	48	66	94	800
2	7	27	18	11	2	0	0	0	0	0	2	2	69
3	31	26	29	1	0	0	0	0	1	0	1	5	94
4	23	24	26	13	3	0	0	1	1	6	1	4	102
5	119	140	121	42	15	4	2	4	19	43	47	74	630
6	95	113	136	108	37	13	5	9	30	87	72	77	782
7	1	0	0	0	2	4	5	5	4	2	1	1	24
8	0	1	1	1	3	9	4	3	6	2	2	1	33
9	40	60	66	29	4	0	0	0	2	9	10	17	237
10	1	3	1	1	0	0	0	0	0	1	0	1	8
11	2	3	4	2	0	0	0	0	0	1	1	2	15
12	125	144	125	60	22	12	6	10	41	52	60	84	741

AVERAGE TEMPERATURE (DEGREES CENTIGRADE)

OBS.	JAN.	FEB.	MAR.	APR.	MAY	JUN.	JUL.	AUG.	SEP.	OCT.	NOV.	DEC.	ANNUAL
1	3.7	3.8	4.0	3.6	2.8	1.9	1.1	1.6	2.2	3.0	3.2	3.4	2.9
2	23.5	24.4	24.7	24.2	22.3	20.0	18.6	18.0	18.2	18.7	19.5	21.5	21.1
3	14.1	14.3	14.5	13.4	12.7	11.2	11.5	12.1	12.9	13.4	13.4	13.5	13.1
4	25.3	25.8	26.0	25.8	22.9	21.7	20.3	20.3	21.1	21.3	22.5	23.9	23.1
5	.	.	.	.	.	.	.	.	.	.	.	.	.
6	14.6	14.1	14.2	13.8	13.9	13.2	13.1	14.0	14.5	14.3	14.6	14.3	14.0
7	21.6	22.7	22.5	20.8	18.2	16.4	15.6	15.1	15.4	16.4	17.8	20.0	18.5
8	23.3	24.1	24.1	22.4	19.7	17.9	17.0	16.7	17.2	18.2	19.5	21.4	20.1
9	.	.	.	.	.	.	.	.	.	.	.	.	.
10	23.0	24.1	23.8	22.0	21.0	19.2	18.3	18.0	17.9	18.2	19.0	20.7	20.4
11	23.9	24.9	24.5	22.8	20.6	19.3	18.4	18.3	18.9	19.2	19.9	21.6	21.0
12	.	.	.	.	.	.	.	.	.	.	.	.	.

CLIMATOLOGICAL DATA FOR PERU

GEOGRAPHY

OBS.	LOCATION	YEARS	LATITUDE	LONGITUDE	ELEVATION -M-	LIFEZONE	CCTX	ID	ONMAP
13	CAYALTI	25	7.07 S	79.57 W	149.7	SO	EW93	13	NO
14	CERRO DE PASCO	.	10.75 S	76.17 W	4398.9	PR	EF61	14	YES
15	CHACAPOYAS	11	6.22 S	77.83 W	2434.4	WD	EC61	15	YES
16	CHICLAYO	18	6.78 S	79.83 W	27.7	SO	EW93	16	YES
17	CONOCOCHA	.	10.05 S	77.27 W	3998.7	.	.	17	YES
18	CUZCO	.	13.52 S	71.97 W	3364.1	CM	EC72	18	YES
19	GUADALUPE	21	7.25 S	79.48 W	299.6	SO	EW93	19	YES
20	HUANCABAMBA	10	5.45 S	79.45 W	1956.2	ST	EW81	20	YES
21	HUANCAYO	.	12.12 S	75.33 W	3379.0	CM	EC72	21	NO
22	HUANUCO	10	9.90 S	75.75 W	1911.4	.	.	22	NO
23	HUMANGA	10	13.15 S	74.22 W	2760.3	WD	EC72	23	NO
24	IMATA	.	15.82 S	71.07 W	4403.8	BW	EF72	24	NO

AVERAGE PRECIPITATION (MM)

OBS.	JAN.	FEB.	MAR.	APR.	MAY	JUN.	JUL.	AUG.	SEP.	OCT.	NOV.	DEC.	ANNUAL
13	2	8	9	5	2	0	0	0	1	2	2	3	34
14	134	153	128	70	36	16	11	19	50	68	115	126	926
15	83	135	102	96	40	27	23	28	60	92	73	56	815
16	1	1	3	2	0	0	0	0	0	1	1	0	9
17	74	74	119	55	25	5	0	4	3	16	33	24	432
18	158	128	126	38	9	5	1	8	24	43	81	128	749
19	2	5	3	2	1	0	0	0	0	2	1	2	18
20	83	71	75	29	16	22	5	2	7	25	14	32	381
21	136	126	116	51	22	8	5	11	40	71	69	93	749
22	80	69	60	26	21	5	5	10	16	46	44	52	434
23	106	100	78	38	16	2	5	10	31	103	51	79	619
24	148	147	105	41	15	5	3	2	20	20	23	90	620

AVERAGE TEMPERATURE (DEGREES CENTIGRADE)

OBS.	JAN.	FEB.	MAR.	APR.	MAY	JUN.	JUL.	AUG.	SEP.	OCT.	NOV.	DEC.	ANNUAL
13	24.8	25.9	26.1	24.6	22.5	20.4	19.3	19.0	19.6	20.1	21.0	22.7	22.2
14	2.8	2.4	3.3	2.1	1.6	2.1	1.3	2.6	2.2	2.6	3.6	3.1	2.5
15	15.9	15.7	15.6	15.4	15.8	15.2	14.5	15.1	15.2	15.9	16.5	16.2	15.6
16	24.2	25.5	25.7	23.9	22.4	20.5	19.6	19.3	19.4	19.8	20.7	22.1	21.9
17	.	.	.	.	.	.	.	.	.	.	.	.	.
18	12.0	11.4	11.9	11.4	10.2	9.5	8.8	10.2	11.5	12.3	12.9	12.4	10.9
19	24.6	25.5	25.9	24.4	22.3	20.1	18.8	18.6	19.3	19.9	20.6	22.4	21.9
20	19.7	19.3	19.3	19.3	19.4	19.2	19.4	19.4	20.3	19.5	20.1	20.2	19.6
21	12.1	12.0	11.7	11.6	10.2	9.8	9.3	10.2	11.9	12.4	12.7	12.2	11.2
22	.	.	.	.	.	.	.	.	.	.	.	.	.
23	15.7	15.2	14.6	14.9	14.2	12.4	13.0	14.1	15.1	16.0	16.5	16.1	14.8
24	4.5	4.7	4.7	3.7	1.9	1.4	1.6	1.0	2.2	2.7	3.6	4.5	2.9

CLIMATOLOGICAL DATA FOR PERU

GEOGRAPHY

OBS.	LOCATION	YEARS	LATITUDE	LONGITUDE	ELEVATION -M-	LIFEZONE	CCTX	ID	ONMAP
25	IQUITOS	10	3.75 S	73.25 W	116.7	.	.	25	YES
26	JAUJA	.	11.78 S	75.50 W	3386.0	CM	EC72	26	NO
27	JUANJU	10	7.22 S	76.72 W	349.6	TD	EH51	27	NO
28	JULIACA	.	15.48 S	70.15 W	3823.7	.	.	28	NO
29	LAMBAYEQUE	27	6.70 S	79.90 W	27.7	SO	EW93	29	NO
30	LIMA	11	12.03 S	77.03 W	136.9	SO	EC93	30	NO
31	LIMATAMBO	11	12.10 S	77.03 W	129.8	SO	EW93	31	YES
32	LOBITOS	26	4.45 S	81.15 W	19.8	.	.	32	NO
33	MOLINA (LA)	31	12.08 S	76.93 W	250.9	SO	EW92	33	NO
34	MOYOBAMBA	10	6.03 S	76.97 W	859.5	.	.	34	NO
35	NEGRITOS	13	4.65 S	81.32 W	4.9	.	.	35	NO
36	PACHACHACA	.	11.50 S	75.92 W	3629.9	.	.	36	YES

AVERAGE PRECIPITATION (MM)

OBS.	JAN.	FEB.	MAR.	APR.	MAY	JUN.	JUL.	AUG.	SEP.	OCT.	NOV.	DEC.	ANNUAL
25	260	200	270	314	257	172	177	138	204	219	270	246	2727
26	107	109	101	42	11	6	2	3	19	53	54	85	592
27	92	159	141	203	127	63	80	60	122	168	153	108	1476
28	96	101	100	37	13	3	4	4	25	35	50	86	554
29	1	7	7	3	1	1	0	0	1	1	1	2	25
30	1	0	1	0	1	3	5	6	5	2	1	1	26
31	2	1	1	0	3	4	6	6	7	5	2	2	39
32	.	.	.	.	.	.	.	.	.	.	.	.	.
33	1	1	1	1	2	3	3	3	2	1	1	1	20
34	148	149	174	185	154	76	78	79	140	180	142	96	1601
35	.	.	.	.	.	.	.	.	.	.	.	.	.
36	102	96	115	58	28	9	25	29	46	69	59	83	719

AVERAGE TEMPERATURE (DEGREES CENTIGRADE)

OBS.	JAN.	FEB.	MAR.	APR.	MAY	JUN.	JUL.	AUG.	SEP.	OCT.	NOV.	DEC.	ANNUAL
25	.	.	.	.	.	.	.	.	.	.	.	.	.
26	11.3	11.1	11.0	11.0	10.4	9.8	9.4	10.0	10.9	11.2	11.8	11.3	10.8
27	27.4	26.6	26.5	26.4	26.0	25.6	25.6	26.3	26.6	26.7	26.9	27.5	26.5
28	.	.	.	.	.	.	.	.	.	.	.	.	.
29	25.1	26.4	26.2	24.6	22.8	20.9	19.6	19.3	19.7	20.2	21.1	22.8	22.9
30	21.4	22.2	22.4	19.8	17.5	15.6	15.2	15.0	15.4	16.2	17.4	19.1	18.0
31	21.8	22.9	22.7	20.9	18.6	16.6	15.8	15.5	15.7	16.7	18.1	20.0	18.8
32	25.8	27.0	27.3	26.3	24.6	22.8	21.4	20.8	20.9	21.2	22.2	23.7	23.7
33	22.8	23.6	23.3	21.4	18.6	16.6	15.9	15.8	16.5	17.6	18.8	20.6	19.3
34	.	.	.	.	.	.	.	.	.	.	.	.	.
35	26.9	28.3	28.6	27.6	25.8	23.7	22.7	22.0	21.5	21.6	22.4	24.0	24.6
36	.	.	.	.	.	.	.	.	.	.	.	.	.

CLIMATOLOGICAL DATA FOR PERU

GEOGRAPHY

OBS.	LOCATION	YEARS	LATITUDE	LONGITUDE	ELEVATION -M-	LIFEZONE	CCTX	ID	ONMAP
37	PARAMONGA	19	10.67 S	77.82 W	14.9	SO	EW93	37	NO
38	PARON	.	8.98 S	77.67 W	4548.5	.	.	38	NO
39	PARQUE DE RESERVA	15	12.07 S	77.03 W	119.8	SO	EW93	39	NO
40	PISCO	10	13.75 S	76.23 E	5.8	SO	EW95	40	YES
41	PIURA	18	5.20 S	80.62 W	52.7	TO	EW93	56	NO
42	PUCALLAPA	10	8.42 S	74.62 W	147.8	TD	EH41	41	YES
43	PUERTO CHICAMA	36	7.68 S	79.43 W	4.0	SO	EW93	42	NO
44	PUERTO MALDONADO	10	12.60 S	69.20 W	199.6	TD	EH42	43	YES
45	PUNO	.	15.83 S	70.02 W	3851.1	CM	EC72	44	YES
46	PUNTA (LA)	11	12.08 S	77.17 W	11.9	SO	EW93	45	NO
47	QUINCEMIL	10	13.27 S	70.67 W	619.7	.	.	46	NO
48	SAN JUAN	10	15.38 S	75.15 W	29.9	.	.	47	YES

AVERAGE PRECIPITATION (MM)

OBS.	JAN.	FEB.	MAR.	APR.	MAY	JUN.	JUL.	AUG.	SEP.	OCT.	NOV.	DEC.	ANNUAL
37	1	0	0	0	1	1	1	3	2	1	0	0	10
38	138	172	166	85	41	16	1	8	36	73	102	85	923
39	1	1	1	1	4	6	7	8	7	3	2	1	42
40	0	0	0	0	0	0	0	0	2	0	0	0	3
41	11	17	19	15	1	0	0	1	0	1	1	6	72
42	142	166	190	190	148	102	36	74	109	190	229	132	1708
43	1	1	1	1	0	0	0	0	0	0	0	0	4
44	287	248	287	176	143	37	60	54	80	112	137	304	1925
45	135	138	109	40	9	6	5	6	24	37	27	88	625
46	1	0	0	0	1	2	2	2	2	1	0	0	11
47	739	301	472	480	406	424	326	240	252	529	531	668	5368
48	0	0	0	0	0	0	0	1	2	1	1	0	5

AVERAGE TEMPERATURE (DEGREES CENTIGRADE)

OBS.	JAN.	FEB.	MAR.	APR.	MAY	JUN.	JUL.	AUG.	SEP.	OCT.	NOV.	DEC.	ANNUAL
37	21.2	22.1	21.8	20.4	18.4	17.3	16.8	16.1	16.1	16.8	18.1	19.5	18.7
38	.	.	.	.	.	.	.	.	.	.	.	.	.
39	23.3	24.0	23.7	21.8	19.3	17.2	16.5	16.3	16.8	17.9	19.3	21.2	19.8
40	22.2	23.2	23.2	21.2	18.7	17.1	16.4	16.3	16.8	17.6	18.7	20.4	19.3
41	27.7	28.9	28.9	27.5	25.1	23.5	22.7	22.9	23.5	23.4	23.9	25.3	25.6
42	27.6	26.8	26.7	26.4	26.8	26.0	26.0	27.1	27.5	27.3	27.3	27.7	26.9
43	23.5	24.7	24.4	23.3	22.0	20.9	20.6	20.1	19.9	20.1	20.6	21.7	21.8
44	26.8	26.7	26.3	26.5	25.8	25.0	24.2	27.2	28.0	27.2	27.2	26.8	26.5
45	8.5	8.5	8.9	8.0	6.6	5.8	6.0	6.8	7.7	9.2	9.5	9.3	7.8
46	22.1	23.1	22.9	21.1	19.2	17.4	16.7	16.5	16.9	18.0	19.4	20.1	19.5
47	.	.	.	.	.	.	.	.	.	.	.	.	.
48	.	.	.	.	.	.	.	.	.	.	.	.	.

CLIMATOLOGICAL DATA FOR PERU

GEOGRAPHY

OBS.	LOCATION	YEARS	LATITUDE	LONGITUDE	ELEVATION -M-	LIFEZONE	CCTX	ID	ONMAP
49	SAN RAMON	10	11.10 S	75.30 W	799.5	.	.	48	NO
50	TACNA	14	18.07 S	70.30 W	452.0	WO	EC93	49	YES
51	TALARA	18	4.57 S	81.25 W	86.9	SO	EW94	50	YES
52	TARAPUTO	10	6.53 S	76.32 W	425.8	.	.	57	NO
53	TICAMPA	.	9.75 S	77.45 W	3558.8	.	.	51	NO
54	TINGO MARIA	12	9.13 S	75.95 W	684.9	SW	EW21	55	YES
55	TRUJILLO	18	8.08 S	79.03 W	25.9	SO	EW93	52	YES
56	VITOR	11	16.45 S	71.80 W	1551.7	SO	EC92	53	YES
57	YURIMAGUAS	10	5.90 S	76.08 W	178.9	.	.	58	NO
58	ZORRITOS	16	3.67 S	80.65 W	6.7	TX	EW93	54	YES

AVERAGE PRECIPITATION (MM)

OBS.	JAN.	FEB.	MAR.	APR.	MAY	JUN.	JUL.	AUG.	SEP.	OCT.	NOV.	DEC.	ANNUAL
49	242	244	243	204	120	54	71	116	128	166	124	219	1942
50	1	1	1	1	4	3	4	8	11	7	1	1	41
51	1	4	8	9	0	0	0	0	0	0	0	0	22
52	92	117	147	133	123	64	55	49	94	136	95	53	1158
53	151	128	135	96	23	4	1	7	31	60	57	81	774
54	424	403	469	351	217	175	157	112	184	341	289	289	3411
55	0	0	1	0	0	0	0	0	0	0	0	0	2
56	8	8	2	1	1	1	0	0	1	1	0	1	22
57	206	198	220	223	194	110	79	93	150	190	175	164	2002
58	18	56	33	17	0	1	1	1	0	0	1	1	129

AVERAGE TEMPERATURE (DEGREES CENTIGRADE)

OBS.	JAN.	FEB.	MAR.	APR.	MAY	JUN.	JUL.	AUG.	SEP.	OCT.	NOV.	DEC.	ANNUAL
49	.	.	.	.	.	.	.	.	.	.	.	.	.
50	20.6	20.7	19.4	17.6	15.1	13.4	12.7	13.0	14.0	14.9	17.0	18.4	16.5
51	25.7	26.6	26.9	25.5	24.6	22.2	21.1	20.6	20.5	21.0	21.6	23.4	23.3
52	.	.	.	.	.	.	.	.	.	.	.	.	.
53	.	.	.	.	.	.	.	.	.	.	.	.	.
54	22.8	22.1	22.4	22.3	22.4	21.8	22.0	22.6	22.5	22.8	23.3	23.4	22.5
55	21.2	22.2	22.1	20.6	19.6	18.4	17.9	17.5	17.3	17.6	18.1	19.3	19.3
56	17.1	17.7	18.3	17.7	18.0	17.8	17.5	17.8	18.1	17.7	17.4	17.5	17.8
57	.	.	.	.	.	.	.	.	.	.	.	.	.
58	26.6	26.9	27.1	26.1	25.6	24.2	23.0	22.7	22.7	23.3	23.4	25.1	24.7

PHILIPPINES*

Agriculture

Of 30,000,000 ha in total area, Philippines was estimated in 1978 to have 8,100,000 ha of arable or permanent cropland, 980,000 ha in permanent pasture, 13,000,000 ha in forest and woodland, and 7,737,000 ha in other terrain categories. In 1978, approximately 1,113,000 ha were irrigated. Of the 1979 population, 46.8% or 8,016,000 individuals of the economically active population were engaged in agriculture. The total population in 1979 was 49,493,000. According to FAO statistics for 1979, there were 325,000 horses, 1,910,000 cattle, 3,018,000 buffalo, 7,300,000 pigs, 30,000 sheep, 1,430,000 goats, 60,000,000 chickens, 5,477,000 ducks, and 200,000 turkeys.[2] Agricultural production is summarized in Table 1. Conventional and energetic yield data on most of these will be found in Appendixes I and II.

Climate

Data from Wernstedt[3] are presented in the Climatological Data Table (see Introduction for explanation). Based on the means for data points cited in Wernstedt, the mean annual precipitation is 25.7 dm (mean of 65 sites) and the mean annual temperature 26.7°C (mean of 65 sites). The averages would suggest that much of the country could be classified as Tropical Moist Forest Life Zone.[4]

Productivity

Land utilization patterns and some major crops are identified in the vegetation map. Since most of the Philippines falls in the Tropical Moist Forest Life Zone, forest productivity should be 18 to 36 MT/ha/yr, assuming reasonable inputs. As one goes into drier or colder regions, productivity would decrease as indicated in the Holdridge life zone map. Commercial forests can produce 25 MT/ha/yr in Tropical Moist Forest Life Zone with good management.

Economics

Food Consumption per capita was 2189 cal/day, energy consumption 342 kg coal equivalent per capita in 1977. In 1970, the caloric supply was 86.0% of requirements. The total population growth rate was 3.0% in 1970. Population density was 123 persons per square kilometer in 1970, with 31.8% of the total population urban. Life expectancy at birth of those born in 1970 was 57 years. In 1977, with the GNP at $460 per capita, agriculture accounted for 27.7% of GDP, manufacturing for 24.6%, other industry for 9.9%, and services for 37.8%. In 1977, food and beverage constituted 50.8% of exports, nonfood agriculture for 7.5%, fuels, minerals, and metal for 16.9%, machines for 1.7%, and other manufacturing for 23.1% of total merchandise exports. Food and beverage constituted 9.5% of imports, nonfood agriculture 3.5%, fuel and lubricants 24.4%, nonfuel minerals and metals 3.5%, machinery and equipment 25.9%, and other manufactured goods constituted 33.3% of total merchandise imports.[10]

Energetics

The Philippines has 25 million b.o.e in oil 1375 million ton coal, and a hydroelectric potential of 7504 MW. Already with nuclear facilities on line, the Philippines is expected to be producing geothermal energy in 1990. Importing more than 75% of its commercial energy demands as oil, the Philippines is expected to be experiencing fuelwood problems by the year 2000. However, the country is becoming a petroleum pro-

* Maps and tables follow text.

Table 1
AGRICULTURAL PRODUCTION IN PHILIPPINES

Crop	Area planted (ha)	Yield (kg/ha)	Total production (MT)
Rice paddy	3,500,000	2,000	7,000,000
Corn	3,276,000	1,007	3,300,000
Potatoes	3,000	6,349	20,000
Sweet potatoes	228,000	4,556	1,037,000
Cassava	182,000	9,802	1,782,000
Dry beans	63,000	653	41,000
Soybeans	9,000	772	7,000
Peanuts in shell	49,000	761	38,000
Castor beans	7,000	1,538	10,000
Seed cotton	5,000	567	3,000
Cottonseed			2,000
Coconuts			8,918
Copra			1,890
Palm kernels			2,400
Palm oil			11,000
Cabbages	9,000	7,798	73,000
Tomatoes	19,000	8,149	153,000
Eggplants	17,000	5,494	95,000
Dry onions	13,000	6,709	85,000
Garlic	5,000	3,283	17,000
Watermelon	12,000	16,508	205,000
Sugar cane	483,000	42,411	20,480,000
Sugar (centrifugal, raw)			2,355,000
Sugar (noncentrifugal)			67,000
Oranges			20,000
Tang. mand. clement. satsuma			35,000
Lemons and limes			40,000
Grapefruit and pomelo			31,000
Avocados			28,000
Mangoes			338,000
Pineapples			479,000
Bananas			2,430,000
Plantains			270,000
Papayas			88,000
Cashew nuts			4,200,000
Green coffee	123,000	695	86,000
Cocoa beans	4,000	727	3,000
Tobacco leaves	74,000	775	57,000
Cotton lint			1,000
Fiber crops nes[a]			286,000
Natural rubber			55,000

From FAO 1979 Production Yearbook, Vol. 33, Food and Agriculture Organization, Rome, 1980.

[a] Nes, not elsewhere specified.

ducer and might come closer to self sufficiency, or reduce their dependency significantly.[10] The Philippines is installing a large-scale alcohol distillery on Luzon and plans others there and in Luzon. Until 1978, when oil was discovered on one of the southern islands, they were totally dependent on imported petroleum.[13]

Soils

The soils in the Philippines are shown on the Soils Map in four categories: (1) Alluvial soils, (2) Grumusols, (3) Latosols of plains and hills, and (4) Latosols of mountains and steep hills.

Alluvial soils are the most important soils for agriculture. They occur not only on broad plains, but also on narrow valley floors along streams, in combination with other soils, in areas too small to show on the map. Alluvial soils vary greatly in texture and drainage conditions, and in many places are subject to prolonged flooding. As a group, they are more fertile and more intensively cultivated than other Philippine soils. The dominant crop on these soils is wetland rice, but on the better drained areas coconuts, sugar cane, tobacco, and abaca are generally grown; near large towns, fruits and vegetables are commonly the dominant crops.

Second in agricultural importance, Grumusols occupy small areas of plains and hills throughout the western and central parts of the island group; some areas are too small to show on the map. Grumusols are fertile, dark-gray to black clays with high organic matter content and slowly permeable subsoils. Often wet and sticky in the wet season, hard and deeply cracked during the dry season, they are difficult to manage for most crops, except wetland rice.

Latosols on the plains and hills are moderately well suited for agriculture, although inherently low in fertility. They are acid, red to reddish-brown, well-drained friable clays with a structure favorable to plant growth. Because of their granular structure and friable consistence, they can be worked with machinery almost immediately after rains. Latosols respond well to fertilizers, but have the undesirable property of rapidly fixing phosphate, making it unavailable to plants; consequently, these soils need frequent applications of fertilizers for satisfactory yields. Only a small proportion of these soils is used for crops, chiefly under shifting cultivation. The dominant crops are upland rice, abaca, and coconuts. Fruits, vegetables, and root crops are grown for local consumption.

Latosols of the mountains and steep slopes, the most extensive soils in the Philippines, are little used for agriculture. The steep slopes generally preclude cultivation; however, locally on Luzon, valley slopes are terraced for wetland rice production.

Lithosols, not shown on the small scale map, are common among the Latosols, on steeper slopes and rugged surfaces of the coral islands. These shallow, stony soils are too droughty for most crops.

Regosols, even less extensive than Lithosols, occur on and near volcanic cones scattered throughout the island group and also on coral islands. Generally sandy and containing cobbles and boulders, they are usually droughty, but would support drought-resistant crops on moderate slopes. On coral sand, coconut palms do well.

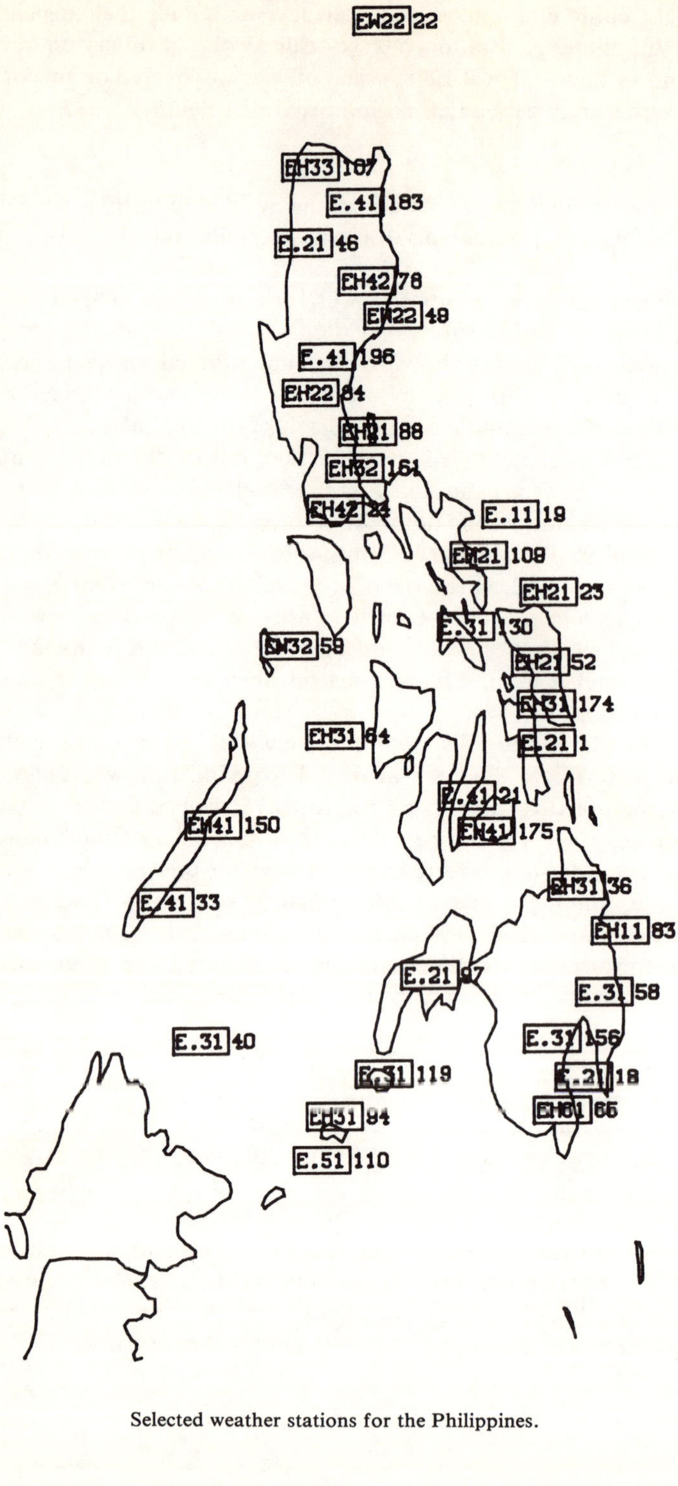

Selected weather stations for the Philippines.

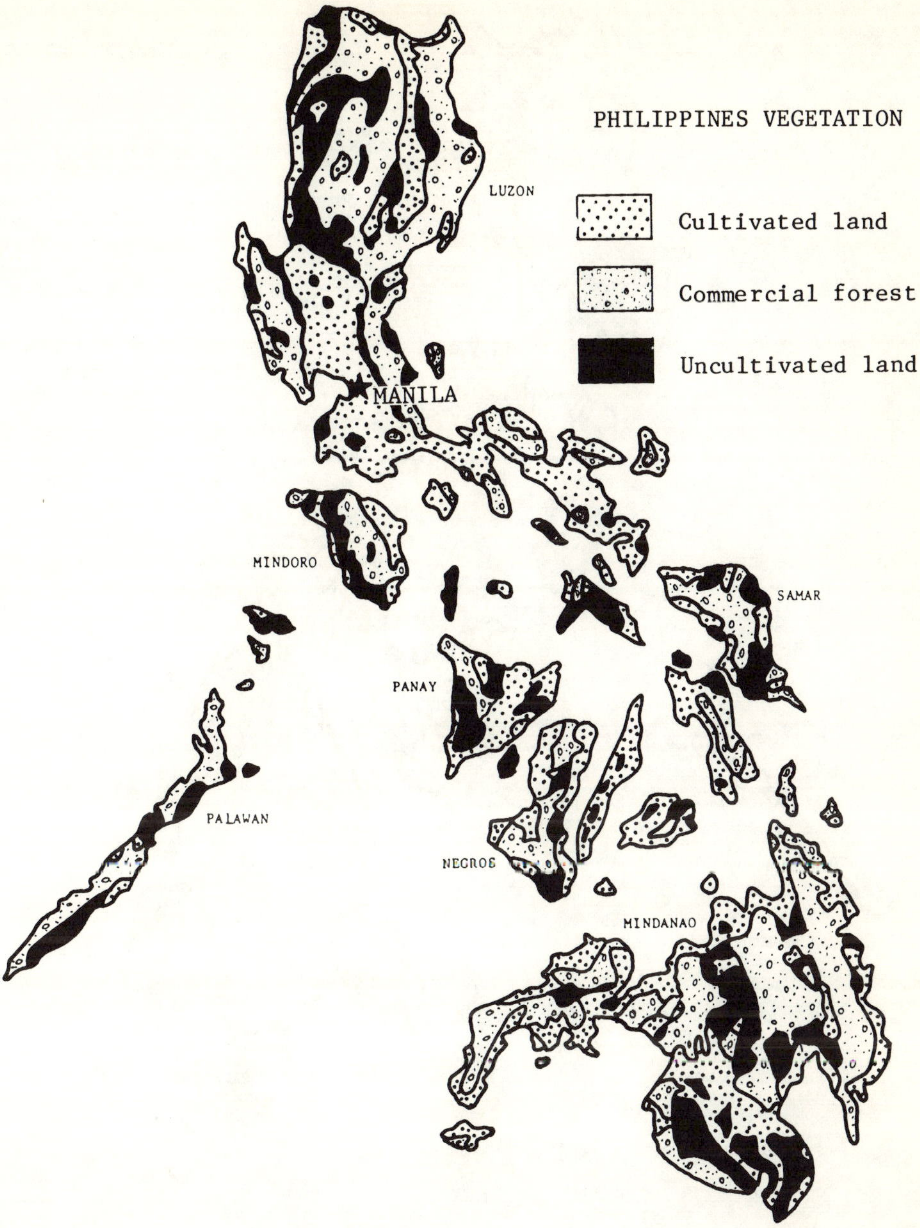
PHILIPPINES VEGETATION
Cultivated land
Commercial forest
Uncultivated land
LUZON
MANILA
MINDORO
SAMAR
PANAY
PALAWAN
NEGROS
MINDANAO

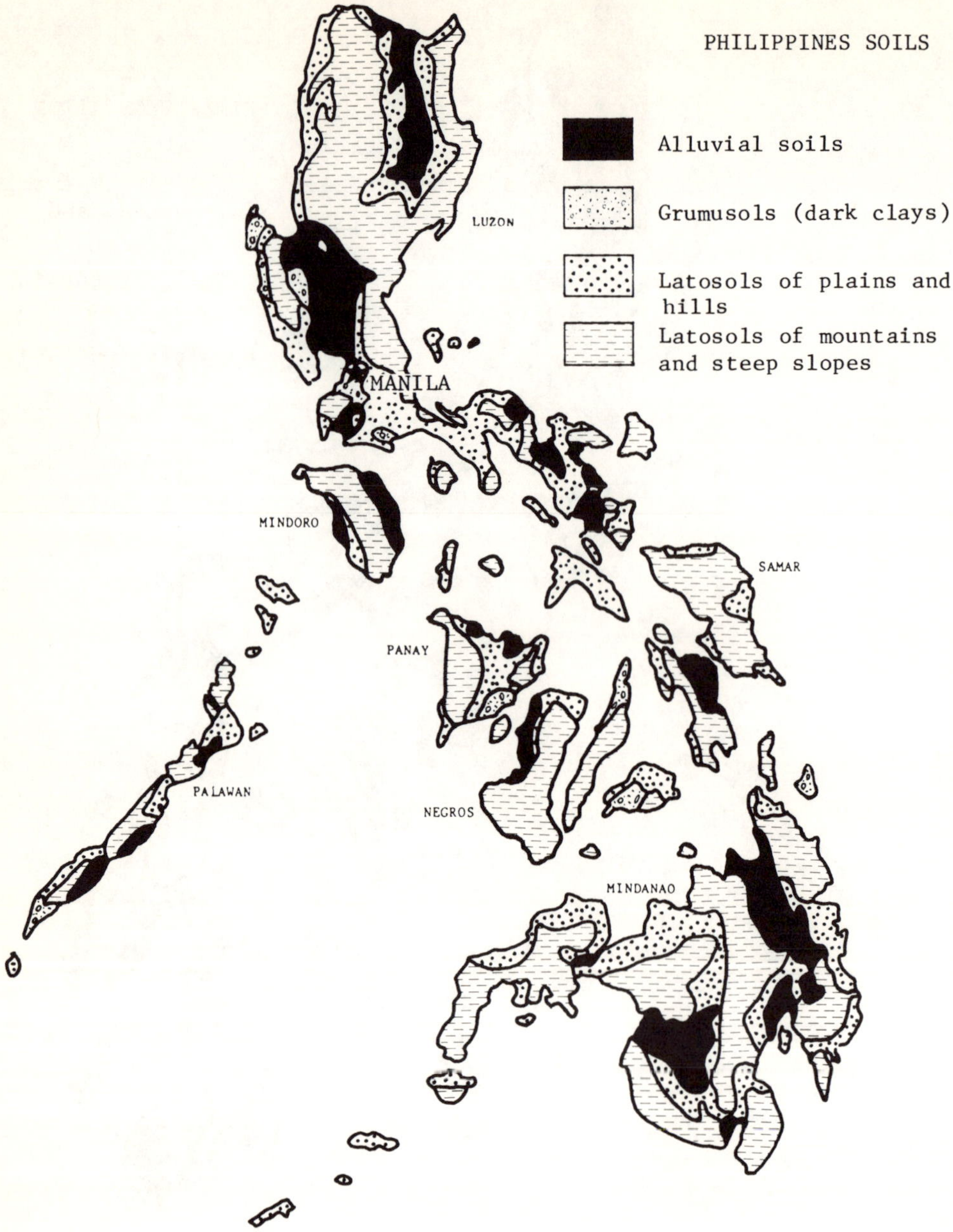
PHILIPPINES SOILS
Alluvial soils
Grumusols (dark clays)
Latosols of plains and hills
Latosols of mountains and steep slopes
LUZON
MANILA
MINDORO
SAMAR
PANAY
PALAWAN
NEGROS
MINDANAO

CLIMATOLOGICAL DATA FOR PHILIPPINES

GEOGRAPHY

OBS.	LOCATION	YEARS	LATITUDE	LONGITUDE	ELEVATION -M-	LIFEZONE	CCTX	ID	ONMAP
1	ABUYOG LEYTE	14	10.75 N	125.00 E	.	.	.	1	YES
2	AGGUNETAN CAGAYAN	14	18.07 N	121.65 E	.	.	.	2	NO
3	AMBUKLAO MOUNTAIN	11	16.50 N	120.75 E	670.6	.	.	3	NO
4	AMBULONG BATANGAS	41	14.08 N	121.05 E	.	TD	EH42	4	NO
5	ANTIPOLO RIZAL	32	14.60 N	121.17 E	.	TM	EH22	5	NO
6	APARRI CAGAYAN	52	18.37 N	121.63 E	.	TM	EH32	6	NO
7	ARINGAY LA UNION	12	16.40 N	120.35 E	.	.	.	7	NO
8	ATIMONAN QUEZON	38	14.00 N	121.92 E	.	TM	EH22	8	NO
9	AURORA QUEZON	12	15.43 N	121.58 E	.	TD	EH41	9	NO
10	BACOLOD NEGROS OCC	25	10.67 N	122.92 E	.	TM	EH31	10	NO
11	BAGABAG NUEVA VIZ	13	16.60 N	121.25 E	243.8	.	.	11	NO
12	BAGO NEGROS OCC	26	10.50 N	122.83 E	.	.	.	12	NO

AVERAGE PRECIPITATION (MM)

OBS.	JAN.	FEB.	MAR.	APR.	MAY	JUN.	JUL.	AUG.	SEP.	OCT.	NOV.	DEC.	ANNUAL
1	581	464	388	254	285	193	162	155	148	294	399	648	3972
2	40	29	14	37	33	66	72	97	115	133	200	98	932
3	4	15	44	87	166	236	333	438	329	165	86	24	1928
4	26	14	21	47	147	200	306	339	299	232	168	111	1910
5	25	17	18	42	189	326	555	591	462	263	199	116	2804
6	144	90	55	49	111	172	189	234	295	367	336	218	2260
7	4	6	7	21	197	392	583	718	325	156	52	20	2481
8	205	113	79	85	147	187	202	163	283	381	447	384	2676
9	58	21	44	30	94	154	242	213	212	262	202	167	1700
10	100	59	41	37	177	227	375	306	286	261	236	172	2277
11	45	199	72	104	137	175	183	217	334	247	209	207	1977
12	120	44	47	85	277	301	459	387	384	352	263	132	2852

AVERAGE TEMPERATURE (DEGREES CENTIGRADE)

OBS.	JAN.	FEB.	MAR.	APR.	MAY	JUN.	JUL.	AUG.	SEP.	OCT.	NOV.	DEC.	ANNUAL
1	.	.	.	.	.	.	.	.	.	.	.	.	.
2	.	.	.	.	.	.	.	.	.	.	.	.	.
3	.	.	.	.	.	.	.	.	.	.	.	.	.
4	25.5	26.1	27.3	28.5	28.7	27.8	27.1	27.0	26.8	26.5	26.4	25.6	26.9
5	25.0	25.5	27.0	28.4	28.8	28.0	26.4	26.2	26.1	26.2	26.1	25.5	26.6
6	23.2	23.8	25.3	27.1	28.1	28.5	28.1	27.9	27.5	26.5	25.2	23.9	26.3
7	.	.	.	.	.	.	.	.	.	.	.	.	.
8	25.2	25.3	26.4	27.7	28.2	27.9	27.4	27.5	27.0	26.9	26.6	25.8	26.8
9	25.4	25.7	26.3	27.3	27.9	27.8	27.3	27.3	27.1	26.8	26.5	25.7	26.8
10	25.9	26.1	26.8	27.9	28.3	27.5	26.6	26.6	26.7	26.8	26.4	26.3	26.8
11	.	.	.	.	.	.	.	.	.	.	.	.	.
12	.	.	.	.	.	.	.	.	.	.	.	.	.

CLIMATOLOGICAL DATA FOR PHILIPPINES

GEOGRAPHY

OBS.	LOCATION	YEARS	LATITUDE	LONGITUDE	ELEVATION -M-	LIFEZONE	CCTX	ID	ONMAP
13	BAGUIO MOUNTAIN	52	16.42 N	120.58 E	1524.0	SR	EW12	197	NO
14	BALANGA BATAAN	13	14.67 N	120.50 E	.	.	.	13	NO
15	BALAOAN LA UNION	12	16.83 N	120.40 E	.	.	.	14	NO
16	BALER QUEZON	50	15.75 N	121.57 E	.	TM	EH21	15	NO
17	BALETE AKLAN	14	11.58 N	122.42 E	.	.	.	16	NO
18	BALINGASAG MIS OR	6	8.75 N	124.73 E	.	.	.	17	NO
19	BANGA COTABATO	4	6.53 N	125.62 E	.	.	.	18	YES
20	BARAS CATANDUANES	13	13.67 N	124.37 E	.	.	.	19	YES
21	BARBAZA ANTIQUE	14	11.17 N	122.08 E	.	.	.	20	NO
22	BARILI CEBU	11	10.08 N	123.50 E	.	.	.	21	YES
23	BASCO, BATANES	51	20.13 N	121.98 E	.	TM	EW22	22	YES
24	BATAG SAMAR	16	12.67 N	125.07 E	.	TM	EH21	23	YES

AVERAGE PRECIPITATION (MM)

OBS.	JAN.	FEB.	MAR.	APR.	MAY	JUN.	JUL.	AUG.	SEP.	OCT.	NOV.	DEC.	ANNUAL
13	20	19	48	122	352	422	922	1043	650	362	163	56	4177
14	9	13	10	32	228	367	700	910	403	230	122	39	3058
15	3	6	22	21	181	407	625	630	330	130	55	9	2418
16	202	177	221	285	287	231	258	199	289	375	404	378	3305
17	266	125	219	104	230	267	268	198	222	410	438	358	3107
18	75	52	26	21	48	178	241	179	274	226	253	86	1659
19	72	92	92	136	310	334	322	276	341	225	262	121	2583
20	457	405	384	148	205	225	358	219	371	802	1097	812	5482
21	28	13	19	59	224	444	570	812	574	513	250	138	3644
22	127	83	98	61	154	158	180	139	153	172	287	230	1845
23	229	130	128	98	205	242	308	402	363	367	346	324	3142
24	515	317	211	136	148	223	183	182	213	393	504	492	3518

AVERAGE TEMPERATURE (DEGREES CENTIGRADE)

OBS.	JAN.	FEB.	MAR.	APR.	MAY	JUN.	JUL.	AUG.	SEP.	OCT.	NOV.	DEC.	ANNUAL
13	16.9	17.4	18.3	18.9	19.2	18.9	18.4	18.0	18.2	18.3	18.1	17.5	18.2
14	.	.	.	.	.	.	.	.	.	.	.	.	.
15	.	.	.	.	.	.	.	.	.	.	.	.	.
16	24.5	24.7	25.6	26.9	27.9	28.2	28.0	28.1	27.7	27.0	26.1	25.2	26.7
17	.	.	.	.	.	.	.	.	.	.	.	.	.
18	.	.	.	.	.	.	.	.	.	.	.	.	.
19	.	.	.	.	.	.	.	.	.	.	.	.	.
20	.	.	.	.	.	.	.	.	.	.	.	.	.
21	.	.	.	.	.	.	.	.	.	.	.	.	.
22	.	.	.	.	.	.	.	.	.	.	.	.	.
23	22.3	22.7	23.9	25.1	27.6	28.3	28.3	27.9	27.5	26.3	24.8	23.1	25.7
24	24.9	25.3	26.1	26.9	27.6	27.2	27.4	27.4	27.2	26.8	26.4	26.1	26.6

CLIMATOLOGICAL DATA FOR PHILIPPINES

GEOGRAPHY

OBS.	LOCATION	YEARS	LATITUDE	LONGITUDE	ELEVATION -M-	LIFEZONE	CCTX	ID	ONMAP
25	BATANGAS BATANGAS	32	13.75 N	121.05 E	.	TD	EH42	24	YES
26	BAYOMBONG NVA VIZ	25	16.50 N	121.15 E	.	.	.	25	NO
27	BINALBAGAN NEGROS	18	10.17 N	122.83 E	.	.	.	26	NO
28	BOAC MARINDUQUE	16	13.45 N	121.83 E	.	.	.	27	NO
29	BOGO CEBU	14	11.08 N	124.00 E	.	.	.	28	NO
30	BOJEADOR ILOCOS N	17	18.50 N	120.57 E	.	.	.	29	NO
31	BOLINAO PANGASINAN	38	16.38 N	119.88 E	.	TM	EH22	30	NO
32	BONTOC MOUNTAIN	20	17.10 N	120.98 E	792.5	.	.	31	NO
33	BORONGAN SAMAR	52	11.62 N	125.60 E	.	TW	EH11	32	NO
34	BROOKES POINT PAL	14	8.75 N	117.83 E	.	.	.	33	YES
35	BUGUIAS MOUNTAIN	11	16.75 N	120.83 E	1219.2	.	.	34	NO
36	BULUAN COTABATO	18	6.73 N	124.80 E	.	.	.	35	NO

AVERAGE PRECIPITATION (MM)

OBS.	JAN.	FEB.	MAR.	APR.	MAY	JUN.	JUL.	AUG.	SEP.	OCT.	NOV.	DEC.	ANNUAL
25	25	17	11	34	112	156	303	265	271	188	189	115	1684
26	48	30	53	107	179	107	190	163	229	176	160	99	1540
27	59	41	46	54	210	204	393	309	322	285	197	98	2218
28	90	66	44	71	158	208	235	185	198	288	282	262	2088
29	124	68	67	52	118	174	173	142	195	196	224	161	1696
30	12	9	12	16	139	272	496	416	248	169	107	45	1943
31	8	9	11	35	203	391	615	649	416	155	50	9	2551
32	13	25	58	124	249	236	387	336	322	214	140	47	2151
33	626	437	359	270	275	229	192	157	181	340	543	654	4262
34	70	44	71	58	123	165	163	194	146	164	181	198	1575
35	13	6	59	102	134	208	241	458	292	153	156	43	1864
36	79	75	90	84	194	234	198	287	172	164	138	93	1807

AVERAGE TEMPERATURE (DEGREES CENTIGRADE)

OBS.	JAN.	FEB.	MAR.	APR.	MAY	JUN.	JUL.	AUG.	SEP.	OCT.	NOV.	DEC.	ANNUAL
25	25.8	26.2	27.8	29.0	29.0	28.6	27.6	27.7	27.2	27.6	26.6	26.1	27.4
26	.	.	.	.	.	.	.	.	.	.	.	.	.
27	.	.	.	.	.	.	.	.	.	.	.	.	.
28	.	.	.	.	.	.	.	.	.	.	.	.	.
29	.	.	.	.	.	.	.	.	.	.	.	.	.
30	.	.	.	.	.	.	.	.	.	.	.	.	.
31	25.9	26.1	27.2	28.7	28.8	27.9	26.9	26.8	26.8	27.3	27.0	26.5	27.2
32	.	.	.	.	.	.	.	.	.	.	.	.	.
33	25.9	25.8	26.4	27.1	27.6	27.6	27.6	27.8	27.8	27.1	26.8	26.3	27.0
34	.	.	.	.	.	.	.	.	.	.	.	.	.
35	.	.	.	.	.	.	.	.	.	.	.	.	.
36	.	.	.	.	.	.	.	.	.	.	.	.	.

CLIMATOLOGICAL DATA FOR PHILIPPINES

GEOGRAPHY

OBS.	LOCATION	YEARS	LATITUDE	LONGITUDE	ELEVATION -M-	LIFEZONE	CCTX	ID	ONMAP
37	BUTUAN AGUSAN	49	8.93 N	125.53 E	.	TM	EH31	36	YES
38	CABANATUAN NVA ECI	15	15.50 N	121.08 E	.	TD	EH43	37	NO
39	CADIZ NEGROS OCC	15	11.00 N	123.33 E	.	.	.	38	NO
40	CAGAYAN MISAMIS OR	45	8.43 N	124.63 E	.	TD	EH41	39	NO
41	CAGAYAN SULU	12	7.00 N	118.50 E	.	.	.	40	YES
42	CALAPAN MINDORO OR	36	13.42 N	121.18 E	.	TM	EH31	41	NO
43	CALAUAG CAMARINES	11	14.00 N	122.25 E	.	.	.	42	NO
44	CALBAYOG SAMAR	37	12.07 N	124.60 E	.	TM	EH21	43	NO
45	CALIVO CAPIZ	12	11.72 N	122.37 E	.	.	.	44	NO
46	CAMP OVERTON LANAO	10	8.17 N	124.17 E	.	.	.	45	NO
47	CANDON ILOCOS SUR	31	17.20 N	120.45 E	.	.	.	46	YES
48	CANTILAN SURIGAO	17	9.33 N	126.00 E	.	.	.	47	NO

AVERAGE PRECIPITATION (MM)

OBS.	JAN.	FEB.	MAR.	APR.	MAY	JUN.	JUL.	AUG.	SEP.	OCT.	NOV.	DEC.	ANNUAL
37	304	189	181	150	176	167	152	110	145	189	247	314	2325
38	6	6	26	53	125	211	276	410	304	176	130	77	1799
39	151	108	75	53	126	150	144	123	161	307	287	198	1884
40	90	54	45	35	128	212	234	193	217	193	135	124	1661
41	254	71	87	103	170	183	138	165	213	174	266	291	2116
42	111	70	59	110	177	215	239	223	193	278	289	199	2162
43	214	87	89	85	98	167	200	250	180	348	377	573	2669
44	203	173	115	136	167	224	271	204	254	280	338	314	2680
45	161	88	70	89	189	199	273	207	227	408	397	225	2533
46	89	84	90	86	149	215	252	385	719	1048	313	211	3641
47	4	7	9	14	197	361	706	727	380	169	51	8	2633
48	856	576	596	411	330	156	155	131	142	205	494	937	4991

AVERAGE TEMPERATURE (DEGREES CENTIGRADE)

OBS.	JAN.	FEB.	MAR.	APR.	MAY	JUN.	JUL.	AUG.	SEP.	OCT.	NOV.	DEC.	ANNUAL
37	25.1	25.3	26.1	27.0	28.1	27.7	27.6	27.6	27.4	26.9	26.1	25.6	26.7
38	24.9	26.6	27.8	29.3	29.8	28.7	28.1	27.6	27.7	27.6	26.7	26.0	27.7
39	.	.	.	.	.	.	.	.	.	.	.	.	.
40	25.9	26.2	26.8	27.6	28.0	27.7	27.2	27.4	27.2	27.2	26.3	26.2	27.0
41	.	.	.	.	.	.	.	.	.	.	.	.	.
42	25.5	25.8	26.8	27.9	28.1	27.7	27.3	27.2	27.2	27.0	26.6	25.9	26.9
43	.	.	.	.	.	.	.	.	.	.	.	.	.
44	24.9	24.8	25.7	26.5	27.0	27.0	27.1	27.5	27.1	26.5	25.7	25.4	26.3
45	.	.	.	.	.	.	.	.	.	.	.	.	.
46	.	.	.	.	.	.	.	.	.	.	.	.	.
47	.	.	.	.	.	.	.	.	.	.	.	.	.
48	.	.	.	.	.	.	.	.	.	.	.	.	.

CLIMATOLOGICAL DATA FOR PHILIPPINES

GEOGRAPHY

OBS.	LOCATION	YEARS	LATITUDE	LONGITUDE	ELEVATION -M-	LIFEZONE	CCTX	ID	ONMAP
49	CARMEN BOHOL	16	9.83 N	124.20 E	182.9	.	.	48	NO
50	CASIGURAN QUEZON	14	16.25 N	122.17 E	.	TM	EW22	49	YES
51	CASTILLA SORSOGON	12	12.92 N	123.92 E	.	.	.	50	NO
52	CATARMAN SAMAR	12	12.50 N	124.58 E	.	TM	EH21	51	NO
53	CATBALOGAN SAMAR	39	11.78 N	124.90 E	.	TM	EH21	52	YES
54	CATEEL DAVAO	15	7.80 N	126.43 E	.	.	.	53	NO
55	CAVINTI LAGUNA	12	14.33 N	121.50 E	.	.	.	54	NO
56	CEBU CEBU	52	10.30 N	123.90 E	.	TD	EH41	55	NO
57	CLARIN MISAMIS OCC	14	8.20 N	123.85 E	.	.	.	56	NO
58	CLAVERIA CAGAYAN	14	18.62 N	121.08 E	.	.	.	57	NO
59	COMPOSTELA DAVAO	18	7.60 N	126.02 E	198.1	.	.	58	YES
60	CORON PALAWAN	11	12.00 N	120.17 E	.	TM	EH32	59	YES

AVERAGE PRECIPITATION (MM)

OBS.	JAN.	FEB.	MAR.	APR.	MAY	JUN.	JUL.	AUG.	SEP.	OCT.	NOV.	DEC.	ANNUAL
49	216	110	110	48	148	198	207	201	200	221	286	263	2208
50	231	225	268	166	201	258	216	245	323	377	579	529	3618
51	231	141	144	132	119	145	198	235	260	358	435	396	2795
52	419	233	259	161	136	144	176	183	189	447	584	516	3447
53	281	184	159	136	165	201	248	214	256	319	364	375	2902
54	841	614	557	359	235	104	104	85	82	182	309	739	4210
55	183	82	82	112	184	285	261	309	293	402	419	393	3004
56	105	71	55	53	121	177	197	153	187	201	163	138	1620
57	431	204	336	217	331	490	407	455	414	551	822	776	5435
58	473	319	175	185	141	218	197	310	341	486	545	657	4046
59	400	233	185	160	237	169	132	118	148	186	163	296	2429
60	38	7	8	30	169	367	376	496	452	290	140	128	2499

AVERAGE TEMPERATURE (DEGREES CENTIGRADE)

OBS.	JAN.	FEB.	MAR.	APR.	MAY	JUN.	JUL.	AUG.	SEP.	OCT.	NOV.	DEC.	ANNUAL
49	.	.	.	.	.	.	.	.	.	.	.	.	.
50	23.7	24.0	24.9	26.4	27.2	27.6	27.4	27.2	27.0	26.5	25.5	24.3	26.0
51	.	.	.	.	.	.	.	.	.	.	.	.	.
52	25.4	25.4	25.9	26.6	27.4	27.6	27.3	27.6	27.4	26.8	26.5	25.8	26.6
53	25.9	26.1	26.7	27.6	28.1	28.1	27.9	28.2	27.9	27.5	26.8	26.3	27.3
54	.	.	.	.	.	.	.	.	.	.	.	.	.
55	.	.	.	.	.	.	.	.	.	.	.	.	.
56	26.3	26.4	27.2	28.2	28.4	28.0	27.6	27.7	27.6	27.5	27.1	26.7	27.4
57	.	.	.	.	.	.	.	.	.	.	.	.	.
58	.	.	.	.	.	.	.	.	.	.	.	.	.
59	.	.	.	.	.	.	.	.	.	.	.	.	.
60	27.1	27.3	28.0	28.9	28.6	27.7	26.7	26.9	27.0	27.4	27.4	27.0	27.5

CLIMATOLOGICAL DATA FOR PHILIPPINES

GEOGRAPHY

OBS.	LOCATION	YEARS	LATITUDE	LONGITUDE	ELEVATION -M-	LIFEZONE	CCTX	ID	ONMAP
61	CORREGIDOR CAVITE	30	14.38 N	120.58 E	.	.	.	60	NO
62	COTABATO COTABATO	43	7.22 N	124.25 E	.	TM	EH31	61	NO
63	CULASI ANTIQUE	26	11.43 N	122.07 E	.	.	.	62	NO
64	CULION PALAWAN	12	11.85 N	120.00 E	.	.	.	63	NO
65	CUYO PALAWAN	52	10.85 N	121.02 E	.	TM	EH31	64	YES
66	DADIANGAS COTABATO	12	6.12 N	125.23 E	.	TV	EH61	65	YES
67	DAET CAMARINES NOR	34	14.12 N	122.95 E	.	TM	EH21	66	NO
68	DAGOHOY BOHOL	21	9.90 N	124.27 E	.	.	.	67	NO
69	DAGUPAN PANGASINAN	52	16.05 N	120.33 E	.	TM	EH32	68	NO
70	DAO ANTIQUE	14	10.50 N	121.92 E	.	.	.	69	NO
71	DAPITAN ZAMBOANGA	30	8.67 N	123.42 E	.	.	.	70	NO
72	DAVAO DAVAO	52	7.02 N	125.58 E	.	TD	EH41	71	NO

AVERAGE PRECIPITATION (MM)

OBS.	JAN.	FEB.	MAR.	AFR.	MAY	JUN.	JUL.	AUG.	SEP.	OCT.	NOV.	DEC.	ANNUAL
61	11	4	5	21	96	334	583	638	433	211	103	53	2491
62	81	85	84	150	222	240	267	243	225	281	184	104	2165
63	118	202	67	79	278	467	676	742	611	512	403	252	4255
64	29	19	11	53	207	351	670	604	518	291	254	101	3107
65	17	8	8	33	198	336	424	387	368	284	152	75	2290
66	44	57	49	52	113	111	100	106	76	78	82	67	934
67	382	250	212	152	163	166	225	198	299	541	660	576	3824
68	213	108	112	68	142	198	232	199	193	205	260	245	2175
69	9	13	25	85	203	315	535	541	436	171	81	26	2439
70	26	9	15	32	128	318	358	404	373	260	152	73	2149
71	141	91	72	132	155	205	199	149	137	251	370	303	2208
72	118	103	119	142	236	217	176	162	176	193	143	145	1928

AVERAGE TEMPERATURE (DEGREES CENTIGRADE)

OBS.	JAN.	FEB.	MAR.	AFR.	MAY	JUN.	JUL.	AUG.	SEP.	OCT.	NOV.	DEC.	ANNUAL
61	.	.	.	.	.	.	.	.	.	.	.	.	.
62	27.1	27.4	27.8	28.3	28.0	27.7	27.2	27.3	27.4	27.6	27.5	27.2	27.5
63	.	.	.	.	.	.	.	.	.	.	.	.	.
64	.	.	.	.	.	.	.	.	.	.	.	.	.
65	26.9	27.0	27.9	28.8	28.7	28.0	27.4	27.3	27.3	27.6	27.6	27.4	27.7
66	26.6	27.0	27.5	27.8	27.4	26.7	26.3	26.2	26.5	26.7	26.9	26.8	26.9
67	25.1	25.2	25.9	26.9	27.7	27.9	27.4	27.4	27.1	26.6	26.3	25.5	26.6
68	.	.	.	.	.	.	.	.	.	.	.	.	.
69	25.8	26.4	27.7	29.0	29.0	28.2	27.4	27.3	27.4	27.5	26.8	26.1	27.4
70	.	.	.	.	.	.	.	.	.	.	.	.	.
71	.	.	.	.	.	.	.	.	.	.	.	.	.
72	26.3	26.6	27.2	27.7	27.6	27.0	26.8	26.9	27.0	27.1	27.0	26.6	27.0

CLIMATOLOGICAL DATA FOR PHILIPPINES

GEOGRAPHY

OBS.	LOCATION	YEARS	LATITUDE	LONGITUDE	ELEVATION -M-	LIFEZONE	CCTX	ID	ONMAP
73	DEL CARMEN PAMPANG	27	15.00 N	120.50 E	.	.	.	72	NO
74	DEL MONTE BUKIDNON	13	8.42 N	124.83 E	609.6	.	.	73	NO
75	DINGRAS ILOCOS NOR	13	18.08 N	120.67 E	.	.	.	74	NO
76	DIPOLOG ZAMBCANGA	7	8.60 N	123.35 E	.	TM	EH21	75	NO
77	DUMAGUETE NEGROS	43	9.33 N	123.33 E	.	TD	EH51	76	NO
78	DUMARAO CAPIZ	20	11.27 N	122.70 E	.	.	.	77	NO
79	ECHAGUE ISABELA	32	16.70 N	121.67 E	.	TD	EH42	78	YES
80	GINGOOG MISAMIS OR	5	8.82 N	125.12 E	.	.	.	79	NO
81	GUBAT SORSOGCN	14	12.92 N	124.13 E	.	TM	EH22	80	NO
82	GUINOBATAN ALBAY	27	13.17 N	123.58 E	.	.	.	195	NO
83	GUIUAN SAMAR	38	11.03 N	125.73 E	.	TM	EH21	81	NO
84	HAC REFUGIO NEGROS	19	10.47 N	123.38 E	.	.	.	82	NO

AVERAGE PRECIPITATION (MM)

OBS.	JAN.	FEB.	MAR.	APR.	MAY	JUN.	JUL.	AUG.	SEP.	OCT.	NOV.	DEC.	ANNUAL
73	11	15	19	43	181	305	544	666	340	208	109	57	2500
74	92	75	112	93	284	370	386	288	322	268	210	203	2704
75	5	1	14	36	214	408	447	592	408	132	88	13	2358
76	136	54	110	102	189	279	283	290	230	350	400	300	2723
77	98	70	55	47	110	149	148	113	136	189	176	138	1429
78	171	99	73	65	234	263	243	225	257	267	292	240	2430
79	60	37	61	85	173	112	198	209	236	228	261	171	1832
80	272	238	246	186	167	236	206	134	213	215	402	463	2977
81	313	235	172	89	104	134	172	100	193	273	395	506	2685
82	196	108	117	83	143	223	327	263	320	327	420	320	2847
83	539	360	293	208	211	211	176	122	172	275	454	521	3540
84	134	91	68	51	109	115	123	115	120	220	241	163	1548

AVERAGE TEMPERATURE (DEGREES CENTIGRADE)

OBS.	JAN.	FEB.	MAR.	APR.	MAY	JUN.	JUL.	AUG.	SEP.	OCT.	NOV.	DEC.	ANNUAL
73	.	.	.	.	.	.	.	.	.	.	.	.	.
74	.	.	.	.	.	.	.	.	.	.	.	.	.
75	.	.	.	.	.	.	.	.	.	.	.	.	.
76	26.7	27.0	27.5	28.1	28.0	27.7	27.3	27.4	27.2	27.4	27.3	26.8	27.4
77	26.4	26.5	27.2	27.9	28.0	27.7	27.4	27.7	27.5	27.3	27.1	26.8	27.3
78	.	.	.	.	.	.	.	.	.	.	.	.	.
79	23.7	24.2	26.7	28.1	29.1	29.2	28.3	28.2	27.8	26.6	25.2	24.3	26.7
80	.	.	.	.	.	.	.	.	.	.	.	.	.
81	25.8	25.9	26.6	27.7	28.4	28.4	28.0	28.4	28.1	27.4	26.9	26.3	27.3
82	.	.	.	.	.	.	.	.	.	.	.	.	.
83	26.2	26.3	26.5	27.1	27.7	28.0	28.0	28.3	28.6	27.8	27.4	27.0	27.6
84	.	.	.	.	.	.	.	.	.	.	.	.	.

CLIMATOLOGICAL DATA FOR PHILIPPINES

GEOGRAPHY

OBS.	LOCATION	YEARS	LATITUDE	LONGITUDE	ELEVATION -M-	LIFEZONE	CCTX	ID	ONMAP
85	HINATUAN SURIGAO	22	8.37 N	126.33 E	.	TW	EH11	83	YES
86	IBA ZAMBALES	45	15.25 N	120.58 E	.	TM	EH22	84	YES
87	ILAGAN ISABELA	25	17.13 N	121.90 E	.	.	.	85	NO
88	ILOILO ILOILO	52	10.70 N	122.57 E	.	TM	EH31	86	NO
89	IMPALUTAO BUKIDNON	13	8.25 N	125.00 E	944.9	.	.	87	NO
90	INFANTA QUEZON	28	14.75 N	121.67 E	.	TM	EH21	88	YES
91	INOPACAN LEYTE	14	10.50 N	124.75 E	.	.	.	89	NO
92	ISABELA ZAMBOANGA	20	6.70 N	121.97 E	.	.	.	90	NO
93	ITOGON MOUNTAIN	13	16.37 N	120.68 E	1219.2	.	.	91	NO
94	IWAHIG PALAWAN	30	9.75 N	118.67 E	.	.	.	92	NO
95	JANIUAY ILOILO	18	10.95 N	122.50 E	.	.	.	93	NO
96	JOLO SULU	57	6.05 N	121.00 E	.	TM	EH31	94	YES

AVERAGE PRECIPITATION (MM)

OBS.	JAN.	FEB.	MAR.	APR.	MAY	JUN.	JUL.	AUG.	SEP.	OCT.	NOV.	DEC.	ANNUAL
85	724	427	460	322	295	220	218	199	195	210	371	665	4303
86	6	5	18	44	257	569	905	985	687	232	85	33	3825
87	59	45	42	60	124	138	195	184	248	287	308	166	1855
88	59	38	36	52	153	265	390	370	294	263	207	121	2246
89	111	86	139	107	268	321	354	274	367	308	203	209	2747
90	404	263	190	200	213	208	229	232	281	502	556	519	3797
91	176	121	96	76	136	157	246	213	244	274	232	269	2241
92	78	77	58	89	140	209	200	200	201	255	160	152	1817
93	7	15	41	120	211	340	466	649	443	222	129	69	2711
94	59	33	41	53	168	198	221	195	241	267	275	236	1986
95	119	57	66	87	183	203	266	219	240	263	250	125	2078
96	105	106	107	143	217	223	178	170	188	226	215	169	2048

AVERAGE TEMPERATURE (DEGREES CENTIGRADE)

OBS.	JAN.	FEB.	MAR.	APR.	MAY	JUN.	JUL.	AUG.	SEP.	OCT.	NOV.	DEC.	ANNUAL
85	25.6	25.5	25.9	26.4	26.9	26.8	26.8	27.1	27.0	26.9	26.7	25.9	26.5
86	25.4	25.7	26.8	28.1	28.4	27.7	27.0	26.7	26.8	27.0	26.6	25.9	26.8
87	.	.	.	.	.	.	.	.	.	.	.	.	.
88	25.8	26.1	26.9	28.2	28.3	27.7	27.2	26.9	27.1	26.9	26.6	26.1	27.0
89	.	.	.	.	.	.	.	.	.	.	.	.	.
90	24.6	25.0	25.9	27.1	27.9	28.4	28.0	28.0	27.6	26.9	26.3	25.3	26.8
91	.	.	.	.	.	.	.	.	.	.	.	.	.
92	.	.	.	.	.	.	.	.	.	.	.	.	.
93	.	.	.	.	.	.	.	.	.	.	.	.	.
94	.	.	.	.	.	.	.	.	.	.	.	.	.
95	.	.	.	.	.	.	.	.	.	.	.	.	.
96	26.2	26.0	26.3	26.7	27.0	26.9	26.9	27.0	26.9	26.6	26.4	26.3	26.6

CLIMATOLOGICAL DATA FOR PHILIPPINES

GEOGRAPHY

OBS.	LOCATION	YEARS	LATITUDE	LONGITUDE	ELEVATION -M-	LIFEZONE	CCTX	ID	ONMAP
97	JUBAN SORSOGON	14	12.83 N	124.00 E	.	.	.	95	NO
98	KABAKAN COTABATO	11	7.17 N	124.83 E	304.8	.	.	96	NO
99	KABASALAN ZAMBOANG	13	7.82 N	122.77 E	.	.	.	97	YES
100	KABAYAN MOUNTAIN	11	16.58 N	120.82 E	1066.8	.	.	98	NO
101	KABUGAO MOUNTAIN	12	18.00 N	121.17 E	243.8	.	.	99	NO
102	KIDAPAWAN COTABATO	18	7.02 N	125.05 E	304.8	.	.	100	NO
103	LA CARLOTA NEGROS	32	10.42 N	122.92 E	.	.	.	101	NO
104	LA CASTELLANA NEG	13	10.33 N	123.00 E	.	.	.	102	NO
105	LA UNION DAVAO	10	6.70 N	126.08 E	.	.	.	103	NO
106	LAKE BUHI CAMARINE	11	13.42 N	123.50 E	.	.	.	104	NO
107	LAMITAN ZAMBOANGA	15	6.67 N	122.17 E	.	.	.	105	NO
108	LANGANGILAN ABRA	13	17.62 N	120.73 E	.	.	.	106	NO

AVERAGE PRECIPITATION (MM)

OBS.	JAN.	FEB.	MAR.	APR.	MAY	JUN.	JUL.	AUG.	SEP.	OCT.	NOV.	DEC.	ANNUAL
97	415	245	304	175	155	153	193	249	249	424	564	557	3682
98	96	87	102	137	237	207	190	159	194	180	148	133	1870
99	133	117	162	198	267	322	340	327	304	322	275	237	3004
100	14	25	51	123	246	264	340	586	375	250	180	46	2497
101	66	66	73	105	185	227	246	276	338	457	449	215	2702
102	127	110	92	131	200	197	244	181	234	253	189	169	2128
103	68	55	67	99	253	258	456	383	395	358	268	144	2804
104	64	29	43	57	191	240	344	404	488	427	295	223	2803
105	229	127	133	150	113	482	263	234	228	217	103	162	2448
106	209	105	115	94	157	183	212	279	273	385	454	422	2888
107	96	102	83	111	167	220	166	149	207	157	160	89	1707
108	8	17	54	169	306	495	533	716	533	227	139	31	3228

AVERAGE TEMPERATURE (DEGREES CENTIGRADE)

OBS.	JAN.	FEB.	MAR.	APR.	MAY	JUN.	JUL.	AUG.	SEP.	OCT.	NOV.	DEC.	ANNUAL
97	.	.	.	.	.	.	.	.	.	.	.	.	.
98	.	.	.	.	.	.	.	.	.	.	.	.	.
99	.	.	.	.	.	.	.	.	.	.	.	.	.
100	.	.	.	.	.	.	.	.	.	.	.	.	.
101	.	.	.	.	.	.	.	.	.	.	.	.	.
102	.	.	.	.	.	.	.	.	.	.	.	.	.
103	.	.	.	.	.	.	.	.	.	.	.	.	.
104	.	.	.	.	.	.	.	.	.	.	.	.	.
105	.	.	.	.	.	.	.	.	.	.	.	.	.
106	.	.	.	.	.	.	.	.	.	.	.	.	.
107	.	.	.	.	.	.	.	.	.	.	.	.	.
108	.	.	.	.	.	.	.	.	.	.	.	.	.

CLIMATOLOGICAL DATA FOR PHILIPPINES

GEOGRAPHY

OBS.	LOCATION	YEARS	LATITUDE	LONGITUDE	ELEVATION -M-	LIFEZONE	CCTX	ID	ONMAP
109	LAOAG ILOCOS NORTE	46	18.20 N	120.58 E	.	TM	EH33	107	YES
110	LAOANG SAMAR	11	12.58 N	125.00 E	.	.	.	108	NO
111	LEGASPI ALBAY	52	13.15 N	123.75 E	.	TM	EH21	109	YES
112	LEPAC SULU	18	5.50 N	120.75 E	.	.	.	110	YES
113	LIBOG ALBAY	13	13.25 N	123.75 E	.	.	.	111	NO
114	LIMAY BATAAN	10	14.42 N	120.58 E	.	.	.	112	NO
115	LOS BANOS LAGUNA	11	14.17 N	121.25 E	.	.	.	199	NO
116	LOS BANOS LAGUNA	11	14.17 N	121.25 E	.	.	.	198	NO
117	LUCENA QUEZON	33	13.92 N	121.58 E	.	TD	EH42	113	NO
118	MAASIN LEYTE	38	10.13 N	124.83 E	.	TM	EH21	114	NO
119	MABINI PANGASINAN	13	16.08 N	119.92 E	.	.	.	115	NO
120	MALABANG LANAO	15	7.60 N	124.07 E	.	.	.	116	NO

AVERAGE PRECIPITATION (MM)

OBS.	JAN.	FEB.	MAR.	APR.	MAY	JUN.	JUL.	AUG.	SEP.	OCT.	NOV.	DEC.	ANNUAL
109	4	5	5	11	178	313	532	628	388	168	48	13	2294
110	412	398	312	59	170	158	219	138	256	362	536	652	3671
111	366	265	218	158	178	194	235	209	252	313	479	503	3383
112	83	92	99	94	145	189	124	92	120	157	161	96	1452
113	440	242	301	210	213	158	165	220	242	412	628	592	3823
114	10	9	9	18	248	321	692	835	430	240	130	61	3002
115	69	34	34	38	166	184	286	244	231	266	197	182	1930
116	69	34	34	38	166	184	286	244	231	266	197	182	1930
117	113	55	50	57	107	169	161	182	185	267	311	257	1913
118	260	168	135	76	125	165	257	200	258	254	293	318	2511
119	8	11	61	83	236	441	598	787	469	211	81	21	3008
120	116	110	189	200	403	390	564	507	443	379	264	131	3696

AVERAGE TEMPERATURE (DEGREES CENTIGRADE)

OBS.	JAN.	FEB.	MAR.	APR.	MAY	JUN.	JUL.	AUG.	SEP.	OCT.	NOV.	DEC.	ANNUAL
109	24.7	25.3	26.9	28.3	28.6	27.9	27.3	27.0	27.2	26.9	26.2	25.2	26.8
110	.	.	.	.	.	.	.	.	.	.	.	.	.
111	25.7	25.9	26.7	27.7	28.2	28.1	27.4	27.4	27.3	27.1	26.7	26.2	27.0
112	.	.	.	.	.	.	.	.	.	.	.	.	.
113	.	.	.	.	.	.	.	.	.	.	.	.	.
114	.	.	.	.	.	.	.	.	.	.	.	.	.
115	.	.	.	.	.	.	.	.	.	.	.	.	.
116	.	.	.	.	.	.	.	.	.	.	.	.	.
117	25.3	25.8	26.7	28.0	28.7	28.3	27.8	27.7	27.6	27.0	26.5	25.5	27.1
118	25.8	26.1	26.5	27.3	28.0	28.0	27.5	27.4	27.1	27.2	26.8	26.4	27.0
119	.	.	.	.	.	.	.	.	.	.	.	.	.
120	.	.	.	.	.	.	.	.	.	.	.	.	.

CLIMATOLOGICAL DATA FOR PHILIPPINES

GEOGRAPHY

OBS.	LOCATION	YEARS	LATITUDE	LONGITUDE	ELEVATION -M-	LIFEZONE	CCTX	ID	ONMAP
121	MALAYBALAY BUKIDNO	12	8.17 N	125.08 E	670.6	TM	EW21	117	NO
122	MALITA DAVAO	14	6.40 N	125.60 E	.	.	.	118	NO
123	MALUSO ZAMBOANGA	9	6.58 N	121.92 E	.	.	.	119	YES
124	MAMBAJAO MISAMIS	29	9.23 N	124.72 E	.	.	.	120	NO
125	MANAPLA NEGROS OCC	13	10.92 N	123.08 E	.	.	.	121	NO
126	MANGATAREM PANGASI	11	15.75 N	120.25 E	.	.	.	122	NO
127	MANILA CITY	90	14.52 N	121.00 E	.	TM	EH32	123	NO
128	MANITO ALBAY	14	13.17 N	123.83 E	.	.	.	124	NO
129	MANTALONGAN CEBU	17	9.75 N	123.42 E	487.7	.	.	125	NO
130	MAPAIT ZAMBALES	12	15.50 N	119.97 E	.	.	.	126	NO
131	MARAWI LANAO DEL S	16	8.00 N	124.30 E	670.6	.	.	127	NO
132	MASBATE MASBATE	50	12.37 N	123.62 E	.	TD	EH41	128	NO

AVERAGE PRECIPITATION (MM)

OBS.	JAN.	FEB.	MAR.	APR.	MAY	JUN.	JUL.	AUG.	SEP.	OCT.	NOV.	DEC.	ANNUAL
121	107	68	92	96	244	339	358	340	370	331	195	195	2736
122	161	142	142	171	181	130	102	75	100	108	167	178	1659
123	99	74	118	156	200	233	212	243	221	270	239	161	2225
124	352	173	150	61	107	161	202	162	97	261	281	375	2381
125	174	102	122	106	179	223	234	218	220	426	407	305	2717
126	12	18	68	112	220	289	334	523	360	217	80	45	2277
127	23	11	17	32	128	252	414	437	353	195	138	68	2069
128	150	93	108	54	78	67	81	109	118	196	273	260	1586
129	169	98	127	60	176	198	247	166	201	249	271	235	2196
130	5	10	25	42	157	500	497	929	476	301	72	16	3029
131	165	138	138	186	303	363	263	229	330	273	230	209	2828
132	186	114	81	39	102	154	180	171	176	192	224	259	1879

AVERAGE TEMPERATURE (DEGREES CENTIGRADE)

OBS.	JAN.	FEB.	MAR.	APR.	MAY	JUN.	JUL.	AUG.	SEP.	OCT.	NOV.	DEC.	ANNUAL
121	22.9	23.1	23.6	24.2	24.4	24.1	23.5	23.5	23.6	23.7	23.6	23.1	23.6
122	.	.	.	.	.	.	.	.	.	.	.	.	.
123	.	.	.	.	.	.	.	.	.	.	.	.	.
124	.	.	.	.	.	.	.	.	.	.	.	.	.
125	.	.	.	.	.	.	.	.	.	.	.	.	.
126	.	.	.	.	.	.	.	.	.	.	.	.	.
127	25.1	25.6	26.9	28.3	28.6	28.0	27.2	27.0	26.9	26.8	26.0	25.3	26.8
128	.	.	.	.	.	.	.	.	.	.	.	.	.
129	.	.	.	.	.	.	.	.	.	.	.	.	.
130	.	.	.	.	.	.	.	.	.	.	.	.	.
131	.	.	.	.	.	.	.	.	.	.	.	.	.
132	26.3	26.8	27.4	28.6	29.3	29.1	28.7	28.4	28.3	28.2	27.5	26.7	27.9

CLIMATOLOGICAL DATA FOR PHILIPPINES

GEOGRAPHY

OBS.	LOCATION	YEARS	LATITUDE	LONGITUDE	ELEVATION -M-	LIFEZONE	CCTX	ID	ONMAP
133	MIDSAYAP COTABATO	14	7.17 N	124.50 E	.	.	.	129	NO
134	MILAGROS MASBATE	14	12.23 N	123.50 E	.	.	.	130	YES
135	MINGLANILLA CEBU	13	10.25 N	123.83 E	.	.	.	131	NO
136	MOUNT DATA MOUNTAI	11	16.87 N	120.85 E	1828.8	.	.	132	NO
137	MUNOZ NVA ECIJA	20	15.72 N	120.90 E	.	.	.	196	YES
138	NAGA CAMARINES SUR	37	13.62 N	123.18 E	.	TM	EH31	133	NO
139	NAGANAGA ZAMBOANGA	9	7.50 N	122.92 E	.	.	.	134	NO
140	NAGUILIAN LA UNION	13	16.50 N	120.42 E	.	.	.	135	NO
141	NORZAGARAY BULACAN	27	14.92 N	121.08 E	.	.	.	136	NO
142	ODIONGAN ROMBLON	15	12.42 N	122.00 E	.	.	.	137	NO
143	OLONGAPO ZAMBALES	22	14.82 N	120.25 E	.	.	.	138	NO
144	ORAS SAMAR	13	12.15 N	125.43 E	.	.	.	139	NO

AVERAGE PRECIPITATION (MM)

OBS.	JAN.	FEB.	MAR.	APR.	MAY	JUN.	JUL.	AUG.	SEP.	OCT.	NOV.	DEC.	ANNUAL
133	66	115	101	178	268	233	217	200	169	208	220	113	2087
134	186	104	89	43	144	199	280	240	285	279	297	310	2456
135	96	81	72	110	174	199	292	212	236	235	253	233	2193
136	24	44	84	176	336	307	395	563	404	256	272	99	2959
137	3	7	19	29	207	232	409	399	274	193	87	29	1888
138	115	81	69	86	156	202	274	187	291	292	309	290	2353
139	116	101	154	170	325	282	325	281	288	225	187	188	2643
140	6	4	20	58	193	420	508	769	539	209	101	32	2858
141	35	24	31	65	249	461	608	648	533	202	151	123	3129
142	66	25	22	51	194	265	450	353	288	345	222	176	2459
143	5	4	3	29	328	524	986	903	687	251	118	44	3881
144	371	198	252	207	206	182	183	149	195	318	493	557	3311

AVERAGE TEMPERATURE (DEGREES CENTIGRADE)

OBS.	JAN.	FEB.	MAR.	APR.	MAY	JUN.	JUL.	AUG.	SEP.	OCT.	NOV.	DEC.	ANNUAL
133	.	.	.	.	.	.	.	.	.	.	.	.	.
134	.	.	.	.	.	.	.	.	.	.	.	.	.
135	.	.	.	.	.	.	.	.	.	.	.	.	.
136	.	.	.	.	.	.	.	.	.	.	.	.	.
137	.	.	.	.	.	.	.	.	.	.	.	.	.
138	25.1	25.0	25.9	27.0	28.2	28.2	27.9	28.1	27.7	27.0	26.3	25.6	26.8
139	.	.	.	.	.	.	.	.	.	.	.	.	.
140	.	.	.	.	.	.	.	.	.	.	.	.	.
141	.	.	.	.	.	.	.	.	.	.	.	.	.
142	.	.	.	.	.	.	.	.	.	.	.	.	.
143	.	.	.	.	.	.	.	.	.	.	.	.	.
144	.	.	.	.	.	.	.	.	.	.	.	.	.

CLIMATOLOGICAL DATA FOR PHILIPPINES

GEOGRAPHY

OBS.	LOCATION	YEARS	LATITUDE	LONGITUDE	ELEVATION -M-	LIFEZONE	CCTX	ID	ONMAP
145	ORMOC LEYTE	38	11.00 N	124.60 E	.	TM	EH32	140	NO
146	PALANAS MASBATE	14	12.15 N	123.92 E	.	.	.	141	NO
147	PALPALAN MOUNTAIN	11	16.48 N	120.80 E	883.9	.	.	142	NO
148	PAMPLONA NEGROS	11	9.50 N	123.58 E	.	.	.	143	NO
149	PANDAN ANTIQUE	12	11.73 N	122.10 E	.	.	.	144	NO
150	PANTUKAN DAVAO	10	7.17 N	125.92 E	.	.	.	145	NO
151	PASACAO CAMARINES	13	13.52 N	123.05 E	.	.	.	146	NO
152	PILAR CAPIZ	22	11.50 N	123.00 E	.	.	.	147	NO
153	PINGKIAN NVA VIZ	13	16.33 N	120.92 E	487.7	.	.	148	NO
154	PORT LEBAC COTABAT	12	6.53 N	124.03 E	.	.	.	149	NO
155	PUERTO PRINCESA	14	9.73 N	118.73 E	.	TD	EH41	150	YES
156	ROMBLON ROMBLON	51	12.58 N	122.27 E	.	TM	EH31	151	NO

AVERAGE PRECIPITATION (MM)

OBS.	JAN.	FEB.	MAR.	APR.	MAY	JUN.	JUL.	AUG.	SEP.	OCT.	NOV.	DEC.	ANNUAL
145	176	114	99	81	109	195	278	225	277	256	488	212	2282
146	162	80	113	80	97	123	157	191	206	295	289	287	2080
147	51	85	58	163	472	544	914	1163	1060	545	376	171	5603
148	185	73	84	88	131	146	245	196	163	247	203	179	1939
149	335	161	106	89	344	351	754	604	584	578	614	407	4928
150	161	112	114	154	200	186	170	179	171	182	155	138	1922
151	171	85	99	75	182	233	292	158	340	276	313	218	2442
152	239	113	141	81	233	273	272	308	311	408	384	308	3069
153	10	23	52	100	166	139	207	307	250	156	123	84	1606
154	103	117	135	190	237	278	374	282	308	388	279	122	2812
155	38	28	60	42	136	171	181	49	221	207	204	118	1607
156	117	78	64	72	142	210	277	228	230	321	297	234	2271

AVERAGE TEMPERATURE (DEGREES CENTIGRADE)

OBS.	JAN.	FEB.	MAR.	APR.	MAY	JUN.	JUL.	AUG.	SEP.	OCT.	NOV.	DEC.	ANNUAL
145	25.2	25.2	25.8	26.4	26.8	26.7	27.6	26.8	26.6	26.1	25.8	25.7	26.2
146	.	.	.	.	.	.	.	.	.	.	.	.	.
147	.	.	.	.	.	.	.	.	.	.	.	.	.
148	.	.	.	.	.	.	.	.	.	.	.	.	.
149	.	.	.	.	.	.	.	.	.	.	.	.	.
150	.	.	.	.	.	.	.	.	.	.	.	.	.
151	.	.	.	.	.	.	.	.	.	.	.	.	.
152	.	.	.	.	.	.	.	.	.	.	.	.	.
153	.	.	.	.	.	.	.	.	.	.	.	.	.
154	.	.	.	.	.	.	.	.	.	.	.	.	.
155	26.9	27.2	27.7	23.5	28.5	27.4	27.2	27.1	27.1	27.2	27.2	26.9	27.4
156	26.4	26.5	27.6	23.8	29.0	28.7	28.2	28.0	28.1	27.8	27.4	26.7	27.8

CLIMATOLOGICAL DATA FOR PHILIPPINES

GEOGRAPHY

OBS.	LOCATION	YEARS	LATITUDE	LONGITUDE	ELEVATION -M-	LIFEZONE	CCTX	ID	ONMAP
157	ROXAS CAPIZ	52	11.58 N	122.75 E	.	TM	EH31	152	NO
158	SAN FERNANDO LA UN	38	16.62 N	120.32 E	.	TM	EH33	153	NO
159	SAN FERNANDO PAMP	13	15.03 N	120.70 E	.	.	.	154	NO
160	SAN ISIDRO NVA ECI	32	15.32 N	120.92 E	.	TD	EH42	155	NO
161	SAN JOAQUIN ILOILO	13	7.02 N	125.05 E	.	.	.	156	YES
162	SAN JOSE ANTIQUE	31	10.73 N	121.92 E	.	TM	EH22	157	NO
163	SAN JOSE MINDORO	16	12.37 N	121.08 E	.	.	.	158	NO
164	SANTA BARBARA PANG	13	16.00 N	120.40 E	.	.	.	159	NO
165	SANTA CRUZ DAVAO	12	7.10 N	125.42 E	.	.	.	160	NO
166	SANTA CRUZ LAGUNA	44	14.28 N	121.42 E	.	TM	EH32	161	YES
167	SANTA CRUZ MARIN	14	13.48 N	122.02 E	.	.	.	162	NO
168	SANTA FE NVA VIZ	13	16.17 N	120.92 E	548.6	.	.	163	NO

AVERAGE PRECIPITATION (MM)

OBS.	JAN.	FEB.	MAR.	APR.	MAY	JUN.	JUL.	AUG.	SEP.	OCT.	NOV.	DEC.	ANNUAL
157	135	82	54	58	177	262	269	230	249	369	290	226	2401
158	8	7	9	22	180	330	604	671	403	143	49	19	2446
159	16	8	17	41	225	283	467	431	238	144	133	32	2036
160	12	7	16	43	187	238	391	362	298	189	103	57	1900
161	94	69	93	141	248	258	276	271	307	265	232	157	2412
162	33	23	19	25	223	370	616	531	490	352	183	54	2939
163	14	36	7	35	195	465	743	604	499	305	212	69	3184
164	2	15	28	88	174	292	292	411	276	115	55	19	1767
165	99	79	84	157	321	282	249	252	195	214	167	121	2221
166	54	29	33	48	137	216	260	310	271	250	259	165	2032
167	103	62	67	46	115	183	197	167	154	282	230	197	1803
168	31	33	58	100	172	212	329	448	317	239	179	111	2229

AVERAGE TEMPERATURE (DEGREES CENTIGRADE)

OBS.	JAN.	FEB.	MAR.	APR.	MAY	JUN.	JUL.	AUG.	SEP.	OCT.	NOV.	DEC.	ANNUAL
157	26.0	26.2	26.9	27.9	28.4	27.8	27.4	27.4	27.1	27.2	27.0	26.4	27.1
158	25.0	25.1	26.8	28.6	29.1	28.4	27.6	27.5	27.3	27.2	26.3	25.8	27.1
159	.	.	.	.	.	.	.	.	.	.	.	.	.
160	24.7	25.2	26.7	28.3	28.3	27.7	26.5	26.5	26.5	26.4	25.7	25.2	26.5
161	.	.	.	.	.	.	.	.	.	.	.	.	.
162	26.2	26.4	27.3	28.2	28.3	27.5	26.8	26.9	26.8	27.1	27.0	26.8	27.1
163	.	.	.	.	.	.	.	.	.	.	.	.	.
164	.	.	.	.	.	.	.	.	.	.	.	.	.
165	.	.	.	.	.	.	.	.	.	.	.	.	.
166	24.9	25.2	26.5	27.7	28.4	28.1	27.5	27.5	27.0	26.8	26.2	25.7	26.8
167	.	.	.	.	.	.	.	.	.	.	.	.	.
168	.	.	.	.	.	.	.	.	.	.	.	.	.

CLIMATOLOGICAL DATA FOR PHILIPPINES

GEOGRAPHY

OBS.	LOCATION	YEARS	LATITUDE	LONGITUDE	ELEVATION -M-	LIFEZONE	CCTX	ID	ONMAP
169	SIATON NEGROS	14	9.08 N	123.00 E	.	.	.	164	NO
170	SIBUKO ZAMBOANGA	13	7.33 N	122.08 E	.	.	.	165	NO
171	SILANG CAVITE	24	14.23 N	120.97 E	304.8	TM	EW32	166	NO
172	SILAY NEGROS	31	10.75 N	123.00 E	.	.	.	167	NO
173	SINAIT ILOCOS SUR	13	17.87 N	120.45 E	.	.	.	168	NO
174	SINDANGAN ZAMBOANG	19	8.17 N	122.98 E	.	.	.	169	NO
175	SIOCON ZAMBOANGA	12	7.33 N	122.08 E	.	.	.	170	NO
176	SORSOGON SORSOGON	24	12.97 N	124.00 E	.	.	.	171	NO
177	SURIGAO SURIGAO	52	9.80 N	125.48 E	.	TM	EH21	172	NO
178	TABUK MOUNTAIN	14	17.42 N	121.33 E	.	.	.	173	NO
179	TACLOBAN LEYTE	51	11.25 N	125.00 E	.	TM	EH31	174	YES
180	TAGBILARAN BOHOL	38	9.63 N	123.85 E	.	TD	EH41	175	YES

AVERAGE PRECIPITATION (MM)

OBS.	JAN.	FEB.	MAR.	APR.	MAY	JUN.	JUL.	AUG.	SEP.	OCT.	NOV.	DEC.	ANNUAL
169	20	13	31	39	95	235	342	392	416	348	263	118	2311
170	70	59	31	31	152	181	189	164	203	264	168	75	1588
171	41	22	29	38	199	273	455	422	329	187	190	108	2293
172	176	103	123	95	182	205	299	269	251	325	436	266	2731
173	7	5	8	13	232	366	665	615	228	108	36	15	2298
174	84	79	73	66	228	342	307	293	323	320	267	141	2524
175	75	53	84	101	217	312	384	431	296	431	314	222	2921
176	490	293	275	125	147	174	219	229	280	412	597	546	3789
177	554	382	368	256	176	122	173	137	169	279	426	634	3676
178	20	36	39	72	126	159	218	259	288	210	170	63	1660
179	319	209	178	134	163	172	161	133	151	207	288	353	2468
180	130	92	91	123	116	143	198	132	168	225	209	158	1785

AVERAGE TEMPERATURE (DEGREES CENTIGRADE)

OBS.	JAN.	FEB.	MAR.	APR.	MAY	JUN.	JUL.	AUG.	SEP.	OCT.	NOV.	DEC.	ANNUAL
169	.	.	.	.	.	.	.	.	.	.	.	.	.
170	.	.	.	.	.	.	.	.	.	.	.	.	.
171	24.2	24.4	25.3	26.3	25.9	25.2	24.4	24.7	24.8	24.9	24.7	24.1	24.9
172	.	.	.	.	.	.	.	.	.	.	.	.	.
173	.	.	.	.	.	.	.	.	.	.	.	.	.
174	.	.	.	.	.	.	.	.	.	.	.	.	.
175	.	.	.	.	.	.	.	.	.	.	.	.	.
176	.	.	.	.	.	.	.	.	.	.	.	.	.
177	25.6	25.6	26.1	26.7	27.2	27.4	27.4	27.7	27.5	27.1	26.4	25.8	26.7
178	.	.	.	.	.	.	.	.	.	.	.	.	.
179	25.7	25.8	26.4	27.3	27.8	27.6	27.5	27.7	27.6	27.2	26.7	26.0	26.9
180	25.7	25.6	26.3	27.0	27.6	27.3	27.4	27.4	27.3	26.9	26.5	26.1	26.8

CLIMATOLOGICAL DATA FOR PHILIPPINES

GEOGRAPHY

OBS.	LOCATION	YEARS	LATITUDE	LONGITUDE	ELEVATION -M-	LIFEZONE	CCTX	ID	ONMAP
181	TAGUM DAVAO	11	7.38 N	125.80 E	.	.	.	176	NO
182	TALA RIZAL	11	13.42 N	122.67 E	.	.	.	177	NO
183	TALACOGON AGUSAN	15	8.53 N	125.67 E	.	.	.	178	NO
184	TARLAC TARLAC	32	15.50 N	120.58 E	.	TD	EH42	179	NO
185	TILIK MINDORO	13	13.82 N	120.20 E	.	.	.	180	NO
186	TIWI ALBAY	25	13.50 N	123.67 E	.	.	.	181	NO
187	TOLONG NEGROS OR	11	9.33 N	122.83 E	.	.	.	182	NO
188	TUAO CAGAYAN	14	17.73 N	121.45 E	.	.	.	183	YES
189	TUBAO LA UNION	13	16.33 N	120.42 E	.	.	.	184	NO
190	TUBIGON BOHOL	14	9.92 N	124.00 E	.	.	.	185	NO
191	TUBURAN CEBU	22	10.75 N	123.83 E	.	TD	EH41	186	NO
192	TUGBOK DAVAO	11	7.08 N	125.50 E	.	.	.	187	NO

AVERAGE PRECIPITATION (MM)

OBS.	JAN.	FEB.	MAR.	APR.	MAY	JUN.	JUL.	AUG.	SEP.	OCT.	NOV.	DEC.	ANNUAL
181	202	145	139	179	264	229	215	191	237	173	214	190	2380
182	18	24	32	64	143	346	358	598	411	288	156	98	2536
183	478	273	296	190	237	240	235	252	253	249	195	383	3282
184	7	11	22	60	191	260	408	401	291	165	83	32	1932
185	13	8	5	29	97	324	388	626	472	275	76	54	2367
186	306	218	205	174	252	169	197	173	191	353	519	446	3203
187	35	15	25	66	102	280	435	439	423	459	218	111	2607
188	42	37	50	72	147	191	185	206	244	230	204	126	1732
189	2	1	14	30	94	190	272	425	284	70	58	29	1470
190	193	88	96	48	114	130	185	144	159	181	186	180	1704
191	155	83	61	34	102	132	147	84	155	210	225	153	1541
192	120	107	114	256	281	250	224	219	248	228	202	130	2380

AVERAGE TEMPERATURE (DEGREES CENTIGRADE)

OBS.	JAN.	FEB.	MAR.	APR.	MAY	JUN.	JUL.	AUG.	SEP.	OCT.	NOV.	DEC.	ANNUAL
181	.	.	.	.	.	.	.	.	.	.	.	.	.
182	.	.	.	.	.	.	.	.	.	.	.	.	.
183	.	.	.	.	.	.	.	.	.	.	.	.	.
184	25.7	26.3	27.9	29.3	29.3	28.7	27.5	27.5	27.4	27.4	26.6	26.1	27.5
185	.	.	.	.	.	.	.	.	.	.	.	.	.
186	.	.	.	.	.	.	.	.	.	.	.	.	.
187	.	.	.	.	.	.	.	.	.	.	.	.	.
188	.	.	.	.	.	.	.	.	.	.	.	.	.
189	.	.	.	.	.	.	.	.	.	.	.	.	.
190	.	.	.	.	.	.	.	.	.	.	.	.	.
191	25.5	25.4	26.1	27.5	28.4	28.0	27.8	27.7	27.4	27.2	26.7	26.1	27.0
192	.	.	.	.	.	.	.	.	.	.	.	.	.

CLIMATOLOGICAL DATA FOR PHILIPPINES

GEOGRAPHY

OBS.	LOCATION	YEARS	LATITUDE	LONGITUDE	ELEVATION -M-	LIFEZONE	CCTX	ID	ONMAP
193	TUGUEGARAO CAGAYAN	51	17.60 N	121.72 E	.	TD	EH42	188	NO
194	VALDERRAMA ANTIQUE	14	11.00 N	122.17 E	.	.	.	189	NO
195	VERUELA AGUSAN	15	8.08 N	125.92 E	.	.	.	190	NO
196	VICTORIAS NEGROS	13	10.92 N	123.08 E	.	.	.	191	NO
197	VIGAN ILOCOS SUR	52	17.57 N	120.38 E	.	TM	EH22	192	NO
198	VIRAC CATANDUANES	45	13.58 N	124.23 E	.	TM	EH21	193	NO
199	ZAMBOANGA CITY	54	6.90 N	122.08 E	.	TD	EH51	194	NO

AVERAGE PRECIPITATION (MM)

OBS.	JAN.	FEB.	MAR.	APR.	MAY	JUN.	JUL.	AUG.	SEP.	OCT.	NOV.	DEC.	ANNUAL
193	31	25	34	65	132	156	233	206	236	237	269	139	1763
194	35	11	30	88	311	403	553	675	607	498	295	134	3640
195	547	368	318	273	251	246	227	230	300	283	244	427	3714
196	171	86	120	92	173	171	253	256	199	508	413	285	2703
197	3	6	8	20	187	369	651	710	401	140	44	13	2552
198	241	191	161	126	174	201	229	167	246	366	480	426	3006
199	52	53	38	53	90	113	126	113	123	155	121	88	1124

AVERAGE TEMPERATURE (DEGREES CENTIGRADE)

OBS.	JAN.	FEB.	MAR.	APR.	MAY	JUN.	JUL.	AUG.	SEP.	OCT.	NOV.	DEC.	ANNUAL
193	23.4	24.4	26.4	28.2	29.0	28.9	28.9	27.9	27.5	26.4	25.1	24.0	26.6
194	.	.	.	.	.	.	.	.	.	.	.	.	.
195	.	.	.	.	.	.	.	.	.	.	.	.	.
196	.	.	.	.	.	.	.	.	.	.	.	.	.
197	25.4	25.7	27.0	28.3	28.7	28.0	27.3	26.9	27.1	27.3	26.7	26.0	27.0
198	25.9	25.9	26.4	27.2	27.8	28.1	27.8	28.1	27.8	27.3	26.9	26.4	27.1
199	26.3	26.4	26.8	27.1	27.1	26.8	26.6	26.8	26.7	26.6	26.6	26.4	26.7

REFERENCES

1. FAO, 1975 Production Yearbook, Vol. 29, Food and Agriculture Organization, Rome, 1976.
2. FAO, 1979 Production Yearbook, Vol. 33, Food and Agriculture Organization, Rome, 1980.
3. Wernstedt, F. L., *World Climatic Data,* Climatic Data Press, Lemont, Pa., 1972.
4. Holdridge, L. R., Determination of world plant formations from simple climatic data, *Science,* 105(2727), 367, 1947.
5. Duke, J. A., The quest for tolerant germplasm, in *Crop Tolerance to Suboptimal Land Conditions,* ASA Special Symp. 32, American Society of Agronomy, Madison, Wis., 1978, chap 1.
6. Duke, J. A. and Terrell, E. E., Crop diversification matrix: introduction, *Taxon,* 23(5/6), 759, 1974.
7. Zeven, A. C. and Zhukovsky, P. M., *Dictionary of Cultivated Plants and Their Centres of Diversity,* Centre for Agricultural Publishing and Documentation, Wageningen, Netherlands, 1975.
8. Federov, A. A., Ed., *Chromosome Numbers of Flowering Plants,* Academy of Science, U.S.S.R., 1969.
9. Griffiths, J. F., Ed., World survey of climatology, in *Climates of Africa,* Vol. 10, Elsevier, Amsterdam, 1972.
10. World Bank, *World Tables,* 2nd ed., Johns Hopkins University Press, Baltimore, 1980.
11. United Nations, World Energy Supplies 1973-1978, Statistical Papers Series J, Number 22, Department of International Economic and Social Affairs, Statistical Office, New York, 1979.
12. Spring, J. D., Biomass Crop Substitution Feasibility Study, Tech. Rep. MTR. 80W327, MITRE Corporation, McLean, Virginia, 1981.
13. NAS, *Calliandra. A Versatile Small Tree for the Humid Tropics,* National Academy of Sciences, Washington, D.C., 1983.
14. Neary, J., Pickleweed, Palmer's grass, and saltwort, *Science 81,* June, 39, 1981.
15. Williams, L., Forest and agricultural resources of Dahomey, West Africa, *Econ. Bot.,* 23(4), 352, 1969.
16. TPI, *Oil Palm News,* Tropical Products Institute 21, Oil Palm Advisory Bureau, London, 1976.
17. Van der Zon, A. P. M. and Grubben, G. J. H., Les Legumes, Feuilles Spontanes et Cultives du Sud Dahomey, Comm. 65, Dept. Rech. Agron. Koninklijk, Institut voor de Tropen, Amsterdam, Holland, 111, 1976.
18. Vergara, W. and Pimentel, D., Fuels from biomass. Comparative study of the potential in five countries: U.S., Brazil, India, Sudan, and Sweden, in *Advances in Energy Systems and Technology,* Auer, P., Ed., Academic Press, New York, 1978.
19. Calvin, M., Oil from Plants, Lecture at Beltsville, Maryland, September 8, 1982.
20. Hammond, A. L., Alcohol: a Brazilian answer to the energy crisis, *Science,* 195, 564, 1977.
21. Duke, J. A., Pitfalls in Simplistic Climatic Analogs, Unpublished, 1984.
22. Sanchez, F. and Duke, J. A., La Papa de Nadi, *El Campesino,* 115(4), 15, 1984.
23. Espinal, L. S. and Montenegro, M. E., Formaciones Vegetales de Colombia. Instituto Geografico "Agustin Codazzi", with the collaboration of Tosi, J. A., Jr. and Holdridge, L.R., Departamento Agrologico, Bogota, 1963.
24. Holdridge, L. R., Grenke, W. C., Hatheway, W. H., Liang, T., and Tosi, J. A., Jr., *Forest Environments in Tropical Life Zones. A Pilot Study,* Pergamon Press, Oxford, 1971.
25. Koppen, W., *Grundriss der Klimakunde — Die Klimate der Erde,* 2nd Ed. Walter de Gruyter & Co., Berlin, 1931.
26. Thornthwaite, C. W., An approach toward a rational classification of climate, *Georgr. Rev.,* 38(1), 55, 1948.
27. Standley, P. C. and Steyermark, J. A., *Flora of Guatemala, Fieldiana: Botany* 34(V), Field Museum of Chicago, Chicago, Ill., 1946, 502.
28. NAS, *Leucaena: Promising Forage and Tree Crop for the Tropics,* National Academy of Sciences, Washington, D.C., 1977.
29. Curtis, C. R. and Duke, J. A., *An Assessment of Land Biomass and Energy Potential for the Republic of Panama,* Vol. 3, Institute of Energy Conversion, University of Delaware, Newark, 1982.
30. Atchley, A. A. and Mathis, E. S., A climate classification for crop introduction and germplasm collection, in preparation.
31. SAS Institute, SAS User's Guide: Basics, SAS Institute, Cary, N. C., 1982, 114.
32. Integrated Software Systems Corporation, Display Integrated Software System and Plotting Language User's Manual, version 9.0, Section D, Integrated Software Systems Corporation, San Diego, Calif., 1981.
33. NAS, *Firewood Crops, Shrub and Tree Species for Energy Production,* National Academy of Sciences, Washington, D.C., 1980.
34. Holdridge, L. R. and Budowski, G., Report of an ecological survey of the Republic of Panama, *Caribb. For.,* 1956.

35. Golley, F. B., McGinnis, J. R., Clements, R. G., Child, G. I., and Duever, M. J., *Mineral Cycling in a Tropical Moist Forest Ecosystem,* University of Georgia Press, Athens, 1975.
36. AID, Panama Project Paper, LAC/DR 79-15, Alternative Energy Sources, Panama, Republic of Panama, 1979.
37. Bille, J. C., Woody Forage Species in the Sahel: Their Biology and Use, Proc. 1st Int. Rangeland Congr. 1978, p. 392, 1978.
38. Duke, J. A., The Gene Revolution, Paper No. 1, in Background Papers for Innovative Biological Technologies for Lesser Developed Countries, Office of Technology Assessment, U.S. Government Printing Office, Washington, D.C., 1981, 89.
39. Duke, J. A., *Handbook of Energy Species,* CRC Press, Boca Raton, Fla., 1984, in press.
40. Rhind, J. M. L. C. and Goodenough, D. C. W., *Acroceras macrum* Stapf. (Nile Grass) — a review, *Proc. Grassland Soc. S. Afr.,* 14, 27, 1976.
41. Gohl, B., Tropical Feeds, Feed Information Summaries and Nutritive Values, FAO Animal Production and Health Ser. No. 12, Food and Agriculture Organization, Rome, 1981.
42. Palit, S., Trials of *Albizia falcataria* (L.) Fosberg and *Leucaena leucocephala* (Lam.) De Wit in North Bengal, *Indian For.,* 106(7), 456, 1980.
43. Kawahara, T., Kanazawa, Y., and Sakura, S., Biomass and net production of man-made forests in the Phillipines, J., *Jpn. For. Soc.,* 63(9), 320, 1981.
44. Webb, D. E., Wood, P. J., and Smith, J., A Guide to Species Selection for Tropical and Sub-Tropical Plantations, Tropical Forestry Papers, No. 15, Commonwealth Foresty Institute, Oxford, 1980.
45. Fenton, R., Roper, R. E., and Watt, G. R., Lowland Tropical Hardwoods, External Aid Division, Ministry of Foreign Affairs, Wellington, New Zealand, 1977.
46. Krishnamurti, B. G., Rain tree fruit — a new raw material for alcohol, *Curr. Sci.,* 43, 700, 1974.
47. Roskoski, J. P., Gonzalez, G. C., Dias, M. I. F., Tejeda, E. P., and Vargas-Mena y Amezcua, A., Woody tropical legumes: potential sources of forage, firewood, and soil enrichment, in SERI: Tree Crops for Energy Co-production on Farms, SERI/CP-622-1086, U.S. Government Printing Office, Washington, D.C., 1980. 135.
48. Page, J., Sunflower power, *Science 81,* July/Aug., 92, 1981.
49. Spurling, A. T. and Spurling, D., Effect of various organic and inorganic fertilizers on the yield of Montana tung *(Aleurites montana)* in Malawi, *Trop. Agric.,* 51(1), 1, 1974.
50. Wu Leung, Woot-Tsuen, Butrum, R. R., and Chang, F. H., I. Proximate composition mineral and vitamin contents of East Asian foods, in Food Composition Table for Use in East Asia, Food and Agriculture Organization, Rome, 1972.
51. Dennington, V. N., Chadwick, M. J., and Chase, D. S., Energy cropping on derelict and waste land, *J. Environ. Manage.,* 16(3), 241, 1983.
52. Kestemont, P., Biomasse, Necromasse et Productivites Aeriennes Ligneuses de Quelques Peuplements Forestiers en Belguique, Thesis, Faculty of Sciences, Free University of Brussels, 1975; as cited in Cannell, M. G. R., *World Forest Biomass and Primary Production Data,* Academic Press, New York, 1982.
53. Cannell, M. G. R., *World Forest Biomass and Primary Production Data,* Academic Press, New York, 1982.
54. NAS, *Tropical Legumes: Resources for the Future,* National Academy of Sciences, Washington, D. C., 1979.
55. Little, E. L., Jr., *Common Fuelwood Crops: A Handbook for Their Identification,* McClain Printing, Parsons, W.Va., 1983.
56. Van Adrichem, M. C. J., Performance of Meadow Foxtail *(Alopecurus pratensis), J. Forage Notes,* 19(2), 50, 1974.
57. Nambiar, M. C. and Haridasan, M., Fertilizing cashew for higher yields, *Indian Farming,* 28(12), 16, 1979.
58. Johnson, D., Cashew cultivation in Brazil, *Agron. Mocamb. Lourenco Marques,* 7(3), 119, 1973.
59. Gopalakrishnan, C. and Kasturi, P., The economics of biomass energy: a study of two agricultural wastes, *Agric. Wastes,* 2, 83, 1980.
60. Kohls, D., Production d'ethanol a partir de sous-produits vegetaux en zone tropical humide:cas de l'alcool medicinal a partir de dehectes d'ananas, Entropie 98 *(Valorisation Energ.* 1; pp. 37-38), ca. 1981.
61. Stewart, G. A., Gartside, G., Gifford, R. M., Nix, H. A., Rawlins, W. H. M., and Siemon, J. R., *The Potential for Liquid Fuels from Agriculture and Forestry in Austrialia,* Commonwealth Scientific and Industrial Research Organization, Alexander Brothers, Mentone, Victoria, Australia, 1979.
62. Marzola, D. L. and Bartholemew, D. P., Photosynthetic pathway and biomass energy production, *Science,* 205, 555, 1979.
63. Haggar, R. J., The effect of quantity, source, and time of application of nitrogen fertilizers on the yield and quality of *Andropogon gayanus* at Shika, Nigeria, *J. Agric. Sci. Cambr.,* 84, 529, 1974.

64. Bogdan, A. V., *Tropical Pasture and Fodder Plants,* Longman, London, 1977.
65. Keyser, H. H., personal communication, 1983.
66. Devlin, R. M., personal communication, 1984.
67. Woodroof, J. G., Peanuts *Arachis hypogaea,* in *CRC Handbook of Biosolar Resources,* Vol. 2, McClure, T. A. and Lipinsky, E. S., Eds., CRC Press, Boca Raton, Fla. 1981.
68. C.S.I.R., *The Wealth of India,* 11 vols., Council of Scientific and Industrial Research, New Delhi, 1948-1976.
69. Anon., Most oil crops potential fuel source, *Crops Soils,* 33(6), 30, 1981.
70. Harwood, H. J., Vegetable Oils as an On the Farm Diesel Fuel Substitute: The North Carolina Situation, RTI Final Rep. FR-41U-1671-4, Research Triangle Park, North Carolina, 1981.
71. Westlake, D. F., Comparisons of plant productivity, *Biol. Rev.,* 38, 385, 1963.
72. Buchanan, R. A. and Duke, J. A., Botanochemical Crops, in *CRC Handbook of Biosolar Resources,* Vol. 2, McClure, T. A. and Lipinsky, E. S., Eds., CRC Press, Boca Raton, Fla. 1981, 157.
73. Carlsson, R., Quantity and quality of leaf protein concentrates from *Atriplex hortensis* L., *Chenopodium quinoa* Willd., and *Amaranthus caudatus* L., grown in southern Sweden, *Acta Agric. Scand.,* 30, 418, 1980.
74. NAS, *Methane Generation from Human, Animal, and Agricultural Wastes,* National Academy of Sciences, Washington, D.C., 1977.
75. Schoner, C. A., Kearney, T. E., and George, M. R., Update: oat hay variety trials, *Calif. Agric.,* Jan/Feb, 20, 1982.
76. Schrickel, D. J. and Scantland, D. A., Oats, in *CRC Handbook of Biosolar Resources,* Vol. 2, McClure, T. A. and Lipinsky, E. S., Eds., CRC Press, Boca Raton, Fla., 1981, 51.
77. Weaver, S. H., Energy use in the production of oats, in *Handbook of Energy Utilization in Agriculture,* Pimentel, D., Ed., CRC Press, Boca Raton, Fla., 1980, 85.
78. Palz, W. and Chartier, P., Eds., *Energy from Biomass in Europe,* Applied Science Publishers, London, 1980.
79. Morton, J. F., *Atlas of Medicinal Plants of Middle America, Bahamas to Yucatan,* Charles C Thomas, Springfield, Ill., 1981.
80. McGill, S., Breeding better biomass crops, *Furrow,* 86(5), 38, 1981.
81. Boardman, N. K., Energy from the biological conversion of solar energy, *Philos. Trans. R. Soc. London A,* 295, 477, 1980.
82. Swift-Hook, D. T., Discussion, *Philos., Trans. R. Soc. London A,* 295, 489, 1980.
83. Hills, F. J., Johnson, S. S., Geng, S., Abshahi, A., and Peterson, G. R., Comparison of four crops for alcohol yield, *Calif. Agric.,* 37(3/4), 17, 1983.
84. Miller, C. P., Growth and nitrogen uptake of irrigated grass-legume and nitrogen fertilized grass and pasture in a Queensland tropical highland, *J. Aust. Inst. Agric. Sci.,* 46, 184, 1976.
85. Patel, J. R., Parmar, M. T., and Patel, J. C., Effect of different sowing dates, spacings, and plant populations on yield of mustard, *Indian J. Agron.,* 25(3), 526, 1980.
86. Maity, P. K., Sengupta, A. K., and Jana, P. K., Response of mustard variety varuna *(Brassica juncea)* to levels of irrigation and nitrogen, *Indian Agric.,* 24(1), 43, 1980.
87. Knowles, P. F., Kearney, T. E., and Cohen, D. B., Species of rapeseed and mustard as oil crops in California, in *New Sources of Fats and Oils,* AOCS Monograph No. 9, Pryde, E. H., Princen, L. H., and Mukherjee, K. D., Eds., American Oil Chemists' Society, Champaign, Ill., 1981, 255.
88. Pryde, E. H. and Doty, H. O., Jr., World fats and oils situation, in *New Sources of Fats and Oils,* AOCS Monograph No. 9, Pryde, E. H., Princen, L. H., and Mukherjee, K. D., Eds., American Oil Chemists' Society, Champaign, Ill., 1981, 3.
89. Finlayson, A. J., Krzymanski, J., and Downey, R. K., Comparison of chemical and agronomic characteristics of two *Brassica napus* cultivars, Bronowski and Target, *J. Am. Oil Chemists' Soc.,* 50(10), 407, 1973.
90. Scott, R. K., Ogunremi, E. A., Ivins, J. D., and Mendham, N. J., The effect of sowing date and season on growth and yield of oilseed rape *(Brassica napus), J. Agric. Sci. Camb.,* 81, 277, 1972.
91. Matai, S., Bagchi, D. K., and Chandra, S., Optimal seed rate and fertilizer dose for maximum yield of extracted protein from the leaves of mustard *(Brassica nigra* Koch) and turnip *(Brassica rapa* L.), *Indian J. Agric. Sci.,* 43(2), 165, 1973.
92. Reed, C. F., Information Summaries on 1000 Economic Plants, Typescripts submitted to the U.S. Department of Agriculture, 1976.
93. Duke, J. A., *Handbook of Legumes of World Economic Importance,* Plenum Press, New York, 1981.
94. Aronson, J. A. and Zur, M., Bioenergy research in Israel: current assessment, *Arid Lands Newsl.,* No. 16, 11, 1982.
95. Peoples, T. R. and Lee, C. W., Dry matter and hydrocarbon yields of *Calotropis procera, Biomass,* 2(2), 153, 1982.

96. Erdman, M. D. and Erdman, B. A., *Calotropis procera* as a source of plant hydrocarbons, *Econ. Bot.*, 35(4), 467, 1981.
97. Magambo, M. J. S. and Cannell, M. G. R., Dry matter production and partition in relation to yield of tea *Camellia sinensis*, *Exp. Agric.*, 17(1), 33, 1981.
98. Murthy, G. V. K. and Natarajan, C. P., Papaya — a commercial crop for the production of papain and pectin, in *Cultivation and Utilization of Medicinal Plants*, Atal, C. K. and Kapur, B. M., Eds., Council of Scientific and Industrial Research, Jammu-Tawi, 1982, 436.
99. Morton, J. F., *Major Medicinal Plants*, Charles C Thomas, Springfield, Ill., 1977.
100. Sharma, V. C. and Ogbeide, O. N., Pawpaw as a renewable energy resource for the production of alcohol fuels, *Energy*, 7(10), 871, 1982.
101. Khoshoo, T. N., *Energy from Plants: Problems and Prospects*, Reprinted from Part II of the Proc. 69th Session, Indian Sci. Congr. Lucknow Printing House, India, 1982.
102. Raghu, J. S. and Sharma, S. R., Response to irrigation and fertility levels of safflower, *Indian J. Agron.*, 23(2), 93, 1978.
103. N.R.C., Innovations in Tropical Reforestation VI: Casuarinas (Draft, National Research Council), National Academy of Sciences, Washington, D.C., 1982.
104. Aspiras, R. B., Nitrogen fixation in nodulated non-legumes growing in the Philippines, *Canopy Int.*, 7(7), 3, 1981.
105. Vaisman, I., Kipnis, T., Shalhevet, I., Feigin, A., Sharabani, N., and Bnei-Moshe, S., Dry matter production of Rhodes grass grown on sand dunes and irrigated with secondary sewage effluents, *Herbage Abstr.*, 51, 2165, 1980.
106. Farnsworth, J., A Comparison of Tropical Grasses Grown on a Reclaimed Clay Loam Soil at Hofuf, Saudi Arabia, Pub. Joint Agric. R. and D. Proj., No. 77, University College of Wales and Ministry of Agriculture and Water, Saudi Arabia, 1977.
107. McLaughlin, S. P. and Hoffman, J. J., Survey of biocrude-producing plants from the Southwest, *Econ. Bot.*, 36(3), 323, 1982.
108. *Chemical Marketing Reporter*, June 15, 1981, Schnell Publishing, New York.
109. Purseglove, J. W., *Tropical Crops*, Vols. 1-4, Longman Group, London, 1968-1972.
110. Duke, J. A., Palms as energy sources: a solicitation, *Principes*, 21(2), 60, 1977.
111. Quick, G. R., Abstract, A Summary of Some Current Research in Australia on Vegetable Oils as Candidate Fuels for Diesel Engines. Seminar II, Peoria, Illinois, U.S. Department of Agriculture, Washington, D.C., 1981.
112. Telek, L. and Martin, F. W., Okra seed: a potential source for oil and protein in the humid lowland tropics, in *New Sources of Fats and Oils*, AOCS Monograph No. 9, Pryde, E. H., Princen, L. H., and Mukherjee, K. D., Eds., American Oil Chemists' Society, Champaign, Ill., 1981, 37.
113. Burkill, I. H., *A Dictionary of Economic Products of the Malay Peninsula*, Vols. 1-2, Art Printing Works, Kuala Lumpur, 1966.
114. U.S. Department of Agriculture Production Research Rep. No. 32, U.S. Government Printing Office, Washington, D.C., 1959.
115. Calvin, M., Hydrocarbons from plants: analytical methods and observations, *Naturwissenschaften*, 67, 525, 1980.
116. ICAR, The Jute Agriculture Research Institute, Annual Report 1970, New Delhi, India, 82, 1973.
117. ICAR, The Jute Agricultural Research Institute, Annual Report 1971-1972, New Delhi, India, 116, 1975.
118. Kolesnikov, V. A., Lange, D. F., and Azamatov, M. A., New regions for the industrial cultivation of hazel, *Vest. Skh. Nauki Kaz.*, 6, 112, 1978.
119. Bengtsson, A. and Olsson, G., *Crambe (Crambe abyssinica Hochst.):* Results from Swedish Cultivation Trials 1976-1979, Rapport Inst. Waxtodling, 90, 31, 1981.
120. Cornelius, J. A. and Simmons, E. A., *Crambe abyssinica* — a new commercial oilseed, *Trop. Sci.*, 11(1), 17, 1969.
121. Lessman, K. J. and Anderson, W. P., *Crambe*, in *New Sources of Fats and Oils*, AOCS Monograph No. 9, Pryde, E. H., Princen, L. H., and Mukherjee, K. D., Eds., American Oil Chemists' Society, Champaign, Ill., 1981, 223.
122. Duke, J. A. and Bagby, M. O., Comparison of oilseed yields: a preliminary review, in *Vegetable Oil Fuels*, ASAE Publication 4-82, American Society of Agricultural Engineers, St. Joseph, Mich., 1982, 11.
123. Gupta, S. K., Effect of clipping on biomass and net productivity of underground parts in grassland ecosytstem, *Indian J. Ecol.*, 7(2), 215, 1980.
124. Almeida, M. R., Productividad primaria de tres praderas de especies tropicales: Para *(Brachiaria mutica)*, grama amarga *(Paspalum conjugatum)*, y pangola *(Digitaria decumbens)*, *Bol. Soc. Bot. Mex.*, Nov. (41), 3, 1981.
125. Hsieh, C. H., Observations on the effect by bulldozer clearing of sloping land for pasture establishment, *J. Taiwan Livestock Res. Inst.*, 12(1), 19, 1979.

126. Holm, L. G., Plunknett, D. L., Pancho, J. V., and Herberger, J. P., *The World's Worst Weeds,* University Press of Hawaii, Honolulu, 1977.
127. Benemann, J. R., Energy from fresh and brackish water aquatic plants, in *Biomass as a Nonfossil Fuel Source,* (ACS Symp. Series 144), Klass, D. L., Ed., American Chemical Society, Washington, D.C., 1981.
128. Smith, W. H. and Dowd, M. L., Biomass production in Florida, *J. For.,* 79(8), 508, 1981.
129. NAS, *Making Aquatic Weeds Useful,* National Academy of Sciences, Washington, D.C., 1976.
130. Wolverton, B. C. and McDonald, R. C., Energy from vascular plant wastewater treatment systems *Eichhornia crassipes, Spirodela lemna, Hydrocotyle ranunculoides, Pueraria lobata,* biomass harvested for fuel production, *Econ. Bot.,* 35(2), 224, 1981.
131. Ueki, K. and Kobayashi, T., Cultivation of new biomass resources, *Energy Dev. Jpn.,* 3(3), 285, 1981.
132. Reddy, K. R. and Tucker, J. C., Productivity and nutrient uptake of water hyacinth *Eichhornia crassipes.* I. Effect of nitrogen source, *Econ. Bot.,* 37(2), 237, 1983.
133. Matai, S. and Bagchi, D. K., Water hyacinth: a plant with prolific bioproductivity and photosynthesis, in *Proc. Int. Symp. Biol. Applications Solar Energy,* Gnanam, A., Krishnaswamy, S., and Kahn, J. S., Eds., MacMillan Co. of India, Madras, 1980, 144.
134. Murry, M. A. and Benemann, J. R., Freshwater plants, in *CRC Handbook of Biosolar Resources,* Vol. 2, McClure, T. A. and Lipinsky, E. S., Eds., CRC Press, Boca Raton, Fla., 1981, 407.
135. Corley, R. H. V., Oil palm, in *CRC Handbook of Biosolar Resources,* Vol. 2, McClure, T. A. and Lipinsky, E. S., Eds., CRC Press, Boca Raton, Fla., 1981, 397.
136. Hodge, W. H., Oil-producing palms of the world: a review, *Principles,* 19(4), 119, 1975.
137. Koeng, W. K., Soft energy from palm oil and its wastes, *Agric. Wastes,* 3(3), 191, 1981.
138. Gaydou, A. M., Menet, L., Ravelojaona, G., and Geneste, P., Vegetable energy sources in Madagascar: ethyl alcohol and oil seeds (French), *Oleagineux,* 37(3), 135, 1982.
139. Sistler, F. E. and Smith, P. A., A total energy model for cotton production, *Louisiana Agric.,* 24(4), 22, 1981.
140. Barker, T. W. and Worgan, J. T., The utilization of palm oil processing effluents as substrates for microbial protein production by the fungus *Aspergillus oryzae, Eur. J. Appl. Microbiol., Biotech.,* 11(4), 234, 1981.
141. Dalrymple, R. J., Weeping Lovegrass Management, The Nobel Foundation, Ardmore, Okla., ca. 1976.
142. Russo, R. O., Resultados Preliminares de Biomasa de la Poda de *Erythrina poeppigiana* (Walpers) O. F. Cook (poro) en Turrialba, Costa Rica, Typescript, CATIE, Turrialba, Costa Rica, 1982.
143. Nemethy, E. K., Otvos, J. W., and Calvin, M., High energy liquid fuels from plants, Preprint chapter for *Fuels from Biomass,* Klass, D. L. and Emert, E. H., Eds., Lawrence Berkeley Laboratory, University of California, Lawrence, Cal., 1980.
144. Sachs, R. M., Low, C. B., MacDonald, J. D., Awad, A. R., and Sully, M. J., *Euphorbia lathyris:* a potential source of petrolum-like products, *Calif. Agric.,* 35(7/8), 29, 1981.
145. *Solarscan, Sun Times,* (Washington, D.C.), October 1980, 15.
146. McLaughlin, S. P., Kingsolver, B. E., and Hoffman, J. J., Biocrude production in arid lands, *Econ. Bot.,* 37(2), 150, 1983.
147. White, G. A., Willingham, B. C., Skrdla, W. H., Massey, J. H., Higgins, J. J., Calhoun, W., Davis, A. M., Dolan, D. D., and Earle, F. R., Agronomic evaluation of prospective new crop species, *Econ. Bot.,* 25(1), 23, 1971.
148. Steinheil, P. de., L'Euphorbe resinifere, *Rev. Gen du Caoutchou,* 18, 55, 1941.
149. Maugh, T. H., II., Unlike money, diesel fuel grows on trees, *Science,* 206, 436, 1976.
150. Leakey, P., Personal communication, September 28, 1981.
151. Sadashivaiah, K. N., Haseeb, A., and Parameswar, N. S., Role of different shade trees foliage as organic manure in Cardamom plantation, *Res. Bull. Marathwada Agric. Univ.,* 4(1), 11, 1980.
152. Freedman, D., Preliminary analysis of the potential for ethanol production from honeylocust pods, in *Tree Crops for Energy Co-Production on Farms,* Nov. 12-14, 1980, SERVCP-622-1086, USGPO, Washington, D.C., 1980, 121.
153. Williams, G., Tree crops for energy production in Appalachia, in *Tree Crops for Energy Co-Production on Farms,* Nov. 12-14, 1980, SERVCP-622-1086, USGPO, Washington, D.C., 1980, 7.
154. Scanlon, D. H., A case study of honey locust in the Tennessee Valley region, in *Tree Crops for Energy Co-Production on Farms,* Nov. 12-14, 1980, SERVCP-622-1086, USGPO, Washington, D.C., 1980, 21.
155. Avgerinos, G. C. and Wang, D. I. C., Utilization of mesquite and honey locust pods as feedstocks for energy production, in *Tree Crops for Energy Co-Production on Farms,* Nov. 12-14, 1980, SERVCP-622-1086, USGPO, Washington, D.C., 1980, 209.
156. Allen, O. N. and Allen, E. K., *The Leguminosae,* University of Wisconsin Press, Oskosh, 1981.

157. Schapaugh, W. T., Jr. and Wilcox, J. R., Relationships between harvest indices and plant characteristics in soybeans, *Crop Sci.*, 20, 529, 1980.
158. Goering, C. E., Vegetable oil as diesel fuel, in Vegetable Oil as Diesel Fuel, Progress Rep., Appendix 10, Seminar II, October 21 and 22, U.S. Department of Agriculture, Peoria, Ill., 1981.
159. Lipinsky, E. S., McClure, T. A., Kresovich, S., Otis, J. L., Wagner, C. K., Trayser, D. A., and Applebaum, H. R., Fats and oils as emergency diesel fuels, in Vegetable Oil as Diesel Fuel, Appendix 13, Seminar II, October 21 and 22, U.S. Department of Agriculture, Peoria, Ill., 1981.
160. USDA, Agricultural Statistics 1981, U.S. Department of Agriculture, Washington, D.C.
161. Whigham, D. K., Soybeans *Glycine max,* in *CRC Handbook of Biosolar Resources,* Vol. 2, McClure, T. A. and Lipinsky, E. S., Eds., CRC Press, Boca Raton, Fla., 1981, 95.
162. Dibb, D. W., Agronomic systems to feed the next generation, *Crops Soils Mag.*, Nov., 5, 1983.
163. Reynolds, L. and Lawson, E. C., A comparison of the fuelwood value of *Gmelina arborea* and *Eucalyptus* spp. from the Bunda Forest, *Res. Bull. Bunda Coll. Agric. Univ. Malawi,* 9, 71, 1978.
164. Akachuku, A. E., Estimation of volume and weight growth in *Gmelina arborea* with X-ray densitometer, in *Kyoto Biomass Studies,* University of Maine at Orono, 1981, 105.
165. Vaing, G. and Delille, V., Design, production, and tests of a low-powered gas-driven prototype tractor for use in tropical countries, *Machinisme Agric. Trop.*, No. 81, 3043, 1983.
166. Parnell, C. B., Jr., Cotton *Gossypium hirsutum,* in *CRC Handbook of Biosolar Resources,* Vol. 2, McClure, T. A. and Lipinsky, E. S., Eds., CRC Press, Boca Raton, Fla., 1981, 115.
167. Rao, T. S. and Rao, M. S. K., Performance of Niger genotypes at Raichur, *Curr. Res.*, 9(5), 78, 1980.
168. Bhardwaj, S. P. and Gupta, R. K., Tilangi, a potential high yielding oil seed crop, *Indian Farm.*, 27(6), 18, 1977.
169. Anon., *S and E Newsmakers,* No. 4, USDA, Beltsville, Md., September 1981.
170. VODF, Vegetable Oil as Diesel Fuel, Seminar II, October 21 and 22, U.S. Department of Agriculture, Peoria, Ill., 1981.
171. USDA, Secretary's Annual Report, U.S. Department of Agriculture, Washington, D.C., 1980.
172. Beard, B. H. and Ingebretsen, K. H., Spring planting is best for oilseed sunflower, *Calif. Agric.*, June 1980, 5, 1980.
173. Chabbert, N., Braun, P., Guiraud, J. P., Arnoux, M., and Galzy, P., Productivity and fermentability of Jerusalem artichoke according to harvesting date, *Biomass,* 3, 209, 1983.
174. Pankhurst, E. S., The prospects for biogas — a European point of view, *Biomass,* 3, 1, 1983.
175. Rogers, T. H., Natural rubber, in *CRC Handbook of Biosolar Resources,* Vol. 2, McClure, T. A. and Lipinsky, E. S., Eds., CRC Press, Boca Raton, Fla., 1981, 387.
176. Campbell, T. A., Personal communication.
177. Whitely, E. L., Kenaf, *Hibiscus cannabinus,* in *CRC Handbook of Biosolar Resources,* Vol. 2, McClure, T. A. and Lipinsky, E. S., Eds., CRC Press, Boca Raton, Fla., 1981, 259.
178. Carlson, K. D., Cunningham, R. L., Garcia, W. J., Bagby, M. O., and Kwolek, W. F., Performance and trace metal content of crambe and kenaf grown on sewage sludge treated stripmine land, *Environ. Pollut.*, A 29(2), 145, 1982.
179. Jiranek, O., Prospects for kenaf *(Hibiscus cannabinus)* cultivation in Tunisia, *Agric. Trop. Subtrop.*, 9, 49, 1976.
180. Crane, J. C., Roselle — a potentially important plant fiber, *Econ. Bot.*, 3, 89, 1949.
181. Morton, J. F., Renewed interest in roselle *(Hibiscus sabdariffa* L.), the long-forgotten "Florida cranberry", *Proc. Fla. State Hortic. Soc.*, 87, 415, 1975.
182. Bukantis, R. and Goodman, N., Energy inputs in barley production, in *Handbook of Energy Utilization in Agriculture,* Pimentel, D., Ed., CRC Press, Boca Raton, Fla., 1980, 59.
183. Sutcliffe, H. M., Ernst, A. J., and Aronovsky, S. I., Hop vines for paper, *The Hopper,* Jan. 1, 1950.
184. Pereira, H., Pequena contribuicao para um diccionario das plantas uteis do estado de S. Paulo, Typographia Brasil de Rothschild and Company, Sao Paulo, 1929.
185. Patterson, D. R., True roots — sweet potatoes, *Texas Agric. Prog.*, Winter, 1979.
186. AVRDC Centerpoint, Asian Vegetable Research and Development Center, Tainan, Taiwan, Winter 1981/1982.
187. Brewbaker, J. L., Ed., Giant Leucaena (Koa Haole) Energy Tree Farm: An Economic Feasibility Analysis for the Island of Molokai, Hawaii, (HNEI-80-06), Hawaii Natural Energy Institute, University of Hawaii, Manoa, 1980.
188. Anon., Oregon's Agriculture Process, Oregon State University, Corvallis, Ore., Spring/Summer 1975.
189. Joliff, G. D., Development and production of meadowfoam *(Limanthes alba),* in *New Sources of Fats and Oils,* AOCS Monograph 9, Pryde, E. H., Princen, L. H., and Mukherjee, K. D., Eds., American Oil Chemists' Society, Champaign, Ill., 1981, 269.
190. Dybing, C. D. and Lay, C., Flax *Linum usitatissimum,* in *CRC Handbook of Biosolar Resources,* Vol. 2, McClure, T. A. and Lipinsky, E. S., Eds., CRC Press, Boca Raton, Fla., 1981, 71.

191. Saleeb, W. F., Yermanos, D. M., Huszer, C. K., Storey, W. B., and Habanauskas, C. K., The Oil and Protein in Nuts of *Macadamia tetraphylla* L. Johnson, *Macadamia integrifolia* Maiden and Betche, and their F_1 Hybrid, *J. Am. Soc. Hortic. Sci.*, 98(5), 453, 1973.
192. Bohan, P. T., Food vs. fuel: the gasohol question, *Vegetarian Times*, No. 51, 1981.
193. Pimentel, D., Ed., *Handbook of Energy Utilization in Agriculture*, CRC Press, Boca Raton, Fla., 1980.
194. Coombs, J. and Vlitos, A. J., An assessment of the potential for biological solar energy utilization using carbohydrates produced by higher plant photosynthesis as chemical feedstock, *Proc. Int. Solar Energy Society Congr.*, Vol. 2 Pergamon Press, New York, 1978.
195. Devendra, C. and Raghaven, G. V., Agricultural by-products in southeast Asia: availability, utilization, and potential value, *World Rev. Anim. Prod.*, 14(4), 11, 1978.
196. Duke, J. A. and Duke, P., Organic energy farms: will they free us?, *Org. Gard.*, September, 135, 1975.
197. Wang, S. L., Huffman, J. B., and Rockwood, D. L., Qualtitative evaluation of fuelwood in Florida — a summary report, *Econ. Bot.*, 36(4), 381, 1982.
198. Lee, D. K., Biomass products as an energy source, *J. Natl. Acad. Sci.*, (Republic of Korea Nat. Sci. Series), 19, 239, No data (ca. 1981 or later).
199. Geoffroy, F., Fabert, V., Calif, E., Saminadin, G., and Varo, H., Value of Banana Leaves and Stems as Material for Fodder. I., Availability and Nutritional Value (in French), *Nouv. Agron. Antilles — Guyane* 4(1), 1, 1978.
200. Markley, K. S., The babassu oil palm of Brazil, *Econ. Bot.*, 25(3), 267, 1971.
201. Pinto, R. W., Babassu — Brazil's wonder nut, *Foreign Commun. Weekly*, XLV(2), 3, 1951.
202. Rutger, J. N., Rice, *Oryza sativa*, in *Handbook of Biosolar Resources*, Vol. 2, McClure, T. A. and Lipinsky, E. S., Eds., CRC Press, Boca Raton, Fla., 1981, 199.
203. Duke, J. A., Hurst, S. J., and Terrell, E. E., Ecological distribution of 1000 economic plants, informacion al dia alerta, *Agronomia*, No. 1, 1, 1975.
204. Han, Y. W. and Anderson, A. W., The problem of rice straw waste a possible feed through fermentation, *Econ. Bot.*, 28, 338, 1974.
205. CIAT, Rice Unit, Centro Internacional de Agricultura Tropical, CIAT 02E1B-77, 1977 Rep., 1978.
206. Rutger, J. N. and Grant, W. R., Energy use in rice producion, in *Handbook of Energy Utilization in Agriculture*, Pimentel, D., Ed., CRC Press, Boca Raton, Fla., 1980, 93.
207. Laughlin, J. C., The effect of band placed nitrogen and phosphorus fertilizer on the yield of poppies (*Papaver somniferum* L.) grown in krasnozem soil, *Acta Hortic.*, 73, 165, 1978.
208. Heltmann, H. and Arslan, O., Comparative Trials of Different Cultivars and Sources of *Papaver Somniferum*, *Acta Hortic.*, 73, 27, 1978.
209. Loof, B., Poppy cultivation (review article), *Field Crop Abstr.*, 19(1), 1, 1966.
210. Duke, J. A., Utilization of papaver, *Econ. Bot.*, 27(4), 390, 1973.
211. Miyagi, E., The effect of planting density on yield of Napier grass (*Pennisetum purpureum* Schumach), *Sci. Bull. Coll. Agric. U. Ryukus*, No. 27, 293, 1980.
212. Kiss, I. E., Experiments with strains of *Typhoides arundinacea* in Hungary, Proc. 13th Int. Grassland Congr., Sec. Paper, Sect. 1-2, 1977, 265.
213. CIAT, Bean Production Program, Annual Reports 1974-1977, 1974-1978.
214. Adams, M. W., Energy inputs in dry bean production, in *Handbook of Energy Utilization in Agriculture*, Pimentel, D., Ed., CRC Press, Boca Raton, Fla., 1980, 123.
215. Sansted, R., Energy inputs in snapbean production, in *Handbook of Energy Utilization in Agriculture*, Pimentel, D., Ed., CRC Press, Boca Raton, Fla., 1980, 127.
216. Kadziulis, L., Three aspects of comparative evaluation of legumes and grasses, *Vaxtodling*, 29, 126, 1974.
217. Cenci, C. A. and Pagiotti, R., Evaluation by grazing of *Phleum pratense* L. and *Hordeum bulbosum* L. ecotypes, *Ann. Fac. Agric. Univ. Degl Studi Perugia*, 33, 707. 1979.
218. Samarawira, I., Date palm *Phoenix dactylifera*, potential source for refined sugar, *Econ. Bot.*, 37(2), 181, 1983.
219. Tackholm, V. and Drar, M., *Flora of Egypt*, Vol. 2, Otto Koeltz Antiquariat, Koenigstein, West Germany, originally published in 1950, reprint 1973.
220. Ford, E. D., High production in a polestage sitka spruce stand and its relation to canopy structure, *Forestry*, 55(1), 1, 1982.
221. Perala, D. A. and Alban, D. H., Biomass, nutrient distribution and litterfall in *Populus*, *Pinus*, and *Picea* stands on two different soils in Minnesota, *Plant Soil*, 64, 177, 1982.
222. Kaul, O. N., Singh, R. P., Srivastava, V. K., and Gurumurti, K., Distribution of organic matter in *Pinus elliottii* plantations, *Indian For.*, 108(1), 39, 1982.
223. Borsuk, V. E. and Glazyrin, G. E., Predicting the yield of *Pistacia vera* by meterological factors, *Lesnoe Khozyaistvo* (Forestry Abstr. 41), (7300), No. 8, 45, 1979.

224. Felker, P., Uses of tree legumes in semiarid regions, *Econ. Bot.*, 35(2), 174, 1981.
225. Veracion, V. P. and Costales, E. F., The bigger, the more, *Canopy Int.*, 7(6), 1981.
226. Anon., Here comes (again), *Canopy Int.*, 7(1), 6, 1981.
227. Anon., Hurov's Tropical Seeds, Catalog 34, 1983.
228. Henry, J. F. and Salo, D. J., Silviculture energy farms, in *CRC Handbook of Biosolar Resources*, Vol. 2, McClure, T. A. and Lipinsky, E. S., Eds., CRC Press, Boca Raton, Fla., 1981, 341.
229. Burkart, A., *Las Leguminosas Argentinas*, Acme Agency, Buenos Aires, 1943.
230. Felker, P., Cannell, G. H., Clark, P. R., Osborn, J. F., and Nash, P., Screening *Prosopis* (mesquite) Species for Biofuel Production on Semiarid Lands, Final Rep. to US Department of Energy, National Technical Information Service, Springfield, Va., 1981.
231. Felker, P., Analyses and Recommendations for Further Development of Tamarugo *(Prosopis tamarugo)* and *Algarrobo* (P. section algarobia) Plantations in the Salar de Pintados of Northern Chile, A report to Sr. Luis Zelada Gonzalez, Programa del Pampa del Tamarugal, CORFO, Casilla 2915, Santiago, Chile, 1982.
232. Simpson, B. B., Ed., *Mesquite, Its Biology in Two Desert Scrub Ecosystems*, Dowden, Hutchinson & Ross, Stroudsburg, Pa., 1977.
233. Shah, F. A., False acacia — a promising bee plant of Kashmir, *Indian Bee J.*, 34(102), 34, 1972.
234. Agriculture Handbook No. 271, Silvics of Forest Trees of the United States, Forest Service, U.S. Department of Agriculture, Washington, D.C., 1965.
235. Irvine, J. E., Sugar cane. Saccharum hybrids, in *CRC Handbook of Biosolar Resources*, Vol. 2, McClure, T. A. and Lipinsky, E. S., Eds., CRC Press, Boca Raton, Fla., 1981, 211.
236. Thring, *Philos. Trans. R. Soc. London A*, p. 487, 1980.
237. Ricaud, R., Energy input and output for sugar cane in Louisiana, in *Handbook of Energy Utilization in Agriculture*, Pimentel, D., Ed., CRC Press, Boca Raton, Fla., 1980, 135.
238. Pryanishnikov, S. N. and Alimaev, I. I., Biological bases of increasing yield of desert pastures on alcareous gray soils in Southeastern Kazakhstan, Proc. 13th Int. Grassland Congr., Leipzig. 1977, Sect. 3-4-5, 209 [CABY 81(10)].
239. Allen, E. B., Water and nutrient competition between *Salsola kali* and two native grass species (*Agropyron smithii* and *Bouteloua gracilis), Ecology*, 63(3), 732, 1982.
240. Karpiscak, M. M., Rawles, R. L., Foster, K. E., Meinel, A. B., and Meinel, M. P., The Feasibility of Using Russian Thistle *Salsola kali* L. for Bioconversion, Fifth Quarterly Rep., Office of Arid Lands Studies, University of Arizona, Tucson, 1980.
241. Foster, K. E., Rawles, R. L., and Karpiscak, M. M., Biomass potential in Arizona, in *Desert Plants*, 2(3), 197, 1980.
242. Karpiscak, M. M., Rawles, R. L., and Foster, K. E., The feasibility of using *Salsola kali* L. (Russian thistle) as a renewable energy resource, *Arizona-Nevada Acad. Sci. J.*, 15 (Abstr.) Proc. Suppl., 1980.
243. Quisumbing, E., Medicinal Plants of the Philippines, Tech. Bull. 16, Philippine Department of Agriculture and Natural Resources, Manila, 1951.
244. Princen, L. H., Potential wealth in new crops: research and development, in *Crop Resources*, Seigler, D. S., Ed., Academic Press, New York, 1977.
245. Scheld, H. W. and Cowels, J. R., Woody biomass potential of the Chinese tallow tree, *Econ. Bot.*, 35(4), 391, 1981.
246. Scheld, H. W., Bell, N. R., Cameron, G. N., Cowels, J. R., Engler, C. R., Krikorian, A. D., and Shultz, E. B., The Chinese tallow tree as a cash and petroleum-substitute crop, in *Tree Crops for Energy Co-Production on Farms*, SERI/CP-622-1086, Solar Energy Research Institute, USGPO, Washington, D.C., 1980.
247. Braun, W. R. H. and Espericueta, B. M., Biomasa y produccion ecologica de jojoba (*Simmondsia chinensis* Link) en el Deserto Costero de Sonora, *Deserta*, 5, 1978; 57, *Deserta*, 1979.
248. Daugherty, P. M., Sineath, H. H., and Wastler, T. A., Industrial Raw Materials of Plant Origin. IV. A Survey of *Simmondsia chinensis* (Jojoba), Eng. Exp. Sta., Bull. No. 17:1-36, Ga. Institute of Technology, Atlanta, Georgia, 1953.
249. NAS, *Products from Jojoba: A Promising New Crop for Arid Lands*, National Academy of Sciences, Washington, D.C., 1975.
250. Yermanos, D. M., Jojoba — a brief survey of the agronomic potential, *Calif. Agric.*, Sept., 1973.
251. Gorelov, E. P., Rasulov, I. R., and Odilov, S. K., Fresh fodders in spring (Russian), *Kormoproizvodstovo*, 5 (Abstr.), 34, 1980.
252. Watt, J. M. and Breyer-Brandwijk, M. G., *The Medicinal and Poisonous Plants of Southern and Eastern Africa*, 2nd ed., E. & S. Livingstone, Edinburgh, 1962.
253. Jarvis, P. G., Production efficiency of coniferous forest in the U.K., in *Physiological Processes Limiting Plant Productivity*, Johnson, C. B., Ed., Butterworths, London, 1981, 81.
254. Bukantis, R., Energy inputs in sorghum production, in *Handbook of Energy Utilization in Agriculture*, Pimentel, D., Ed., CRC Press, Boca Raton, Fla., 1980, 103.

255. Gill, P. S., Tahir, S. M., Al-Younis, A. H., and Younis, M. A., Preliminary studies on the cultivation of sweet sorghum (*Sorghum bicolor* L. Moench) for the production of sugar in Iraq, *Iraqi J. Agric. Sci.*, 12, 3, 1977.
256. Denman, C. E., Sorghum Cultural Practices and Variety-Environment Interaction Studies, OSU No. P-728, Res. Rep., Agricultural Experiment Station, 1975.
257. Mendez, J. and Martinez, J. F., Behaviour of legumes intercropped with forage sorghum, *Pastos Forrajes*, 3(1), 83, 1980.
258. Itnal, C. J., Desai, G. S., Sajjan, G. C., and Parvatikar, S. R., Effect of supplemental nitrogen on the plant characters, grain and fodder yield of rabi sorghum under dryland conditions, *Curr. Res.* 9(2), 24, 1980.
259. Ricaud, R., Martin, F. A., and Cochran, B. J., Sweet sorghum for biomass and alcohol production, *Louisiana Agric.*, 24(4), 18, 1981.
260. Freeman, K. C., Sweet Sorghum for Liquid Fuel, *Agric. Res.*, USGPO, Washington, D.C., July 1981.
261. Pal, U. R. and Malik, H. S., Contribution of Azospirillum Brasilense to the nitrogen needs of grain sorghum (*Sorghum bicolor* L. Moench) in humid sub-tropics, *Plant Soil*, 63(3), 501, 1981.
262. Logsdon, G., Sweet sorghum — four foods in one, *Org. Gardening Farm.*, Aug, 72, 1974.
263. Bennett, H. W., Johnsongrass *Sorghum halepense*, in *CRC Handbook of Biosolar Resources*, Vol. 2, McClure, T. A. and Lipinsky, E. S., Eds., CRC Press, Boca Raton, Fla., 1981, 251.
264. Dyachenko, P. A., Fodder mixtures for the green conveyer, CAB HA 00960 (049), *Korma*, No. 5(Abstr. only seen), 40, 1977.
265. *U. S. Oil Week*, September 17, 1979.
266. Ravantii, S., Winter hardiness and yield of local varieties of Finnish red clover grown in Southern Finland at the Anttila Experimental Farm of the Hankkija Plant Breeding Institute in 1962-6, *Ann. Agric. Fenniae*, 19(2), 142, 1980.
267. Andries, A., The role of red clover in Belgian forage production: changes and perspective, *Fourrages*, No. 90, 2737, 1982.
268. Taylor, N. L. and Smith, R. R., Red clover *Trifolium pratense*, in *Handbook of Biosolar Resources*, Vol. 2, McClure, T. A. and Lipinsky, E. S., Eds., CRC Press, Boca Raton, Fla., 1981, 11.
269. Kvech, O., The importance of crop residues in rotations of an intense farming system, *Rostlinna Vyroba*, 25(10) (Abstr. only), 1013, 1979.
270. USDA, Crop Production Research, Agricultural Research Service 1982 Annual Report, U.S. Department of Agriculture, Washington, D.C., 1983.
271. Briggle, L. W., Introduction to Energy Use in Wheat Production, in *Handbook of Energy Utilization in Agriculture*, Pimentel, D., Ed., CRC Press, Boca Raton, Fla., 1980, 109.
272. Briggle, L. W., Wheat. *Triticum aestivum*, in *CRC Handbook of Biosolar Resources*, Vol. 2, McClure, T. A. and Lipinsky, E. S., Eds., CRC Press, Boca Raton, Fla., 1981, 67.
273. *Alcohol Week*, October 20, 1980.
274. Pratt, D., *Washington Star*, September 4, 1979.
275. Hitzhusen, F. J. and Abdallah, M., Economics of electrical energy from crop residue combustion with high sulfur coal, *Am. J. Agric. Econ.*, 62(3), 416, 1980.
276. Hall, D. O., Biological Solar Energy Conversion, *Nature*, 278, 114, 1979.
277. Bulletin 221, University of Florida, Gainesville, 1935.
278. National Academy of Sciences, Alcohol Fuels — Options for Developing Countries, National Academy Press, Washington, D.C., 52, 1983.

INDEX

A

B

C

D

E

F

G

H

I

J

K

L

M

N

O

P

R

T